# Moral and Political Reasoning in Environmental Practice

This book was set in Sabon by SNP Best-set Typesetter Ltd., Hong Kong.

Printed and bound in the United States of America.

Library of Congress Cataloging-in-Publication Data

Moral and political reasoning in environmental practice / edited by Andrew Light and Avner de-Shalit.
p. cm.
Includes bibliographical references and index.
ISBN 0-262-12252-9 (alk. paper)—ISBN 0-262-62164-9 (pbk. : alk. paper)
1. Environmental sciences—Philosophy. 2. Environmental ethics. I. Light, Andrew, 1966– II. de-Shalit, Avner.

GE40 .M67 2003
179′.1—dc21 2002026349

This book is in memory of Eimear Herbert-Barry.

# Contents

# Introduction: Environmental Ethics—Whose Philosophy? Which Practice?

Andrew Light and Avner de-Shalit

When philosophers talk to one another in conferences, at universities, and through books and articles, they tend to generalize, theorize, and express themselves in the abstract. They often ask questions that are purely hypothetical, about an ideal, theoretical world. Environmental philosophers are no exception.

In this book the editors and authors claim that while traditional philosophical practice has much value, it is not the only way to philosophize and reason about the environment. It may in fact not be the best way to bring philosophy to environmental questions. Coming mainly from a background of pragmatism and communitarian political theory, we claim that philosophy can be expressed as a public event. It can, and perhaps should, aim at changing the world, and it can do so only if it (1) takes seriously arguments that derive from real cases, from practice, and (2) applies itself to searching for novel philosophical tools that can be of use in environmental practice. In other words, environmental philosophers should find a way to become more involved in argumentation that takes place in environmental campaigns and is discussed in the broader environmental literature. Such arguments should be part of what constitutes environmental philosophy, and we believe that they could contribute positively to environmental practice.

But how wide is the scope of environmental practice? Or more precisely, what can be legitimately labeled as environmental practice? Is it only conservation? Is it only activism? Or does it include the production of humanly manipulated landscapes that bring together human and nonhuman communities? Is environmentalism only supportable through biocentric arguments, or can we make human-centered claims as well, such as arguments for obligations to future human generations to protect the environment? If the scope of environmental practice is defined more broadly than it often is, the scope of legitimate philosophical questions about the environment may be much wider than many environmental philosophers have traditionally thought it to be.

The essays in this book confront these two issues: how to philosophize about the environment when "on-the-ground" practice is at stake, and what the scope of environmental practices is that we aim to philosophize about. The authors approach these twin topics in several ways. The first three chapters offer reflections on the general role of political philosophy in environmental reasoning. The next five chapters discuss more specific concepts, or philosophical tools, that could be used as an aid for environmental reasoning. And the last four chapters offer analysis of the role of philosophical argumentation in several case studies. Before moving to these essays, we will first discuss the scope of practices environmental philosophers should attend to, through an investigation of the relationship between environmental philosophy and environmental activism. Second, we will offer our own account of how better to undertake environmental philosophy in light of a claim that it should aim to be helpful to activists and policymakers—those who play a key role in shaping environmental practices. Finally, we will conclude with a summary of the chapters that follow.

You may note that our contribution in this chapter diverges in some ways from the views of the other authors in this book. Nonetheless, it attempts to grapple with the same basic issue as the other essays: how to do environmental philosophy in a broader political and public context.

## Environmental Philosophy and Environmental Activism

If environmental philosophers wish to inspire activists, policymakers, politicians, and the public in general, what should they do?[1] If they wish to make an impact, if they want environmental philosophy to be incorporated into environmental policies, how should they put forward their arguments? What, in the end, should be the focus of their theories?

One way to have an impact on environmental policies and practices is to adjust to and accommodate the ideas and practices already in use. As a late politician once said, "I compromise, I give up, I compromise again, until everybody agrees with me." Of course, this would seem wrong for philosophers to do. Philosophers who work in ethics or political philosophy should only work through the question of whether some view, X, is right. Once they have concluded that X is right then *perhaps*, depending on their views on the role and importance of an understanding of moral psychology in moral reasoning, they should turn to the question of how to persuade others that X is right.[2] At bottom though, the commonly held view is that philosophy and rhetoric are separate projects. According to this model, environmental ethicists should work toward helping societies find out why X is right in relation to the environment. They do so, however, because X is right, not because X is popular or because they themselves desire to be popular.

But if philosophers only see their activities as a search for truth, they may fail to have an impact on policies and practices. After all, the questions asked in everyday life often differ from the questions asked by philosophers. If you want people to move from position A to position B, from one state of consciousness to another, from acquiescing to injustice to embracing justice, from ignoring rights to respecting them, you often have to consider existing questions and not just supply new answers.

Such a view is found in the work of Kate Rawles, a philosopher who has been heavily involved in environmental activism and work with activists. Rawles believes that philosophers can help activists by offering them systematic justification for their policies as a means of both legitimizing and guiding action. But she is cautious about a wholesale endorsement of philosophical methods as they have traditionally been practiced. The responsibility of an environmental philosopher is to consider her motivations and activities more carefully so as not to obstruct the creation of long- and short-term environmental change:

> Consider the question of the metaphysical status of value in nature. The question is of interest in its own right. But from the perspective of improving states of affairs in the world, the complex and convoluted trail one embarks upon in order to try to answer it seems, after a certain point, to become quite unrelated to the starting point. If philosophy is to contribute to change, it needs a constant pulling back to the question, how does this help?[3]

One way of putting the challenge facing environmental philosophy, then, is how to make philosophy, not philosophers, a more relevant and useful tool in reaching public decisions about whether a particular view is right or wrong in relation to environmental issues. A way has to be found for environmental philosophy to overcome the alienation that too often exists between environmental activists and policymakers on the one hand, and environmental philosophy on the other. We must overcome the usual state of affairs in which philosophers and philosophical ideas are simply ignored in the exchange between politicians and environmental advocates.

Both of us have worked with environmental activists on many occasions. Sometimes we have been invited to talk to them; at other times they were invited to participate in academic seminars we either organized or attended. In too many cases, however, the discussion between philosophers and activists ended up in despair, with the latter claiming that environmental philosophy was a far cry from what was needed in practice. "All this is very beautiful, but how can I make the layperson see it?" they asked. Environmental activists have often pointed to a need for an environmental philosophy constructed in language and arguments accessible to wider audiences.[4] Moreover, they maintain, arguments are needed that are relevant to the debates in which they are engaged.

Even more important, if we analyze the gap between environmental philosophers and environmental activists, we see that the two groups sometimes disagree about the causes of environmental problems as well. Environmental philosophers often argue that human chauvinism, or "strong" anthropocentrism, a misguided moral belief about the superior place of humans in the world, is the key cause of environmental problems. This has unfortunately led many of them to discount investigations into other forms of valuing nature for more traditional ethical reasons, including "weak" anthropocentric claims about the positive aesthetic value of nature or about our environmental obligations to future humans.[5] Activists, on the other hand, often think that government policies concerning the environment are determined by a misguided conception of human needs and by a scorn and disregard of human rights. They often claim that this is so because the interests of the wealthy govern politics. Instead, they argue, we should aim at formulating better understandings of genuine human needs: ones that include the needs of future people for environmental protection, as incorporated into the needs of people living today. Some of them argue that we need a new politics that respects rights in a broader sense.

But consistent with their broader rejection of anthropocentric reasoning, many environmental philosophers claim that what is needed is not an expanded anthropocentrism but a change in the predetermined "state of mind," or "worldview," that humans have toward the environment. The claim is that our current state of mind is built on false philosophical assumptions. If people change the way they think about human-nature relations to some form of *non*anthropocentrism, then policies will eventually change as well.

There are many ways to make such a case for preferred philosophical practice. One way is to claim that changing worldviews has always been the traditional work of philosophers in relation to any larger moral, political, or social change. Consistent with this reading of the history of philosophy, J. Baird Callicott argues that the most "lasting and effective" form of environmental activism that philosophers can engage in is simply philosophy itself. As such, environmental philosophers should not be concerned with the sorts of worries we are raising here. Reasons must always precede polices, according to Callicott, and the trajectory of environmental ethics has been primarily aimed at creating better policies insofar as it is "devoted to articulating and thus helping to effect . . . a radical change in [environmental] outlook."[6]

Although nonanthropocentrists are certainly not the only ones who have tended to focus only on changing background worldviews as the focus of environmental ethics, part of changing worldviews according to this line of reasoning is often thought to involve a recognition of the value of nature independent of human judgment. Philosophers pressing this line of argument focus in their work on "intrinsic value" theory (or at least, noninstrumental value) in the context of some form of nonanthropocentrism. Such a move is right in line with Callicott's views on the proper focus of environmental philosophy. Because, according to Callicott, the anthropocentric instrumental values in favor of preserving nature (say, for the potential health benefits of current or future people) can never hope to compete with the anthropocentric instrumental values for developing it, a "*persuasive* philosophical case for the intrinsic value of nonhuman natural entities and nature as a whole would make a huge practical difference [to the resolution of environmental problems]."[7] Callicott approvingly cites Warwick Fox as arguing that the advantage of a claim to the intrinsic value of an entity is that it shifts the burden of proof to those who would want to harm it. Interference requires "sufficient justification" for action under such conditions. A trade-off of one kind of instrumental value for another is then deemed to be insufficient justification for developing some bit of nature. Describing nature as having intrinsic value presumably trumps any claim to its instrumental utility.

But regarding Callicott's suggestion that the case for intrinsic value needs to be "persuasive," we must ask, to whom is it supposed to be persuasive? Only other philosophers? Presumably Callicott's sights are set broader than this. The reason they must be is because his comparison of different anthropocentric accounts of value (some in favor of preservation of nature and others in favor of development) assumes that the anthropocentric case for preservation can never win. But this is surely not because Callicott thinks that the anthropocentric values we find for justifying preservation of nature are not true or not better than the anthropocentric reasons for development. It must be because he thinks that the anthropocentric reasons for preservation can never compete in the court of public inquiry against the anthropocentric case for, as he puts it, "lucrative timber extraction, agricultural conversion, hydroelectric empoundment, mining, and so on."[8]

But if anthropocentric arguments for the preservation of nature fail in this larger arena where persuasion matters, we must also test our intuitions about whether nonanthropocentric arguments for the intrinsic value of nature will also be persuasive in this larger arena. And here we have good reason to be skeptical. The underlying assumption of almost all nonanthropocentrists is that nonanthropocentrism must be developed as an alternative worldview because most people are anthropocentrists. (This is in fact why many environmental philosophers feel there is a philosophical dimension to environmental problems: the history of Western philosophy has been successful in developing the faulty worldviews that assert that only humans have the kind of value that generates moral obligations.) Thus, the nonanthropocentrist advocating the intrinsic value of nature cannot rest after making a persuasive case to other environmental philosophers of the truth of his or her views. The case must also be persuasive to people who do not count themselves as nonanthropocentrists. It must be a case compelling enough to persuade anthropocentrists that they should accept the shift in burden of proof (or burden of protection) that is made manifest through a claim to the nonanthropocentric intrinsic value of nature. But that is surely a higher hurdle of persuasion than starting with an acceptance of an anthropocentric terrain and arguing from within that framework that there are more values at stake in an environmental controversy than just pure economic values in favor of development. Either Callicott only believes that persuasion matters when evaluating the veracity of anthropocentric or instrumental arguments, or he is giving short shrift to the problems that have actually confronted those using nonanthropocentric or intrinsic arguments in public arenas.[9]

In contrast, activists believe that they have no choice but to broaden their horizons as to who needs to be persuaded. They have to persuade developers (be they private or public) to reorient their projects, or at least effectively use policy, law, or protest to stop developers. While philosophers have the luxury of talking to one another, engaging exclusively in an intramural critique of each others' ideas on competing worldviews, most activists cannot afford such a luxury. They must speak to the opposition in the opposition's own largely anthropocentric terms or fail to be granted a hearing at all. Andrew Dobson describes nonanthropocentric environmental thought as follows: "Ecologism . . . appears to want to go beyond human-instrumental reasons for care for the natural world, arguing that the environment has an independent value that should guarantee its "right to life." . . . The private ecologist, in conversation with like-minded people, will most likely place the intrinsic value position ahead of the human-instrumental argument. . . . The public ecologist, however, keen to recruit, will almost certainly appeal first to the enlightened self-interest thesis and only move on to talk about intrinsic value once the first argument is firmly in place."[10]

At the end of the day, the goal of activists (or as Dobson puts it, "public ecologists") is to stop developers from causing harm. They can do so either by persuading politicians to listen to their arguments rather than to those of developers, or by persuading developers to change their plans. As with any policy argument, especially in a democratic context, public support for either of these projects is very important. Activists, therefore, might look for philosophers to help them in encouraging the public to support such policies. Philosophers need to respond by providing a broader array of moral and political tools that can be used to help make these cases, rather than simply demonstrating their ability to joust with one another over the intricacies of value theory.[11]

And yet some philosophers still insist that although they have to be accurate, consistent, and even strive to find the truth about their areas of inquiry, they in no way should be responsible for persuading people to support particular environmental projects. The assumption behind a view like Callicott's, as Rawles puts it, is not that persuasion really matters at all but that "all normative disputes are in the end reducible to disputes about facts. . . . As long as someone is not a psychopath, if she genuinely understands and accepts that from an evolutionary perspective animals are kin and from an ecological perspective the land is a community [which is Callicott's particular worldview], she will agree that she has strong moral obligations toward land and animals and will treat them accordingly."[12] No persuasion is needed, only recognition of the "thick facts" of the matter, as Callicott puts it, which will eventually count as a sufficient reason to change one's ethical and political views once one has accepted a worldview inclusive of these facts.

We are faced then with a strange puzzle. On the one hand there is a common assumption in much philosophical work that there are two spheres of reasoning: one that is purely academic, in every sense of the word, and one that is practical. The former is the realm of philosophy proper, including environmental philosophy; the latter is the realm of activism or advocacy. Such a distinction hinders the ability to make philosophy relevant to environmental activism. But on the other hand, attempts by environmental philosophers such as Callicott to overcome this divide claim that the former kind of activity, philosophy, ought to be understood as activism proper. But such a view is similarly unhelpful. If it were true, all philosophical activity would be a form of activism, because all philosophical activity is aimed at least at discerning why one view of some subject X is better than another view of some subject X, and hence ought to be incorporated to some extent in the worldviews of at least some persons. For after all, at a minimum, any area of philosophy actually engaged in by someone is presumably of some interest at least to the worldview of the philosopher undertaking the philosophical investigation at hand.

But such a conclusion is absurd. It trivializes more formal areas of philosophy by reducing their importance to their measured effect on worldviews. It also undercuts the uniqueness of environmental philosophy, in relation to other areas of philosophical inquiry, by failing to provide reasons for recognizing the difference between philosophizing about a problem immediately at hand that requires a resolution now to avoid causing pain and suffering (such as environmental problems) and philosophizing about issues that we can afford to discuss among a very small set of people without resolution for thousands of years. The subject matter of environmental philosophy begs the question of whether it should have an advocacy component distinct from other forms of philosophical activity. Our claim is that environmental philosophy must at least assert the gap between philosophy and activism even if only to eventually overcome it. Environmental philosophy is not, or at least ought not be, just another branch of academic philosophy.

But the uniqueness of environmental philosophy should not only be understood in terms of the object of its concern. We also think that it ought to be different in the way it is practiced. And it is in finding a specific and uniquely justifiable practice that environmental philosophy will overcome the divide between academic and practical activity, not in simply asserting that there is no divide. At least one reason environmental philosophy should be practiced differently is that the original grounding intuition of environmental philosophy, when it became organized as a formal subdiscipline in philosophy, was that philosophers should do it so as to make a contribution to the resolution of environmental problems in philosophical terms. But if those terms produce only arcane discussions by a few theorists of issues such as the intrinsic value of nature, we will have failed in our aspirations to make a contribution to the resolution of environmental problems. While it is conceivable that eventually our theories of value could filter down to the broader environmental community and to policymakers, the importance of environmental problems warrants taking seriously a more practical and pragmatic set of tasks for the field that might make a more immediate contribution to the solution of these problems. This is not to say that environmental philosophy is only practical or worth doing if it has policy relevance. But if our work has no relevance in such spheres, we must ask ourselves why we are doing this kind of philosophy at all.[13] Are we doing environmental philosophy simply because the problem of natural value is philosophically interesting? Such a conclusion is counterintuitive to the original grounding intuitions that shaped the field. Environmental philosophy ought to be a form of applied philosophy, or, as some would insist, "practical philosophy": its argumentation should be inspired by problems in the real world and by the need to solve them. We therefore want to suggest that there is a way to do environmental philosophy that is more relevant to real-life cases, and more helpful to society in its efforts to solve moral dilemmas.[14]

What is this way? The two key ideas we will present here are: (1) to reason from and within a community, in this case, broadly speaking, the environmental community, and (2) to broaden the scope of the objects of reasoning. Environmental philosophers must take the questions that disturb the environmental community seriously; they must consider arguments raised by activists and developers and analyze them, in order to be relevant. This is not to say that purely theoretical questions are not important in environmental philosophy. Far from it. It means that the community and the discourse taking place in the community must be part of what constitutes the philosophical debate in the field. But how to do this?

## Environmental Philosophy as Public Reflective Equilibrium

We want to suggest that the way to incorporate community issues and concerns into environmental philosophical debates is to widen the notion of the "philosophical text." Why shouldn't the philosopher regard written works of activists, and indeed, the arguments used in real-life cases, as texts? Of course, there is good philosophy and poor philosophy, and one should not pursue proving that a bad argument is valid. But a careful and professional scrutiny of arguments is one thing; ignoring these arguments is another. Excluding from environmental philosophy arguments and theories that are raised in real-life cases will only fortify the sense of alienation that activists sometimes have toward environmental philosophers. And although it is true that many, if not most, environmental philosophers have appealed to cases in their work, rarely have the cases been more than merely illustrative of the general philosophical argument being pursued. While philosophers criticizing restoration have illustrated their arguments with cases, rarely have they used the controversies unfolding in the cases themselves as the orienting point for their philosophical work.[15]

Now we come to the difficult questions: If environmental philosophers are to consider a broader variety of issues and questions, and if they are to take activists' arguments seriously, how should they do it? Let us begin with the question of what to do with these texts. We suggest that the mechanism to be applied to helping to integrate the activities and experiences of the broader environmental community into environmental philosophy is "public reflective equilibrium."[16]

It is widely accepted that moral exploration involves a process of "reflective equilibrium." This concept—which is often attributed to John Rawls—means "that we 'test' various parts of our system of moral beliefs against other beliefs we hold, seeking coherence among the widest set of moral and non-moral beliefs."[17] We find the coherence, which involves "more than logical consistency," by going backward and forward, continuously revising and modifying our theories and intuitions. These revisions are very important because failing to change our intuitions or theories in cases of serious contradictions would imply dogmatism.

But a question immediately arises: Whose intuitions and whose theories count in the process of reflective equilibrium? There are three possible answers. One is that the *private reflective equilibrium* of the philosopher counts: the philosopher considers his or her intuitions and theories. Often she will also consider her colleagues' theories, but the actual process of reasoning is private, and in theory can be detached from any community or any other individual. The philosopher can simply sit in her armchair and reflect. The second possible answer to this question is that although the intuitions considered should reflect the various intuitions present in the community, the role of the philosopher is to hold up a mirror and interpret society's intuitions, and then to offer the theory that meets or criticizes these intuitions.[18] This reasoning can be called *contextual reflective equilibrium*, because the philosopher exercises his or her reflection within a given cultural or moral context. The third answer, and the one we endorse, is called *public reflective equilibrium*, because the philosopher considers not only the public's intuitions but also the public's theories. Many theories—explanations and justifications, arguments and consistent methods of reasoning—are put forward by activists, politicians, and the public in general. The role of the philosopher, according to this model of reflective equilibrium, is to engage in the public's discourse, refine the arguments coming out of this discourse, and help to foment public debate about important issues. But before elaborating on this model, let us say a few words about why the private and contextual models are not enough for environmental philosophy.

The model of private reflective equilibrium assumes that the best way to philosophize is in detachment. It regards philosophy as not far from science, and both philosophy and science as "neutral." Moreover, because philosophy is the search for wisdom, there is no need to consult the layperson. In fact, it is often the case that the more remote and detached the process of reasoning is, the better it is: better in the sense that the theory constructed is more accurate, less biased, and therefore more "professional." The criterion for a theory to be successful is that the philosopher can offer a consistent and coherent theory, which ties in with the philosopher's declared intuitions. Sometimes these intuitions happen to be shared by many readers, but the point of private reflective equilibrium is not to convince the reader that a theory fits the *reader's* intuition, but rather to convince the reader that the philosopher has managed to develop a very accurate theory in which the principles of morality offered live in harmony with philosophical intuitions that in themselves are reasonable. When will the theory also meet the intuitions of the reader? According to the private reflective equilibrium model, this will happen when the theory is taken from a view that is as objective as possible. Then it is more likely that anyone looking over the philosopher's shoulder will be able to recognize his or her way of reasoning as correct.[19]

However, it is precisely this dislocation, this placelessness of the philosopher that is the first difficulty in applying the method of private reflective equilibrium to environmental issues. Many environmentalists claim that the physical and spiritual uprooting of humans from place—indeed the very thought that they can and should easily and voluntarily uproot themselves—is what makes them indifferent to the natural and urban environments in which they live. Environmental philosophers should likewise argue from a sense of place, of location, from here.[20] Moreover, the idea of philosophy as supplying neutral arguments contradicts the ideas of "environmentalism" and "environmental lifestyles," which are ideas of the good, namely, about how the community lives with nature and with the urban environment.

Indeed, the idea that a good philosophical theory is one that is not related to the community's ideas of the good, but instead rests on universal claims, will not appeal to activists. By and large they come from local communities. They seek to foster sustainable communities by, among other things, protecting the environment of these communities. They do not wish to ignore the community and its values when reasoning about the environment. Moreover, philosophers who believe in neutral reasoning and wish to apply it to environmental issues will defend their position by arguing that such reasoning is the right way to talk to developers, because it bypasses the question of which form of life is better—one emphasizing development or one emphasizing conservation—and therefore can persuade those resisting environmentally friendly policies by an appeal to a nonbiased argument. But in practice, this is far from reasonable. Debates about development and conservation take place within communities and are closely linked to the community's values. In fact, developers and conservationists both appeal to the community's values to support their respective claims. It is therefore necessary to find out the values and positions of communities in order to make any moral appeal that has a hope of motivating people to take action.

So, is a more contextual reflective equilibrium a better way to practice environmental philosophy? At the end of the day, this model does take the community and its intuitions into account. Indeed, we think it is a step in the right direction, though later we will suggest a way to improve it.

Michael Walzer's philosophy is a very good example of contextual reflective equilibrium. One reason his model suits environmental issues better than does the private reflective equilibrium model is that for him, social criticism is practiced in relation to particular communities and their moral views. The philosopher learns and examines the values of the society and theorizes about them. This process of learning is interpretative rather than inventive.[21] The philosopher examines the behavior and expressed views of the individuals who make up a community and interprets them vis-à-vis what she thinks the community's values are.

This activity is "reiterative" in the sense that the philosopher does not aim at putting forward the last word. Rather, she regards her theory as new input to the moral discourse under reevaluation.[22]

If environmental philosophy is subjected to the debates and opinions of the general public, and the general public will use and apply whatever moral theory is offered in actual cases, then Walzer's contextual reflective equilibrium is a promising philosophy. However, several questions arise: To what extent can the contextual model revise or change the way people think about the environment? What if the theory developed by the philosopher is not relevant to the real cases at hand in a particular community? What if its arguments are consistent with each other and the theory is coherent, but there are "external tensions" between the theory and the way people think, behave, and justify their behavior in real-life cases? Would the process of reflective equilibrium be valid? Will it be accessible to activists? For example, a certain theory of intrinsic value can be consistent and coherent, hence lacking "internal tensions," but it may not be at all relevant to the questions people ask themselves when engaged in environmental activism. Indeed, as we have mentioned, several activists have argued that when they face developers and try to persuade them, it is useless to apply the theory of intrinsic value. This is an external tension between the theory and the actual arguments that people apply.

Our fear is that although contextual reflective equilibrium is a step forward from the private method, it does not go far enough. For a theory in environmental philosophy to avoid such external tensions, and to be relevant to real cases (and the reasoning of activists in those cases), it should also arise from the cases in question. The best way to achieve this is to start with the activists and their dilemmas. Hence, an adequate environmental philosophy should derive from extended sources—that is, not only the "contextual" philosopher or anthropological explorer, but the general public as well. A theory is required that reflects the actual philosophical needs of activists seeking to convince others that their standpoints are morally right by appealing to practical issues, and not necessarily to the philosophical needs of the philosopher, who convinces others by appealing to consistency and simplicity.

At this point two clarifications are needed. First, we do not mean to say that activists are, or should be, interested in instrumental reasoning only. Indeed, if activists are interested in instrumental reasoning alone, perhaps it is a task for the philosopher who practices a more expanded form of reflective equilibrium to see that their arguments are put in a larger philosophical context. Second, it goes without saying that the philosopher should not take the value of activists' claims for granted. Their intuitions, arguments, claims, and theories should be scrutinized. However, the fact that they need to be critically examined does not affect the main point: that the activists' and the public's intuitions, claims, and theories ought to be the starting point for a philosophy aimed at policy change. This is why this model of reasoning is called *public reflective equilibrium*. The role of the philosopher includes finding the balance between the intuitions and theories of the public. This process includes refining the public's theories and representing them in public forums.

Using this method of reasoning, environmental philosophy could become truly public and practical. This is so for two reasons. First, it derives from, and speaks to, people previously denied access to the shaping of the morality of our institutions. Second, environmental philosophy that uses public reflective equilibrium will derive from issues and questions previously marginalized by what has been considered the "real" or important issues on the moral agenda. Notice that this critique is pointed at two groups simultaneously. The first group consists of the developers and politicians who maintain that growth, economic considerations, and so on are the only important issues on the environmental agenda. The second group is made up of those environmental philosophers who seem to have ignored what activists have defined as urgent and important—for example, the effectiveness of arguments that try to persuade developers to act more responsibly (rather than simply the truth of those arguments), or moral dilemmas deriving from activism. Instead, many environmental ethicists have decided that the philosophically important and interesting issues are intrinsic value, nonanthropocentrism, and other philosophical concepts. An environmental philosophy based on public reflective equilibrium would not fall into this trap because it would start from questions and arguments that are raised in real-life, public debates.[23]

We should also clarify that by no means should the philosopher who exercises public reflective equilibrium offer his or her services to every activist. We cannot reasonably expect such philosophers to help, for example, a fascist group. There is a preliminary stage in the process of public reflective equilibrium—a stage of private reflective equilibrium whereby the philosopher analyses the theories at stake according to his or her own values. However, this preliminary stage should not imply that public reflective equilibrium is impossible; instead it means that a preliminary stage in which the philosopher "filters" intuitions and theories is inevitable. But even in this stage of the process, the starting point is what the public thinks and argues. In fact, as we will see, many of the contributors to this book (such as Arler, Griffith, and Schlosberg) do just this. We close this section with our own brief example of how such a public philosophy would look at an environmental controversy differently from the more traditional form of philosophical practice.

Our example is the controversies that have arisen over the practice of restoring nature. Restoration ecology is the science and technological practice of restoring damaged ecosystems, most typically ecosystems that have been harmed by human beings. Such projects can range from small-scale urban park reclamations, such as the ongoing restorations in Central Park and Prospect Park in New York City, to huge projects such as the approved $6.8 billion restoration of the Florida Everglades.

In addition to the scientific and technological questions at the heart of restoration work, which have received substantial support and attention, there are also ethical issues involving the actual practice of restoration that have been woefully underexplored by environmental philosophers too often committed to a more abstract form of theorizing. Instead, environmental philosophers looking at this issue have tended to focus on more esoteric questions involving the metaphysical status of restorations, such as the issue of whether restored areas are really part of nature or not. Why is this the case?

As we have suggested, environmental ethics, especially as practiced in North America, is dominated by a concern with abstract questions of value theory, primarily focused on the issue of whether nature has nonanthropocentric value. One effect of this predominant search for a nonanthropocentric normative foundation for natural values has been that restoration, as a human practice, largely falls outside the purview of environmental ethics. When restoration has been taken up by environmental philosophers, the results have been largely deflationary. While there are some notable exceptions, the most influential work by environmental philosophers on this topic, such as that of Eric Katz and Robert Elliot, has argued that ecological restoration does not result in a restoration of nature, and that it may even create a disvalue in nature.[24] In their most famous interventions in this literature, Katz calls restoration "the big lie" of the possibility that humans can restore nature, nonanthropocentrically conceived, and Elliot refers to the practice as "faking nature." While Elliot's more recent work on this topic has found a more positive role for restoration, it is important to note that these criticisms stem directly from the principal concerns and framework of environmental ethics just mentioned. If the goal of environmental philosophy is to describe the non-human-centered value of nature, one assumption of the field is that nature can have a value independent of human appreciation of that value. If nature is to be distinguishable from human appreciation of it, presumably nature cannot be dependent on human creation or manipulation. If nature was dependent on human creation, it would have an irreconcilable anthropocentric (or anthropogenic) component. So, if restorations are human creations, they cannot ever count as containing natural value on nonanthropocentric grounds. Restorations are not natural in this view, and for Katz at least, they are merely technologically produced artifacts. To claim that environmental philosophers should be concerned with ecological restoration is therefore to commit a kind of category mistake: it is to ask that they talk about something that is not part of nature.

But one cannot simply ignore the practice of restoration by pronouncing restored areas to be unnatural. Restoration projects make up a large portion of what counts as on-the-ground environmental activity. The reason is simple: restoration makes sense because, on the whole, it results in many advantages over mere preservation of ecosystems that have been substantially damaged by humans. Through restoration, habitats for endangered species can be created that help stem the tide of the loss of biodiversity. Still further, an environmental philosophy that ignores restoration for overstated theoretical reasons risks losing site of another key facet of this practice: every instance of restoration represents an opportunity to involve local communities more intimately in the nature around them because most restorations present opportunities for public involvement. For example, the cluster of restorations known collectively as the "Chicago Wilderness" project, in the forest preserves surrounding Chicago, have at their height attracted some 2500 to 3000 volunteers annually to help restore 17,000 acres of native oak savanna that have been lost in the area.[25] The final plan for the project is to restore upward of 100,000 acres. And while sometimes controversial, social science research conducted on the volunteers in this cluster of projects shows that participation in these projects has forged a strong and positive bond between the local community and the nature preserves.[26] The goal of the projects has not only been to restore nature per se, but perhaps more important, to reestablish the local cultural relationship with nature. This can be seen as a positive pragmatic outcome of such activity, given that so many restoration efforts allow such participation and given that this participation is the point at which philosophical attention to restoration as a form of public reflective equilibrium can take place.[27]

So, even if it is true that humans cannot reproduce the value of nature as some nonanthropocentrists would claim, it is quite plausible that a moral ground justifying restoration projects can be found in the value of human participation in nature, even a participation mediated by the technology of restoration. Assuming such a case be made, the next interesting ethical issue becomes how to guarantee or further the benefits of participation in restoration projects despite the criticisms of restoration as a practice. The future of restoration as an environmental practice that offers opportunities for the public to become actively involved in the natural areas around them may in fact depend on undertaking a serious inquiry into this set of issues. (The Chicago restorations have sparked many controversies, in part stemming from the high degree of pubic participation in them. Our point, though, is that environmental philosophers should jump into these controversies, rather than standing on the sidelines discussing the metaphysical status of the restored forest preserves.[28])

This brief example helps to demonstrate our earlier conclusion: a more valuable, or at least more pressing, role for environmental ethicists is to engage in the debates that the public places a high priority on. Although the conceptual issues many of our colleagues in the field have focused on are surely crucial to the backgrounds of such public disputes, we would be remiss if we did not involve ourselves more extensively in the ethical dimensions of these real-life arguments as they are unfolding.

## Contents of the Book

The chapters in this book comprise three groups. Those in part I raise general questions about our advocated shift in environmental reasoning from metaethical questions and value theory to a more publicly responsible political theory. The chapters in part II analyze new concepts and methods of argumentation in environmental discourses, most of which derive from activists, nongovernmental organizations (NGOs), and their practices. The chapters in part III adopt something akin to the method of reasoning we have outlined above, arguing from cases though often without explicitly endorsing the methodology of public reflective equilibrium.

We begin with the question of how to work toward a more adequate account of environmental practices. In chapter 1, Michael Freeden—a prominent political theorist—dives into the tumultuous water of environmental reasoning by discussing how to relate environmental philosophy to political theory and in turn to the broader political world. He argues that the first step is to acknowledge that environmental thought is political thought. Freeden claims that "green" thought is a particular way of thinking about politics, not only about the environment. However, for Freeden, there is no single, correct green theory. Hence he maintains that we need to draw a new map of political ideologies which will allow us to better understand arguments raised within the environmental movement.

As we have suggested, reasoning about the environment should involve examining both the intuitions and theories of activists and finding a "reflective equilibrium." However, activists often put forward very preliminary theories, if not simply the beginnings of intuitions. In chapter 2, Mathew Humphrey offers constructive criticism for how we have thought about the role of intuitions in more traditional environmental philosophy. Humphrey examines the role and place of intuition among nonanthropocentrists (specifically, biocentrists), and puts forward a critique of biocentric philosophy's use of intuition, which raises the question of how we know what to do when we have very good reasons supporting contradictory environmental public policies.

If environmental philosophy is to marry more practical political theory, environmental justice is likely to be critical to the success of their union. This notion was first introduced by activists, for example, at the First National People of Color Environmental Leadership Summit and only then taken up by philosophers and theorists.[29] In chapter 3, David Schlosberg provides an analysis of three different concepts of justice in environmental justice theory and in the environmental justice movement. He refines the theories of environmental justice put forward by activists to make them a more accurate reflection of our intuitions regarding equity in environmental matters. This is public reflective equilibrium in practice.

Schlosberg's chapter leads us to part II, in which new philosophical tools for environmental reasoning are put forward and analyzed. The authors in this group of chapters discuss concepts that have been used by the public, yet less so by environmental philosophers. In chapter 4, Tim Hayward looks at "environmental constitutional rights." Constitutional environmentalism has not been a prominent topic of political study, but in practice the phenomenon has been developing apace. Accordingly, after clarifying the scope of potentially feasible environmental rights, Hayward assesses the case for them in the light of four critical questions: whether environment protection can be considered a genuinely fundamental right, whether a new right is necessary for achieving that end, whether such a right is practicable, and whether pursuing environmental ends by means of rights is democratically legitimate. While presenting arguments for an affirmative answer on each score, he also shows that the strength of the case ultimately depends on a number of contextual issues.

Another new concept, already used by activists and politicians, but analyzed next by William B. Griffith in chapter 5, is trusteeship. Griffith makes a case for an increased use of trusteeship for justifying environmental obligations to future generations. He surveys how the concept of trusteeship is used in contemporary U.S. policy on environmental issues and how that concept could be given more power by combining it with notions of sustainability. Once again, this is a refinement of public philosophy.

In chapter 6, Finn Arler takes us to another public debate. He reviews the Danish experience with the practice of creating "ecological utilization spaces." Although focused on one country's experience, the chapter addresses broader themes about the limits of claims to value free scientific assessment. Arler asks how we operationalize notions like "sustainability" so that it becomes a good tool for policy, not merely a theoretical concept. This concern mirrors our claim that theory (as in the case of restoration) should become a tool for policy, rather than remaining theory alone.

If philosophy should be part of practice, so must science be, or so argues Paul B. Thompson in chapter 7. In a way Thompson takes to an extreme the variety of positions and suggestions put forward by all the authors so far. Thompson analyzes arguments applied to the hot questions of genetically modified crops. He examines preservationist and teleological arguments as well as precautionary and risk arguments against genetically modified crops, and argues that in an agricultural context, these arguments do not provide compelling reasons to oppose plant biotechnology. Whether a particular use of plant biotechnology is environmentally beneficial is largely a pragmatic and practical question for Thompson—one that, although it should be informed by philosophy, depends finally on facts and good farming judgment. Thompson is therefore in some ways the most skeptical (or critical) among this book's contributors about the use of pure theory in environmental reasoning.

We end part II with a contribution that, although joining the others in searching for new philosophical tools for environmental practice and reasoning, differs from the rest in finding this tool in basic common sense. In chapter 8, Alan Holland and John O'Neill take us to England, Wales, and Scotland to scrutinize the philosophical language that best suits environmental matters. They suggest that what they call the "old world" historical perspective has considerable explanatory power with respect to new world conservation problems, in particular to cases of conflict over diagnosing environmental loss. The authors argue that this method helps to explain the claim that we feel the natural world has on us. The natural world, like human culture, has a particular history that is part of our history and part of our context, both explaining and giving significance to our lives.

Finding the balance between intuition and theory is one pillar of our proposal for reasoning in environmental practice. The other is the important role for the analysis of case studies. Chapter 9, by Robert Hood, opens part III, with a cautious examination of the advantages and disadvantages of analyzing case studies. Hood does this by comparing environmental philosophy with medical ethics. His claim is that although studying cases may be fruitful, it is also risky. Indeed, he thinks that casuistry came to dominate medical ethics because several mistakes were made in the way cases were analyzed and later used to construct theories. This does not mean we should abandon case studies, only that we should learn from other fields of applied philosophy that have made significant use of them when we come to apply case studies in environmental reasoning.

In chapter 10, Vivian E. Thomson dives more directly into a particular case-study from the United States. She provides a pragmatist analysis of the U.S. federal decision to take lead out of gasoline. She shows that in practice, the way ethics works in the policymaking context does not really correspond to the way we often do applied ethics in an abstract fashion. This may suggest, as we saw with the restoration case, the limited utility of a philosophical method focussing exclusively on metaphysics and metaethics. Such ideas may be profound and interesting, but this eagerness to put forward such views often blinds philosophers from seeing the more pressing moral problems in the sort of cases to which their theories relate. Consistent with this concern, in chapter 11 Clare Palmer and Francis O'Gorman look at fox hunting in the United Kingdom as a case study showing that, surprisingly enough, environmental theorists who have discussed the morality of hunting have overlooked an important political issue: power relations between humans and animals. The authors attempt to reorient us toward this aspect of fox hunting in order to shed light on a set of moral dilemmas largely ignored so far.

If one thinks that the gap between traditional environmental philosophy and environmental activism is limited to the Anglo-Saxon world, chapter 12—by Niraja Gopal Jayal—proves otherwise. She argues that in the study of biodiversity in India, one sees a gap between environmental ethics and environmental practices that emanate from a variety of sources—popular practices, environmental activism, and government regulation—all of which speak quite different languages. In that sense, she confirms our suggestion of the gap between mainstream, professional environmental ethics and environmental activism.

Where will all this end up? We do not wish to devalue the importance of mainstream environmental philosophy. However, we want to suggest that there is an urgent need for a parallel track in environmental reasoning, especially when it comes to environmental practice. We therefore hope that this book will be of interest not only to academics and those doing research in environmental philosophy, but to activists, politicians, members of organizations, and all those to whom the environment is dear.

The essays in this book were originally presented at an international conference that took place at Mansfield College, Oxford, in June 1999. The meeting was the annual conference of the Society for Applied Philosophy held in conjunction with the International Society for Environmental Ethics. We would like to thank the Oxford Centre for Environment, Ethics and Society, at Mansfield College, for its warm hospitality on that occasion. Special thanks to Steve Burwood, Richard Norman, and Brenda Almond of the SAP and to Anne Maclaughlan of OCEES for helping to make the meeting a success. Many thanks to Clay Morgan, our editor at MIT Press, for encouraging us to stick to this project during the last year, despite the difficulties we faced after the tragic events of September 2001. We are also grateful for his valuable comments on this introduction.

## Notes

1. By "activists" we do not simply mean those normally thought of as activists—for example, members of nonprofit environmental organizations such as the Sierra Club, nor only radical environmental groups such as Earth First! We intend by this term to refer to members of these groups as well as those environmental protection from inside government agencies. One might rather claim that those campaigning for stronger environmental protection within governmental organizations are "advocates," rather than "activists" per se. If such a case were made, then whenever we use "activists" in this chapter we would instead insert "advocates and activists," or make a case that the term "advocates" should be used as a blanket term inclusive of those normally thought of as activists. From this perspective, environmental philosophers are most likely advocates as well. Nonetheless, the specialized language and discourse of environmental philosophers can be separated out as different from the language of most other environmental advocates, writ large, and so in this chapter, by "activists" we do not

mean to include philosophers. We include more on this topic below in dicussing Baird Callicott's claim that environmental philosophy is just another form of environmental activism.

2. There is an important distinction to be made between externalists and internalists in ethics. Externalists believe that there are objective reasons for action, not dependent on desires. No antecedent desire to follow a principle need be demonstrated. Internalists such as David Hume demurred that a complete moral theory required an account of how the principles espoused by the theory could be embraced by moral agents. For a helpful discussion see Jennifer Welchman, "The Virtues of Stewardship," *Environmental Ethics* 19, no. 4 (1999): 411–423.

3. Kate Rawles, "The Missing Shade of Green," in *Environmental Philosophy and Environmental Activism*, ed. Don E. Marietta, Jr., and Lester Embree (Lanham, MD: Rowman and Littlefield, 1995), 162.

4. Some environmental philosophers have also called for more attention to such matters. See for example Bryan Norton, "Why I Am Not a Nonanthropocentrist: Callicott and the Failure of Monistic Inherentism," *Environmental Ethics* 17, no. 4 (1995): 341–358. For Callicott's retort, especially on the claim that environmental ethics has not affected the work of activists and policymakers, see his "On Norton and the Failure of Monistic Inherentism," *Environmental Ethics* 18, no. 1 (1996): 219–221.

5. See Tim Hayward, "Anthropocentrism: A Misunderstood Problem," *Environmental Values* 6, no. 1 (1997): 51.

6. J. Baird Callicott, "Environmental Philosophy Is Environmental Activism: The Most Radical and Effective Kind," in *Environmental Philosophy and Environmental Activism*, 21.

7. Callicott, "Environmental Philosophy Is Environmental Activism," 22; our emphasis.

8. Callicott, "Environmental Philosophy Is Environmental Activism," 22.

9. We believe that it is a bit of both. On the one hand, Callicott and many other nonanthropocentrists seem more than content to retreat only into philosophical tests of the strength of their arguments because in part they may think these are the only kinds of arguments that really are ethical in the environmental arena. Anthropocentric and instrumental arguments are prudential and not moral and so have a different set of expectations attached to them. On the other hand, though, Callicott does at times try to point to the power of nonanthropocentric moral philosophy in the development of other areas of environmental inquiry and activism. In several articles he points to the influence of nonanthropocentric environmental thought on the development of the field of conservation biology. Also, in the "Environmental Philosophy Is Environmental Activism" article he argues that Dave Foreman and Earth First! were strongly influenced by academic philosophy with reference to Foreman's own writings on the topic (p. 33). In both of these cases, though, two points are critical: First, the strongest influence on both of these groups was not environmental philosophy per se but deep

ecology in particular. Second, that nonanthropocentric environmental ethicists have influenced either of these groups is no surprise whatsoever and insufficient as a test of the persuasive capacity of such views. With respect to the first point, there are very big differences between the development of deep ecology (which has always been aimed at more practical action through its emphasis on the acceptance of a robust practical pluralism) on the one hand and a more purely philosophical and admittedly monistic view such as Callicott's on the other. On the second point, the goal of an effective environmental philosophy is not only to help persuade like-minded nonanthropocentrists of the relevance of philosophical thought to their work (such as deep ecologists and conservation biologists—the latter a field of science founded expressly on normative environmental assumptions), but to persuade broader audiences who do not count themselves as nonanthropocentrists—that is, most of the people in actual decision-making positions with respect to environmental priorities. In short, we cannot rest by only preaching to the choir. For a comparison of Callicott's monism with the practical pluralism of deep ecology see Andrew Light, "The Case for a Practical Pluralism," in *Environmental Ethics: An Anthology*, ed. Andrew Light and Holmes Rolston III (Cambridge, MA: Blackwell, 2002), 229–247.

10. Andrew Dobson, *Green Political Thought* (London: HarperCollins Academic, 1990), 20.

11. Kate Rawles rightly comments as well that we ought to worry whether a change in worldview really is accompanied by a change in action. In many cases this may not matter, but if philosophers believe that they are making a contribution to the resolution of environmental problems (or doing environmental activism) only by contributing to changes in worldviews, they need to have a good account of how this process works.

12. Rawles, "The Missing Shade of Green," 158.

13. For a very good discussion of how environmental pragmatism must widen its notion of what counts as "practical" and "relevant" environmental philosophy, see David Littlewood, "The Wilderness Years: A Critical Discussion of the Role of Prescribed Newness in Environmental Ethics," unpublished doctoral dissertation, Lancaster University (UK), 2001.

14. For example, Andrew Light and Eric Katz suggest that environmental philosophers embracing some form of pragmatism can be engaged in several philosophical activities relevant to policy, among them "the articulation of practical strategies for bridging gaps between environmental theorists, policy analysts, activists, and the public" and "theoretical investigations into the overlapping normative bases of specific environmental organizations and movements, for the purposes of providing grounds for the convergence of activists on policy choices" and so on (Andrew Light and Eric Katz, "Environmental Pragmatism and Environmental Ethics as Contested Terrain," in *Environmental Pragmatism*, ed. Andrew Light and Eric Katz (London: Routledge, 1996), 5). See also Light, "Taking Environmental Ethics Public," in *Environmental Ethics: What Really Matters? What Really Works?*, ed. David Schmidtz and Elizabeth Willott (Oxford: Oxford University Press, 2001), 556–566.

15. There are some noteworthy exceptions that should not be overlooked, although often, and not surprisingly, they come from theorists who for one reason or another also work in strongly interdisciplinary environments (such as Bryan Norton, a philosopher who holds a position in a School of Public Policy). An impressive example, even though we do not concur with all of her theoretical positions, is found in Greta Gaard's *Ecological Politics: Ecofeminists and the Greens* (Philadelphia: Temple University Press, 1998). Gaard's book is a masterful example of how to do environmental political philosophy with a broader set of texts than most philosophers would ever use, including her own personal interviews with activists. She does not simply pepper her theoretical narrative with a few examples. The raw material of her work is the public record of how activists have shaped their environmental priorities.

16. This notion was suggested in Avner de-Shalit, *The Environment: Between Theory and Practice* (Oxford: Oxford University Press, 2000), 28–36.

17. Norman Daniels, *Justice and Justification* (Cambridge: Cambridge University Press, 1996), 2.

18. Michael Walzer, *Interpretation and Social Criticism* (Cambridge, MA: Harvard University Press, 1987).

19. Thomas Nagel, *The Last Word* (Oxford: Oxford University Press, 1997), 5.

20. See Bryan Norton and Bruce Hannon, "Environmental Values: A Place-Based Approach," *Environmental Ethics* 19, no. 2 (1997): 227–245.

21. See Walzer, *Interpretation and Social Criticism*, 1–33.

22. Michael Walzer, *Thick and Thin: Moral Argument at Home and Abroad* (South Bend, IN: University of Notre Dame Press, 1994), 52–53.

23. This is not to say that no attention whatsoever has been paid to relationships between environmental ethics and environmental policy or activism. Quite the contrary. But we believe that much more can be done in this direction and that the work that has been done was more top-down and insufficiently useful for forming more coherent environmental policies. An example might be Irene Klaver's essay, "The Implicit Practice of Environmental Philosophy," also in the Marietta and Embree volume. Though an admirable and fascinating excursus on the problems with separating theory and practice, going back to Heraclitus, the discussion is one that only other philosophers (and only a fairly small subset of philosophers) could even begin to engage with.

24. Eric Katz, *Nature as Subject: Human Obligation and Natural Community* (Lanham, MD: Rowman and Littlefield, 1997); Robert Elliot, "Faking Nature," *Inquiry* 25 (1982): 81–93, and more recently, Elliot's *Faking Nature* (London: Routledge, 1997). For exceptions to this view that have garnered much less attention by the restoration community, see Alastair Gunn, "The Restoration of Species and Natural Environments," *Environmental Ethics* 13, no. 3 (1991): 291–309; Holmes Rolston III, *Conserving Natural Value* (New York: Columbia University Press, 1994); Donald Scherer, "Evolution, Human Living, and the Practice of Ecological Restoration," *Environmental Ethics* 17, no. 3 (1995): 359–380; William Throop, "The Rationale for Environmental Restoration,"

in *The Ecological Community*, ed. Roger Gottlieb (London: Routledge, 1997), 39–55.

25. William K. Stevens, *Miracle under the Oaks* (New York: Pocket Books, 1995).

26. See Irene Miles, William C. Sullivan, and Frances E, Kuo, "Psychological Benefits of Volunteering for Restoration Projects," *Ecological Restoration* 18, no. 4 (2000): 218–227.

27. Two points are important to note. First, we are not of course offering here the full argument to support the positive view of the value of restoration in terms of its value to help restore the human relationship with nature. For this case see Andrew Light, "Ecological Restoration and the Culture of Nature: A Pragmatic Perspective," in *Restoring Nature: Perspectives from the Social Sciences and Humanities*, ed. Paul Gobster and Bruce Hull (Washington, DC: Island Press, 2000), 49–70; and Andrew Light, "Restoration, the Value of Participation, and the Risks of Professionalization," also in *Restoring Nature*, 163–184. Light makes a strong case for linking the arguments offered in these articles for the positive democratic potential in restoration to the literature on civic republicanism in "Restoring Ecological Citizenship," in *Democracy and the Claims of Nature*, ed. Ben Minteer and Bob Pepperman-Taylor (Lanham, MD: Rowman and Littlefield, 2002), 153–172. Second, an environmental ethicist seeking to make such a positive case for the value of restoration (or some other positive value) ought also to be able to answer the Katz/Elliot brand of philosophical criticisms, which are certainly more extensive than we have had the space to present here. For Light's answer to Katz's views see "Ecological Restoration and the Culture of Nature," just cited. For his answers to Elliot see Light, " 'Faking Nature' Revisited," forthcoming in *The Beauty around Us: Environmental Aesthetics in the Scenic Landscape and Beyond*, ed. Diane Michelfelder and Bill Wilcox (Albany, NY: SUNY Press, 2003).

28. For details on the Chicago restoration controversy, see the introduction to Gobster and Hull, *Restoring Nature*.

29. See David Newton, *Environmental Justice* (Santa Barbara, CA: ABC-Clio, 1996), 23.

# I

# Political Theory and Environmental Practice

# 1

# Political Theory and the Environment: Nurturing a Sustainable Relationship

Michael Freeden

## New Maps for Old?

Consider sitting on a tree. Every year in Oxford hundreds of human beings sit on trees. Most of them are children, often in their back gardens, scrambling over branches, hiding in their tree houses. Some are adults, out for a walk, looking for a view or a place to rest for a while when the ground is wet. Sitting on trees is a recreational activity, and has been so since time immemorial. Not long ago, one group of adults chose to sit on trees on the site of the Oxford Business School to be. Was that a recreational activity? I doubt it. The act was the same, but the human behavior around it was far from routine. So how do we make sense of such acts? We need to fold up the map we usually use when stumbling across people in trees, and obtain a new one. Political maps interpret practices, which are recurring acts (whether deliberate or not) engaged in by groups of people. If sitting on trees in anger is a one-off event, we may deem it insufficiently significant to map. Yet in recent years the innocuous activity of sitting on trees has been redefined as a saliently political practice, a protest. True, Robin Hood's merry men sat on trees in preparation for ambushing the Sheriff of Nottingham's cohorts. But changing military technologies have marginalized that particular tactic. Now once again we require a new theoretical map in order to realize that, rather suddenly, a set of observable actions no longer inhabits the semantic field in which we have been accustomed to find it. Indeed, even if we define sitting on a tree as a political act, further, competing maps may be necessary to decide what *kind* of political act it is. Political maps are never just descriptive of a terrain, but interpretative and organizational; they are themselves a form of scholarly as well as imaginative creativity. We can no longer rest content with Kant, who maintained that "not all activities are called *practice*, but only those realizations of a particular purpose which are considered to comply with certain generally conceived principles of procedure."[1] Practices, to the contrary, may embody and generate theory.

Some of the ecological protesters claimed to act as guardians of those trees' right to life, perhaps even—if this is not overstating the case—to dignified life. This argument by analogy sounds more familiar to advocates of natural law, as well as to procedural rights theorists. According to that argument the tree-sitters were, at different levels of articulation, superimposing a set of fundamental philosophical beliefs drawn from current debates onto their actions and the objects of their actions. They were conferring on trees the status of honorary persons, with the entitlements to respect and consideration that personhood entails. And they were justifying civil disobedience in the name of those fundamental principles, because a moral principle concerning one's obligations toward trees had been violated, and civil disobedience is legitimated whenever a higher moral principle is disregarded by a particular political authority. But in the eyes of other observers, the practice in which they were engaging was one of obstruction. These were tiresome faddists who were flouting the law, at great expense to the welfare of the community, who would therefore have to divert resources from other crucial social objectives in order to remove the protesters.

The map of civil disobedience, however, whether such disobedience is justifiable or not, is a poor one with which to explore the contours of the practice. If it is the only map in town, we may find it difficult to make our way through. The protesters were hugging trees in communion with nature, interacting noninstrumentally with nonhuman objects. Conventional political theory had not adequately prepared us for that. In the realm of civil disobedience, the semantic field has been dominated by concepts such as promising, voluntarism, and consent.[2] But politics is also an arena where strong emotion is expressed, and where group dynamics play a central role. These too need to be incorporated into political theory. One reason they have been excluded from the purview of many political theorists is that hegemonic models of Anglo-American philosophy have become dismissive of nonrational accounts of human conduct. Such accounts, it is asserted, are irrelevant to constructing the versions of best practice, buttressed by requirements of coherence, that the prevalent notion of the rational, purposive agent requires. Another reason is that manifestations of nonrational theory in the experience of the past century have all too frequently been unpalatable. Theorizing about the environment is thus confronted with the choice of extending the category of persons to cover nonhuman objects and consequently incorporating a new domain into mainstream political theory, or seeking alternative conceptual frameworks through which to bestow meaning on its particular foci.

The upshot of this dilemma is to highlight that, if we wish to optimize our understanding of the sociopolitical worlds and the environments in which we are located, political thought needs to take on board a multiplicity of interpretative viewpoints, *many* of which are reasonable, even if not all are equally persuasive. This may upset philosophers looking for consistency, for neatness, and above all, for decisive cogent arguments by means of which to recommend the adoption of certain political practices rather than others. Political thought is not shaped in the closed laboratory of our minds, but is a result of reflecting, in our various and strange ways, on experiences we have undergone, directly or vicariously. Environmental theory and green political thought are, in their current manifestations, relatively new systems of ideas. Unsurprisingly, many of their premises reflect philosophical ideas dominant in their period of inception and early growth. Yet a sustainable political theory needs to follow its own maxims in relation to the methodologies it employs, namely, to engage in a continuously critical assessment of those dominant paradigms both from within and from without. A sustainable *green* political theory needs to identify central values and concepts in the service of the green family of arguments and to decide which theories and methodologies advance them. Pretheories always contain methodological and ideological preferences. There is no view from nowhere because none of us knows where nowhere is.

## Wholes, Links, and Patterns

There is, I believe, an interesting interface between political theory and environmental thought. It is common among recent political philosophers to argue that justice is the first virtue of the state, but it is rare for them to add that well-being is the first virtue of the community. Neither of these statements is incontestable, but they serve as good starting points for broadening our view of political theory. I submit that these statements run in parallel, that to assume societies need to choose between them (on the lines of liberal right versus communitarian good) is a methodological misconception, and that green political thought is particularly well placed to utilize both standpoints. To begin with, green political thought does not have the dual boundary problems constraining much political theory. It does not focus solely on human beings as possessing the only attributes that political theory should consider, nor does it focus solely on political space as constituted by the borders of the nation-state (it has of course other boundary problems, of which more later).

The *substantive* reason for the linking of state and community is the interest green political theorists have in groups, wholes, collectives, organisms, as well as biodiversity and sociodiversity, so that—as political sociologists know—the state is merely one aspect of a society, and—as feminists know—the political cannot be encompassed in the activities of the state alone. The *methodological* reason for linking the two is crucial to the very nature of political theory itself. Let me put this in a grossly oversimplified way to make the point. Anglo-American moral philosophy is profoundly inspired by the power of logic, and its typical manner of arguing is to validate or invalidate particular statements in terms of their consistency with a foundational position, itself formulated in order to promote optimally a moral maxim. Its grammar is cascaded in terms of logical chains: it moves from A to B to C. Thus *if* we wish to pursue justice, *then* we will need veils of ignorance and original positions, and *then* we will arrive at certain distributive rules, and so on. Much of the debate is immanent: Is the path followed coherent, can it be assailed by alternative cascades, and—far more problematic for those who insist that there exists clear blue water between philosophy and ideology—does it conform to our moral intuitions?

But political theory is also constituted by its units, political concepts; and political concepts are signified by words; and words are components of language. And language is a structure of interdependency. Words and the concepts they carry only make sense in complex clusters; by rejigging the pattern of each cluster, we create different messages and apply different meanings to sentences and passages. Political arguments are always configurations of concepts, and competitions over which configuration attains philosophical, or political, or cultural legitimacy.[3] They cannot make sense as isolated concepts linked in linear sequences, unless they serve certain thought experiments, the purpose of which is to test an argument to the point of destruction under insulated "laboratory" conditions. It all depends on what we want political theory to do for us. And it is here that we can learn from the study of ideology, inspired by green concerns. I will table some suggestions in the form of a number of propositions.

**Proposition No. 1**

The first proposition emphasizes the integrity of the environment; it holds that removing political concepts from their natural habitat, from the idea environment of other concepts in which they are located in actual language usage, is a cruel and often unnecessary act. The battery farming of concepts, held in relative isolation and often worked to death in order to gratify the intuitions of human experimenters, should rather proceed with sensitivity and under carefully regulated circumstances, at the end of which the concept should be released back into the sustaining community of language.

Green political thought provides a sympathetic framework for regarding political theory as holistically complex. It also offers an opening for a more elaborate conception of community than the one currently employed by many political philosophers. In general terms, the methodological individualism still predominant in Anglo-American political philosophy might have been more generous in making space for explicit notions of group identity and interests implicit in environmental views. Indeed, political concepts are themselves interconnected, in that their separate meanings are always formed through their interaction with, and proximity to, other concepts. Specifically, community in the green context may be conceptually proximate to a strict species egalitarianism, in which human beings are subsumed within a totalizing view of nature that refuses to accord them special status.[4] Or it may be conceptually proximate to a decentralized bioregionalism that allows for a multitude of distinct and local communities, rather than a monolithic version of community.[5] Thus to subscribe to the interconnectedness of concepts is not to support a single unified view of political theory. Many holisms can be applied to make sense of one series of facts and phenomena; hence there are also many alternative harmonies, not just one. Each holism is constructed through a different internal configuration of the field of concepts it addresses, a different mapping of the same evidence. Some may offer solutions inimical to individual rights and autonomy, but that is certainly not a corollary of organicism per se, as anyone acquainted with social liberal thinking in the early twentieth century knows.

## Proposition No. 2

This proposition involves the biodiversity, or rather morphodiversity, of political concepts. It emphasizes that political concepts exist in multiple as well as indefinite forms, conceptualizations, and configurations. Political theory centrally embraces a study of the complexity of political argument, its evolution, its internal morphology, and the multiple meanings it accrues. Singling out some concepts and arguments as superior—a practice engaged in by many moral philosophers—is a form of discrimination. Let's call it *conceptism*. It has its obvious uses, no doubt, for we wish to make moral judgments, develop our virtues, and improve the world. But it excludes much human thought creation and is often unreasonably disrespectful of the variety and the quality of other conceptual arrangements. Conceptism may even, deliberately or unwittingly, modify the evolutionary process through which unexpected new combinations of political concepts are created and from which future generations may benefit. Moreover, it is quite unjustifiably contemptuous of ideologies, which are vital and necessary communal resources at the disposal of a society without which political decisions cannot be taken.[6] Political ideologies comprise the actual political conceptual arrangements that people, philosophers included, inevitably construct when they map the social world. Anyone who thinks about politics has to select some meanings rather than others from the indeterminate and contested universe of the conceptions a concept embraces. What some philosophers prefer to call moral intuitions, analysts of ideologies may call nonnegotiable, decontested, core beliefs on which a political *Weltanschauung* is anchored. And they may then arrive at the shocking conclusion that more than one moral intuition within the same ethic, or within the same semantic field, may be valid—that is, that political concepts lend themselves to multiple combinations, and that a convincing case can be made for quite a few of those combinations. Even when such ideologies are not at their qualitative best (and most ideological families do display great sophistication in the structure of their conceptual arrangements), that dimension of political thought provides a rich source material from which to distill meaning, to understand the consequences of certain conceptual decontestations that, though logically arbitrary, have crucial cultural significance. If ideologies tend to stress difference, their *analysis* emphasizes connections and sensitizes us to conceptual structure.

Political philosophers who are rude about ideologies and denounce them as bad political theory often fall into a confusion about what we should be studying as scholars. That happens because there is no noticeable disjuncture between the methods of the political or moral philosopher as scholar and the nature of the subject matter she or he explores. The moral philosopher as scholar is indistinguishable from the moral philosophers to which he or she refers. Both present and past philosophers are engaged in a joint discourse, namely, that of morally philosophizing. They often overlook that, by contrast, there is a considerable difference between the *formulators* of ideologies and the *students* of ideologies. The former focus—as do moral philosophers—on the politically desirable, but their justifications rest on widely diverse bases of truth and validity. Their thought practices do not necessarily occupy the same, or even a broadly common, discursive space. They may range from the largely rational to the largely irrational, from the absolutist to the relativist, from the assertive to the tentative. Students of ideologies, exploring as they do ideologies and political language as a thought practice, possess a different starting point. They are concerned neither with advocacy nor with improving the practices and techniques of ideologists. Their interpretations may vary from the functional to the contextual or the morphological. The consequence is the production of a different kind of theorizing about ideologies, which does not participate in the same thought practices it analyses.[7]

Why, then, study "bad" political theory? First, because much of it is not bad at all if we are prepared to be more tolerant about our criteria. Most political thought falls far short of the technical, persuasive, and justificatory standards that philosophers expect it to attain. But that does not make it less significant. Had Rousseau now offered *The Social Contract* to Oxford University Press, it would have been returned post-haste with red pencil marks on every page, and with the following admonition: "The definition of the general will that you employ in book 2, chapter 3 is inconsistent with that in book 4, chapter 2. Clean up your act and resubmit." We might have lost a book that is thought provoking, imaginative, controversial, influential, while indifferent in the technical quality of many of the arguments it marshals.

**Proposition No. 3**

According to this proposition, thinking globally is always a form of thinking locally. Local varieties of political theory require protection from standardizers; indeed, the habitat of political theory is always local. Political theory, like green practice, is invariably the view from somewhere. Here we find both similarities and divergences with respect to green political theory. Its very focus on globalism lends itself to universalism and the prospect of single overarching solutions. But that ignores the fact that the possibility of multiple holisms undermines universality. Put differently, every proposal for a holistic viewpoint constitutes a particular reading of universality. On the other hand, one of the great attractions of environmentalism is its sensitivity to the concrete and to the particular, through its emphasis on action and on practices. The notion of practice is elevated, in similar although certainly not identical fashion to its role in Marxist thought, to the status of a core green principle.[8] Recent trends in political thought argue that the divide between theory and practice is far from clear, that practices are theory rich and decodable in many different ways.

This does not mean that generalizations are impossible to formulate. First, some generalization is of course advantageous, as long as we acknowledge its possible contingency, awaiting contrary argument. Second, some strains of political theory itself may spread uncontrollably, rampantly supplanting other local varieties, as Rawlsian political philosophy has done from its Eastern Seaboard roots, supported not merely by its powerful intellectual appeal but by the capitalist power of American publishing and the Ph.D. factory. The result has been a genetically modified liberalism, in which a spurious neutrality creates the illusion of a suprapolitical stance, though some Rawlsian supporters insist that this fortifies liberalism's immunity to infection by particular conceptions of the good. The losses to green theory are sometimes considerable: Even an eminent and subtle ecotheorist such as Eckersley has attributed a historically quite misleading attribute to liberalism—the autonomy of atomistic individualists, which is at best a mid-nineteenth-century feature of liberalism—and ignored others such as tolerance and the recognition of community rights, both friendly to environmentalism.[9] She has fallen into the common trap of ideologically misrepresenting an ostensible opponent's views and thus excluding them from one's semantic field at considerable theoretical cost.

**Proposition No. 4**

One of the lessons of the above for green political theory is the precautionary principle. According to the fourth proposition, excessive risk aversion in political theory should not crowd out risk taking. Methods adopted by political philosophers and by analysts of ideology often employ parallel perspectives on the same subject matter, and the enthusiasm of either needs to be contained by the legitimate concerns of the other. Some interventions in political thought may be irreversible; some ideas may be irredeemably contaminated and lost to posterity. But some of these losses may be desirable—a view that, incidentally, few environmentalists endorse when it comes to biodiversity, yet is the entire environment equally deserving of blanket preservation just because it is there? Death and decay are just as natural as life and growth. A young discipline is an experimental one and therefore rightly risk-prone, but it should not be risk-averse. Green political theory has developed by provisionally positing certain goods and exploring value systems that may secure them. This trial-and-error process is attractive from the standpoint of a political theory that endorses conceptual polysemy and plural paths of valid or justifiable argumentation and that eschews homogeneous universalism.

The problems of universalization for green political thought are of course those of holding time and space constant. Theories of justice, some of which rely heavily on universalization and immunity to change, run immediately into difficulties when made to intersect with theories whose concerns are grounded in an appreciation of the local, the diverse, and the evolving. All ideologies harbor theories of time and change, and all serious theories of ideology respect the contingency of conceptual permutations. Would a space- and time-sensitive theory be more applicable to green perspectives? Global issues notwithstanding, and recognizing the argument for some generalized obligations, we also need to assess the claims of a political theory that argues the parallel case for our responsibilities for the near over the distant, in both space and time. Not all forms of obligation need to be informed by a strict egalitarianism in terms of its objects—that is, allocated to all potential beneficiaries. In social space the prioritization of the near is commonly practiced, and ethically justifiable, with respect to our attitudes toward the family, as well as toward our interpretations of ethnicity, multiculturalism, and even benevolent forms of nationalism. These put a premium on nonrational aspects of human interaction—bonding, sentiments, pride in group belonging, care—all of which have been underconceptualized in recent political theory, if not eliminated altogether (outside feminism and the revival of interest in nationhood and nationalism). Localized environments and concrete communities are the arenas of these features. Dare we even suggest that localism, or finiteness, in time are equally comprehensible, so that some of our responsibilities to the near future may plausibly be stronger than to the distant? One asset of green political thought is its receptiveness to notions of growth, change, and evolution, which allow for graduated conceptions of time and space, unavailable, say, in the static character of natural rights theory. If green political thought continues to conceptualize such questions in terms of boundary problems and clear-cut categories, as some political philosophers are inclined to do, it may not be able to resolve this kind of issue. Instead, it should begin to build on its own strengths, thus assisting in a broader invigoration of current political theory.

In that light we can begin to analyze green political thought as a particular family of manifestations of thinking about politics. There is no correct green political thought, nor is there a single green political theory, nor *can* there be without totalitarian coercion, nor should there be. That is not to argue that all forms of environmental debate are equally attractive, persuasive, or reasonable. But, like all ideologies, the green family is a multiverse. It is pluralist not only in intent and in its conception of the world, but in the very structure of the political thought patterns and practices it engenders. This leads to the next proposition.

**Proposition No. 5**

The final proposition maintains that human beings must abandon their aspirations to control optimally their cultural as well as physical environments. Green political theory contains the perceptive implication that, all too often, excessive control has serious downsides in its invasiveness and its intellectualization of environmental concerns. Three points follow. First, an agency cum autonomy cum purposive model of human nature is ultimately grounded in the notion of reflective self-control and self-criticism. It is echoed in many philosophers' attempts to control language through logic and persuasive argument. However, Paul Ricoeur's notion of the "surplus of meaning" enables us to pursue alternative routes in exploring political and social thought.[10] The deliberate meanings we intend to convey are always accompanied by unintentional ones; hence, the analysis of political messages imparted by political actors cannot be reduced to examining the purposes of those messages alone. The inevitable absence of complete control over meaning is precisely what a study of ideologies reveals; it opens up a different role for the scholar as one who accepts linguistic usage and linguistic communities as given, though modifiable, rather than threatening or inadequate, just as by analogy ecologists urge us to accept nature and to adjust it in a sustainable way. If we want, contra Marx, to change the world through reinterpreting it, we must, as professional thinkers, descend from the Olympus of a specialized moral language and work with our subject matter, human beings, using their ordinary tools of thought and language as well as ours in our professional capacity. Environmental thinking offers us a good example here.

Second, through the very fact that green political thought, as a family, deals with unreflective as well as with reflective entities, it is well placed to distance itself from the proclivity of some political theorists to dismiss the unreflective altogether. And green political thought, because it crosses the boundary between the human and the nonhuman, and locates human beings in nature, is well placed to appreciate the physical and emotional attributes of human beings. True, these attributes cannot be treated in the same way as the intellectual and the moral attributes, but they are nonetheless ineliminable features of being human and acting as human beings, even though we share some of those features with animals.

Third, we too are passively conditioned by the manifold communities of which we are members, side by side with being purposive agents. No amount of ideal-type liberal theory can wish that away, nor *reduce* us to our capacity for agency alone. Here is where well-being joins autonomy as a twin value of political theory, but one that cannot be catered to entirely through procedural theories of justice designed to maximize our capacity for informed and considered choice. We do not have to choose between humans as moral agents and as possessing other human attributes. The merit—for green political thought—of configurational conceptual analysis, or the appreciation of ideological morphology, lies in its mirroring of the environment's holism and antidualism. As noted above, it rejects the cascades of lexical and logical priority and replaces them with a cluster notion of conceptual interdependence. Indeed, this is also the case with the five propositions offered above. They are not sequential but mutually supportive. You can work your way from any one to the others, much as with Mill's notion of well-being, consisting of the "free development of individuality."[11]

## A Provisional Balance Sheet

Some philosophers are rightly exercised about *resolving* tensions within arguments. Ideologies, to the contrary, suggest alternative routes of *reasonably avoiding* tensions—and avoid them one must if political decisions are to be taken. Otherwise, we get trapped in incommensurabilities or indeterminacy as an argumentational chain unfolds. Studying ideologies helps us to be more modest about our intellectual ambitions and to resign ourselves to the necessity of making path choices on the basis of cultural constraints, emotional attractiveness, as well as conformity to current scholarly practices.

Green ideology in its various manifestations is, however, thin-centered. It lacks adequate conceptual complexity to address the range of issues that mainstream ideologies have addressed over time. As an ideological family, it does not have a full morphology, containing particular interpretations and configurations of all the major political concepts attached to a general plan of public policy that a specific society requires. Unlike liberalism or socialism, its core concepts do not have sufficient pull to constrain the structure of its adjacent concepts. The relation of human beings to nature, the preservation of the integrity of nature, and an appeal to holism are shared by all its variants and hence constitute the core of its actual usage of language, but they are insufficient to prevent loose centrifugal arrangements.[12] Some of its arguments are subsumed in other ideological groupings: ecosocialism, anarchism, or conservatism. Much of its discourse decenters human action and the overriding value, evident in other ideologies, that is attached to human societies. It needs therefore to borrow from other schemes of distributive justice, nor does it have a distinctive approach even to democracy. Flying the colors of a progressive ideology, it nevertheless cannot propose a unique decontestation of concepts central to progressive debate: liberty, equality, or rationality. Does that make it a postmodernist receptacle for all values and none? Far from it. Green thought is imbued with a strong sense of desirable values, among which of course justice is important, but it mirrors the diversity and the interconnectedness of the world it is designed to preserve, and it identifies human beings as possessing attributes irreducible to their mental and moral faculties. For that reason, a political theory tailored to cope with interwoven variance, in method as well as substance, is likely to be a useful tool at its disposal. For political theorists as students of ideology, sitting on trees is more than an expression of moral agency; it offers the elevated vista of multiple routes through the world.

## Notes

1. Immanuel Kant, "On the Common Saying: 'This May Be True in Theory, But It Does Not Apply in Practice,' in *Kant: Political Writings*, ed. H. Reiss (Cambridge: Cambridge University Press, 1991), 61.

2. See, for example, Hannah Pitkin, "Obligation and Consent," in *Philosophy, Politics and Society*, 4th series, ed. Peter Laslett, W. G. Runciman, and Quentin Skinner (Oxford: Blackwell, 1972), 45–85.

3. See Michael Freeden, *Ideologies and Political Theory: A Conceptual Approach* (Oxford: Clarendon Press, 1996), esp. chap. 2.

4. See, for example, Warwick Fox, *Toward a Transpersonal Ecology* (Totnes: Green Books, 1995).

5. For a critical assessment of bioregionalism see John Barry, *Rethinking Green Politics* (London: Sage, 1999), 81–90.

6. See Michael Freeden, "Ideologies as Communal Resources," *Journal of Political Ideologies* 4 (1999): 411–417.

7. For an elaboration of this argument see Michael Freeden, "Practising Ideology and Ideological Practices," *Political Studies* 48 (2000): 302–322.

8. See Freeden, *Ideologies and Political Theory*, 527.

9. Robyn Eckersley, "Greening Liberal Democracy: The Rights Discourse Revisited," in *Democracy and Green Political Thought: Sustainability, Rights and Citizenship*, ed. Brian Docherty and Marius de Geus (London: Routledge, 1996), 212–236.

10. Paul Ricoeur, *Interpretation Theory: Discourse and the Surplus of Meaning* (Fort Worth: Texas Christian University Press, 1976), 55.

11. J. S. Mill, *On Liberty* (London: Dent, [1859] 1910), 115.

12. See Freeden, *Ideologies and Political Theory*, 485–487.

# 2

# Intuition, Reason, and Environmental Argument

Mathew Humphrey

Even a cursory examination of the environmental ethics literature would reveal a wide variety of answers to the question "Why should we preserve natural objects?" Rolston tells us that there are intrinsic values in nature,[1] Paul Taylor that we should have respect for nature,[2] and de-Shalit that we have obligations to posterity grounded in community.[3] Alternatively deep ecologists tell us we need nature in order to fully Self-realize (Naess, Fox),[4] and Goodin[5] or O'Neill[6]—each in their own way—that human beings require a recognizably natural context if they are to flourish. The list could be extended, but this is sufficient to give an indication of the wide variety of reasons environmental literature offers for adopting nature preservation.

In this chapter, I want to ask certain questions of one set of reasons for nature preservation, concerning how we should judge or evaluate them *as* reasons, and moreover as reasons sufficiently strong to justify appropriate action in the environmental policy sphere. First, why should we consider the reasons offered to be valid and compelling reasons for action?[7] Second, by what system of ethical or other principles can we prioritize these preservationist desiderata as against other demands for resource use in society?

These two questions are clearly related. To the extent that we find a reason for environmental action compelling, the greater the weight we will presumably be prepared to give it with respect to other possible uses of the same resources. The questions at hand, then, relate not so much to the abilities of environmental philosophers to offer reasons for the preservation of (some) natural objects in the world when no very obvious immediate human interest is at stake—as we have seen there are a wide variety of arguments to that effect in play. It is rather to ask how those arguments should be judged—first in terms of their epistemological aspects (e.g., how can we *know* that nature has value, or that we owe obligations to future generations?), and second in their potential priority against other distributive principles that are taken to be valid and to some extent compelling. This latter question remains important as long as choices between environmental and other public policy desiderata have to be made. In some (even many) cases it may be that human welfare objectives and preservationist objectives coincide (e.g., there may be more wealth created by whale watching than whale killing). It would, however, be complacency of the highest order to assume that such congruence was to be relied on. As long as choices have to be made between environmental and other valued policy outcomes, each will somehow have to be measured against the other, and the question of whether any general principles can be employed for such comparisons remains a live one.

Questioning the validity of reasons raises questions covering a broad range of disciplines and topics—epistemology, ontology, the philosophy of science, the philosophy of social science, subjectivity, objectivity, foundationalism, and so on. This chapter focuses on the specific field of biocentric environmental political philosophy, which has been very influential in the development of ecological philosophy. How have thinkers in this area sought to deal with the question of validity in the process of developing their own accounts of justifying preservationist policies? Reading the literature with this question in mind highlights an interesting epistemological dichotomy, which I believe merits attention because it can shed light on a particular source of confusion in environmental philosophy. If this confusion can be eliminated, we can bring a welcome degree of clarity to the way environmental philosophy approaches these problems—which is not to claim that these problems necessarily become any easier to solve.

## Intuition in Environmental Philosophy

A significant amount of literature in environmental philosophy seeks to establish its epistemological claims through an appeal to *intuition*, which might refer to a sensory faculty, a mode of hypothesis generation, knowledge gained without the use of conscious reasoning, or a moral belief based on a shared moral culture but not yet fully thought through.[8] I will go on to argue that the conception of intuition at the heart of the epistemology of this type of environmental philosophy is unclear, and to the extent that it is clear, it is unhelpful. The largest claim made for intuition in this field is that *all* fundamental knowledge about the world is intuitive.[9] "Fundamental" knowledge here is knowledge about the order of the natural world. Intuition is thus decontested as a cross-cultural ability to discern important truths about the natural world.

At the opposite end of the epistemological spectrum comes the view that intuition is a completely unreliable instrument for validating normative judgments about the relationship between human beings and the nonhuman natural world: "We cannot use our own, or anyone else's moral intuitions as grounds for accepting or rejecting a theory of environmental ethics."[10] On this view, intuitions are culturally specific, unreasoned feelings about the world that arise from a process of socialization into one's community from an early age. They possess no status independently of that social setting and do no more than reflect a set of existing attitudes regarding, in this case, the correct relationship between human beings and their environment. The grounds for rational acceptability (something that is itself interpreted in different ways by different writers) are here seen as the only acceptable criterion for judging an environmental ethic as valid.

For a significant number of biocentric environmental theorists, the answer to the question "Why should I believe that the ethical status of the natural world is sufficient for us to have good reason to preserve it, as you suggest?" revolves around intuition.[11] The theory we are offered accords with the theorist's deepest intuitions about the subject at hand, and even if it is not conclusive, that is supposed to at least give us good reason to accept their justifications for a biocentric approach. One putative advantage of such a strategy is that no intellectual opponent is likely to disabuse you of the idea that you have a particular intuition. I might try to show you that your intuition will lead to undesirable consequences if enacted, or that its enactment would conflict with other deeply held intuitions that you claim to cleave to, but I am unlikely to convince you that you are in error with respect to holding the intuition itself.

However, such subjective epistemological certitude is bought at the cost of an obvious set of accompanying problems, one of which is the relatively weak possibility, on these grounds, for the discursive persuasion of others. Suppose I intuit differently from you? Your intuition is that the ancient woodland near our home has almost immeasurable intrinsic value, mine is that land only becomes valuable when it is used to directly further human ends, and thus a farm would be a far more valuable use of the land. There we stand, our intuitions about this matter at loggerheads. How do we proceed from here? We could question the bases of each other's intuitions. You might claim that mine arises from being socialized in an anthropocentric, materialist culture, which fails to appreciate the value of natural objects. I might respond that your intuition stems from your growing up in an urban setting with a strongly romanticized image of nature. Such divergent stances would seem to offer support for Taylor's contention that intuitions are culturally specific and not to be trusted. Yet perhaps there is more to the biocentric environmental philosophers' use of intuition than this simple example suggests?

The most explicitly intuitionist environmental philosophers tend to be those of the deep ecology school, such as Arne Naess, David Rothenberg, Warwick Fox, Bill Devall, George Sessions, and Edward Goldsmith. In this literature, *intuition* refers both to an epistemological process and to the status of the knowledge claims made on the basis of that process. That is, *an* intuition is a knowledge claim itself based on the process *of* intuition. Thus, "There is a basic intuition in deep ecology that we have no right to destroy other beings without sufficient reason."[12] Why should anyone else accept that this intuition about the limits of legitimate human behavior is valid? Because (following Spinoza) the highest level of knowledge is direct intuitive knowledge[13] and the intuition-knowledge claims of deep ecologists are direct in that they are experiential: "Ultimate norms of deep ecology cannot be fully grasped intellectually but are ultimately experiential."[14] We will see that the notion of direct experience plays an important part in the justificatory strategy of deep ecologists with regard to their intuitive insights. The important point for our purposes is that intuition is taken here to refer not only to putative knowledge claims that have yet to be subject to the test of reason, but also to a *cognitive process* related to direct experience in a way yet to be determined. This dual role for intuition is also reflected one of Fox's early publications:

> The central intuition of deep ecology . . . is the idea that there is no firm ontological divide in the field of existence.[15]
>
> The central vision of deep ecologists *is* a matter of intuition in Naess' and Worster's sense, that is, it is a matter of trusting one's inner voice in the adoption of a value stance or a view that cannot itself be proven or disconfirmed.[16]

Thus, the idea that there is "no firm ontological divide in the field of existence" both is itself an intuition and is realized through the intuitive process of "trusting one's inner voice."

To explore this idea further, we can turn to the account of the purpose of ecosophy offered by Rothenberg in his introduction to Naess's *Ecology, Community and Lifestyle*: "The intention is to encourage readers to find ways to develop and articulate basic, common intuitions of the absolute value of nature which resonate with their own backgrounds and approaches."[17]

This statement of intent contains a number of interesting features. First, Rothenberg says that intuitions about the "absolute value of nature" are "common"—although it is not clear whether he means "common to us all" or "common to many of us." Second, these "common intuitions" are to be "developed and articulated," suggesting that they are conceived in their initial form to be underdeveloped and inarticulable. Moreover, the development and articulation are to happen in ways that "resonate" with people's own (presumably differing) backgrounds and approaches. This latter intention complements Naess's general foundational pluralism. What is important to Naess is that one ultimately comes to support something like the eight-point platform of deep ecology,[18] not the path one takes in getting there. The reference to backgrounds in particular suggests an acknowledgment not so much that intuitions vary with different cultural backgrounds (which would hardly render them "common"), but rather that their cultural source will vary. If this is so, it in turn suggests a belief that different cultural backgrounds are capable of producing common "basic" intuitions about the natural world.

All this still tells us very little about the basis and content of these intuitions, or what the modes of "development and articulation" of them may be. What "theory" of intuition—if any—underlies the idea that prerational or preverbal emotion and feeling can give access to valid information about the world? Of course the claim may not be that intuition gives access to unchallengeable "truth," but rather that it leads to a sufficiently strong presumption of knowledge that one is justified in taking particular environmental action (e.g., tree spiking or other forms of direct action).

Naess says of his work that it is "primarily intuitions."[19] Where do these intuitions begin? With immediate experience of nature. Says Rothenberg, "If we wish to identify a starting point for a system, spontaneous experience offers itself"[20]—although the system "cannot capture or replace the uniqueness of the original experience."[21] So, we can take as our starting point direct experience of the natural world, from which feelings, emotions, and intuitions about nature are developed. We should, therefore, note that Naess has his own decontestation of "experience," one that, if accepted, would succeed in giving a particular set of individuals privileged access to intuitionist knowledge of nature and the human-nature relationship. *Experience* is held by Naess to carry the correlates of activity and presence. Activity within nature, rather than, say, passive viewing, is the route to experience for Naess:

> To "only look at" nature is extremely peculiar behaviour. Experience of an environment happens by doing something in it, by living in it, meditating and acting. The very concepts of "nature" and "environment/milieu" cannot be delimited in an ecosophical fashion without reference to interactions between elements of which we partake.[22]

The "spontaneous experience" referred to earlier is, then, a product only of *inter*action with a natural environment—"living in it, meditating and acting." It is not clear whether each of these activities is individually sufficient for the relevant kind of experience, or whether they all have to be engaged in. However, a reasonably charitable interpretation would be that for both Rothenberg and Naess, the experience on which valid intuitions about the world are based can only come from time spent in direct, active interaction with the natural environment. If we do not have this interactive experience to refer to, we are not able to delimit the relevant concepts in an appropriate fashion.

In this account, if we do not have interactive experience with nature, we are disabled from doing ecological philosophy; in fact we will not have even reached the prerequisite point for beginning such an exercise. We would be disbarred on these grounds from delimiting, defining, and decontesting concepts in anything other than what would be considered an arbitrary (because unmediated by direct experience) fashion.

One aspect of this approach worth noting is that it offers a resolution to our earlier question about what way forward there might be when deeply held intuitions clash, because now the intuitions that people have do not necessarily share the same epistemological status. If experience is the starting point for valid intuition, then intuitions based on this have a claim to validity that other intuitions (such as one based on passive reflection) do not. The intuition based on experience has a claim here to be *more than merely subjective*. According to Naess,

> An attempt is made [in *Ecology, Community and Lifestyle*] to defend our spontaneous, rich, seemingly contradictory experience of nature as more than just subjective impressions. They make up the concrete contents of our world. This point of view, as every other ontology, is deeply problematic—but of great potential value for energetic environmentalism in its opposition to the contemporary near monopoly of the so-called scientific worldview.[23]

Experience-based intuitions describe to us the "concrete contents" of our world, a fact about them that certainly appears to imply a privileged epistemological status with regard to other intuitions, whatever they may be. The notion of "concrete contents" is an ingredient in Naess's long-running argument against what he sees as a devalued and debased world in the descriptions of natural science, and in particular the attribution of primary, secondary, and tertiary qualities to objects in the world. How are experience-derived intuitions related to concrete contents? Naess holds that the "concrete contents" of the world (for us) refer to constellations of relationships between us and natural objects around us. They are opposed in his ecosophy to "abstract structures," which are the conceptions of the world offered by the natural sciences in general and physics in particular. The mainstream scientific understanding confusedly mistakes the abstract for the concrete, and the identification of primary properties with real objects in the world, which "leads to a conception of *nature without any of the qualities we experience spontaneously*. Now, there is no good reason why we should not look upon such a bleak nature as only a resource."[24]

One of the tasks of environmental philosophy is to put the qualities we experience "back" into nature, by incorporating experiential elements into our descriptions of nature through the use of *gestalts*. Thus "the green tree" does not describe a tree (1) that is (2) green; instead it refers to a single, entire gestalt, "the green tree." Says Naess, "The world of concrete contents has a gestalt character, not an atomic character. I do not know of any better frame of reference than that of gestalts."[25] Of course, according to Naess, most gestalts are not perceptive, as in the example of the green tree, but *apperceptive*, in that they combine descriptive and evaluative components, based on experience, in the one gestalt. Thus a "beautiful green tree" and a "valuable green tree" are as much descriptions of reality, in terms of concrete contents, as the "green tree." Naess says that "environmentalists talk about reality as it is in fact when they talk in terms of feelings."[26]

What about those who see the world differently, who have different experiences and different intuitions about nature? I conjectured earlier that the intuitions of nonenvironmentalists would be of the "wrong sort" because they would not be based on direct experience of nature, and this seems confirmed, if somewhat obliquely, by Naess's diagnosis of a clash between conservationists and developers at the level of ideas. To the developer, the conservationists' appeals are merely subjective, and this feeling is "firmly based on his view of reality."[27] Yet to the environmentalists, "the developer seems to suffer from a kind of radical blindness."[28] Yet we have been told already that it is environmentalists who "see the world as it really is." Now this might appear a throwaway comment, meaning little more than "how it really appears to them." We have also, however, been told that only those who have active experience in nature are qualified to form an ecosophy, and that experience of nature is more than just a subjective impression. It seems, then, that it is indeed true for Naess that the developer suffers a "radical" ontological blindness, even though he sees his own understanding of the world clearly enough.

Thus to the question of whether our intuitions have epistemological validity, whether they actually provide us with any knowledge about the world, we can see that there is a theory of intuitive cognition, based on direct experience, at work in Naess's "ecosophy T." Naess has an answer to the question "Why should your intuitions about nature be more valid and carry more weight than those of a developer?" Valid intuitions are based on direct, active experience, because only direct, active experience allows us to develop meaningful apperceptive gestalts about a place, and these gestalts are what give us access to the concrete contents of the world. The intuitions of the developer or antienvironmentalists are unlikely to include these elements, and, to the extent that they do not, the developer is not qualified to develop an environmental philosophy. This still leaves open a number of difficult questions—apart from what it means to live an active life in nature, it must at least be possible that someone who has led such a life would have strong antienvironmentalist intuitions. Or, more likely, the decision to go and live an active life in nature will itself be the result of, not a cause of, a set of strong intuitions about the value and beauty of nature. The process of living in nature would thus be reinforcing, rather than transforming, a set of already extant intuitions.

A second writer who I would place broadly within the deep ecological tradition is Edward Goldsmith, who also lays great stress on the role of intuition in furnishing access to "fundamental knowledge" about the world. His arguments for this center on a coevolution hypothesis. The most explicit statement of the coevolution hypothesis and its implication for epistemology is in Goldsmith's *The Way: an Ecological Worldview*. In brief, the coevolution hypothesis is a theory about an evolutionary "fit." It claims that any supposed radical dichotomy between the human ability to know about the world, and the way the world is, ignores the common evolutionary history of the human mind and its environment. This common evolutionary history *entails* the conception of a human mind that can ascertain accurate information about its surrounding environment on the reasonable assumption that accurate information about the world increases survival chances in the long term. To borrow a phrase from Rolston, those who did not believe that *lion* referred to a real predator lurking in the grass are extinct. Goldsmith says:

> The perfection of man's cognitive endowment for the purposes of assuring his adaptation to his biological and social environment is an essential principle of the ecological worldview.[29]

> This principle is incompatible with the scientific assumption that subjective knowledge is necessarily imperfect and that only objective scientific knowledge displays sufficient accuracy to provide the basis of rational and hence adaptive behaviour.[30]

> It is encouraging . . . that a number of our more thoughtful scientists have, by implication at least, [realized] that the living brain, after millions of years of being shaped by the environment, is suited to it with an accuracy that is both remarkable and profound. Waddington[31] points to a "congruity between our apparatus for acquiring knowledge and the nature of the things known" and suggests that

the human mind "has been shaped precisely to fit the character of those things with which it has to make contact."[32]

According to Goldsmith, therefore, intuitions have a claim to legitimate cognitive status because they draw on the shaping experiences of evolutionary history and reflect the common evolutionary heritage of mind and environment. This heritage is also reflected in our genes, so that "living things are born with an innate knowledge of the environment . . . [because] each generation of living things inherits genetic information which reflects the experience of its ancestors going back into the mists of time."[33] This knowledge takes the forms of instinct and emotion and is stored in our primitive brains, and it is "the knowledge organized in these brains of which we are unconscious and cannot articulate."[34] The difficulties of bringing this knowledge to conscious scrutiny, and of articulating it to another subject, render such knowledge mysterious and "ineffable" to us. Goldsmith adds that "the process whereby we acquire our ineffable knowledge is usually referred to as "intuition." It is in itself mysterious and ineffable."[35]

Thus does Goldsmith attempt to make a virtue out of apparent necessity. The difficulties inherent in articulating and rationally defending claims about the world made as expressions of a theory of intuitive cognition become, instead of weaknesses for that theory, demonstrations of its links with our evolutionary history. They are thus legitimated through the coevolutionary hypothesis that our primitive brains are optimally integrated with our environment. Therefore, information about the world gained through intuition has good cognitive standing.

Goldsmith gives us a theory of intuitive cognition that seeks to explain why intuitions about the value of nature, or the ecologically positive forms of the human-nature relationship, should have sound cognitive standing. It may have some explanatory value in allowing us to understand why intuitive thought processes sometimes do apparently lead to valid knowledge. It does not, in and of itself, despite Goldsmith's anti-empiricist rhetoric, tell us why intuitive knowledge claims should be considered any more than hypotheses, possible truths awaiting empirical testing. Insofar as he does object to such a position, he suggests that the attempt to test intuitions rationally and consciously can be dysfunctional or can cause dysfunctionality. He raises the psychological condition of hyperreflexivity, a condition whereby we try to govern, via the neocortex, behavior that, if it is to function correctly, should be governed by the cerebellum. This leads to dysfunctionalism, as in the well-known example of the piano player who begins to concentrate on what his fingers are doing. The implication Goldsmith is drawing is that if we as a society seek to rationalize our ecological thinking and behavior, we will become ecologically dysfunctional—we have to trust "our" ecological intuitions. This raises the obvious problem of whether the analogy between individual and society here in terms of causes of dysfunctionality actually works.

As to why this theory of intuitive cognition should be associated with a preservationist principle, or why intuitions about conservation should carry the day, this is clear enough to Goldsmith. His reasons are not in themselves dependent on a theory of natural value, although they are used to generate a theory of value in two ways. The coevolution hypothesis tells us that we are uniquely adapted to achieve cognitive certainty about the environment in which we have evolved. In what environment have we evolved? In the natural environment that has surrounded our evolutionary development for millennia—the biosphere, not what Goldsmith calls the "technosphere" of the modern world. It is this environment that we instinctively understand and are adapted to. It is also in this, and only this, environment that we can expect our behavior to be "homeotelic"—that is, having the crucial ecological property of contributing to the stability of the biosphere, which itself constitutes Goldsmith's strictly consequential test of the rightness of an action. The environment in which our hunter-gatherer ancestors lived would be, says Goldsmith, the "optimal" environment for us. Thus the preservation of the natural world has value for us because (1) it provides the conditions for us to engage in homeotelic behavior, and (2) similarly to Naess, Goldsmith holds that an evaluative component is an inevitable part of our intuitions about the natural world. Given that these intuitions arise as a result of coevolution with the natural world, they have as much claim to epistemological validity as intuitive, experience-based knowledge of the physical qualities of nature—the tree is not just green, it is also valuable.

Again we can see that the set of "acceptable" intuitions about the world has been delimited, here by the coevolution hypothesis and functionalist ethics that Goldsmith uses to frame his theory. Adaptation to a particular kind of environment is "read off" from the experience of coevolution. This environment is, as said, that which allows homeotelic behavior to flourish. However, the "modernist ethic," which holds that only human-made things are of value, has become dominant in the present day, as once-useful evolutionary traits have become ecologically dysfunctional. Intuitions about the world based on the "modernist" viewpoint are seen as misguided because they fail to account for coevolution and lead to heterotelic behavior patterns.

Both Naess and Goldsmith are biocentrists at the level of environmental philosophy.[36] Naess cleaves to a principle of biospherical egalitarianism, and Goldsmith operates with a strict consequentialist ethic based on the flourishing of the biosphere. We might, then, come to expect a general connection between biocentrism and intuitionist epistemology. This, however, is not the case. Paul Taylor, another self-avowed biocentrist, argues that far from being the universal product of a coevolutionary history, or the outcome of an unmediated experience, intuitions are no more than socially constructed feelings about the world. These will vary immensely from culture to culture, and they represent an unreliable guide to questions of nature preservation or normative questions about human attitudes toward the natural world. What Taylor has to say runs so contrary to the approach of Rothenberg, Naess, and Goldsmith that it is worth quoting at some length:

> A second point that needs to be made here has to do with our "intuitive" moral judgements. . . . It seems perfectly obvious to us, something self-evident to our intuitive moral sense, that there can be nothing wrong in the way we treat plants or those animals that cannot feel pain. Thus the appeal to practically everybody's moral intuition is enough to refute the idea, central to a life-centered theory of environmental ethics, that every living thing is a moral subject.
>
> It is important to see why this sort of argument is not sound. . . . Our moral intuitions regarding how the living things of the natural world should be treated are psychologically dependent on certain basic attitudes toward nature that we were imbued with in childhood. What attitudes were given to us early in life reflect the particular outlook on animals and plants of our social group. Extreme variations are to be found in the outlooks of different groups even within western culture, to say nothing of other cultures throughout the world.[37]
>
> Since our intuitive judgements in matters of ethics are in this way strongly affected by our early moral conditioning and since different societies will imbue children with different attitudes and feelings about the treatment of animals and plants, we cannot use either our own or anyone else's moral intuitions as grounds for accepting or rejecting a theory of environmental ethics.[38]

> Reliance on moral intuitions is at bottom an appeal to our most strongly held inner convictions. Such an appeal has no relevance to the truth or falsity of what is felt and believed so deeply. Indeed, the search for truth in these matters is *seriously hindered* by the tendency to rely on our intuitive judgements. We want to think that the moral beliefs we hold with certainty are correct and that the opposite beliefs of others are mistaken. But they hold their beliefs with as much certainty as we do ours. Unless we can give good, adequate reasons that justify our beliefs, we cannot simply reject the intuitive convictions of others as erroneous. Reference to what seems intuitively to be so is no substitute for thinking things through to their foundations.[39]

Taylor makes this anti-intuitionist argument in order to defend a biocentric environmental ethics against a putative intuitionist "sentience" critique. That is, Taylor thinks that many of us will make a "commonsense" intuitive assumption that an ability to feel pain marks the distinguishing characteristic for when a creature can be considered a moral subject—an argument often made in animal rights discourse. Because Taylor wants to argue that nonsentient entities, including plants, have moral standing, he takes his own argument to be counterintuitive and this, at least in part, explains his anti-intuitionist position.

However, this anti-intuitionist validatory standard for ethical argument has, if it is to be applied consistently, to apply equally to intuitionist ecological arguments as it does to skepticism about environmental ethics. If our intuitions about sentience and moral considerability are not to be trusted as standards of judgment for purposes of environmental argument, we should not trust intuitions about the intrinsic value of nature, or the identification and interrelatedness that we (according to Nacss and Goldsmith) have with the world around us.

Taylor accepts that intuitions may be, as Rothenberg suggests above, common, indeed they may be held by "practically everybody." This, however, is consistent with his accompanying claim that "extreme variations" occur between different social groups as to the value of, and human action toward, nonhuman nature. Our culture imbues us in childhood with certain attitudes toward nature, and, we are told, the "search for truth" is seriously hindered by references back to these childhood-centered prejudices. Intuitions can thus never be foundational for an ethics, and adequacy of reason is the only standard by which arguments for nature preservation should be judged. We must establish the "rational acceptability"[40] of an environmental ethics. "Only the most careful and critical reflection, carried through with total openness and honesty," Taylor says, "can bring us nearer to whatever truth the human mind can attain in this domain of thought."[41] Further, "We must strive for objectivity, and this requires a certain detachment from our immediate intuitions in this area so that we can consider without prejudice the merits of the case for a life-centered view."[42]

Suggesting that we must keep a "certain detachment" from our intuitions is clearly not the same as arguing for their complete rejection, which the earlier passages seem to suggest. How then, if at all, do intuitions "fit" in Taylor's validatory strategy for environmental ethics? Taylor argues that our ethics have to be rendered compatible with our beliefs about the nature of the world, but what validates our beliefs about the nature of the world? Do we not merely push the epistemological problem back a stage here? To answer this, we need to assess the relationship between Taylor's conception of belief systems and intuitionism. Taylor believes that, similarly to systems of ethics, worldviews should be held on objective and rational grounds, or at least that in order to be acceptable worldviews must satisfy a set of rational criteria. Specifically, a worldview must (and, Taylor holds, biocentrism does)

1. Satisfy well-established criteria for the acceptability of a philosophical worldview.
2. Be of such a character that, "fully rational, enlightened, and reality-aware" moral agents will want to adopt this worldview as their own.[43]

It is somewhat off the point of this essay, and anyway I do not have the space, to engage in an extended analysis of Taylor's use of these arguments, worthy of detailed consideration though they are. Relevant to our purposes is his rejection of the idea that intuition is adequate to the task of giving validity to either an environmental ethics, or a worldview, thus contradicting both the ethical and ontological epistemology of mainstream biocentrism and deep ecology.

The one aspect of Taylor's alternative I do want to pick up on is his contention that in both deciding about worldviews and deciding about ethics, we have to strive for objectivity. This, at least prima facie, stands in contrast to both the clearly subjective elements of intuitionism, and the claim made by both Goldsmith and Naess that the pursuit of objectivity is a chimera, and that anyway the subjective-objective dichotomy is outmoded and nonecological in its epistemological and ontological assumptions.

So, how, according to Taylor, can we hope to achieve objectivity? This is claimed as an integral part of the rationality of thought. It is

> The capacity and disposition to take into account only reasons that are relevant to the decision to be made, and only reasons that are open to all inquirers for consideration. Objectivity means being disposed to consider only the merits of an argument, regardless of who offers it and regardless of their motives. In examining the reasons for and against any world view, the objective evaluator avoids all ad hominem appeals.[44]

Thus for Taylor, objective, rational canons of argument constitute the only philosophically valid form of discourse available for establishing, and judging the adequacy of, a system of environmental ethics and, indeed, a worldview. In neither of these fields does he consider "any appeal to pre-theoretical convictions, however deeply held, to be philosophically relevant."[45]

So, the intuitions that we have can never be "philosophically relevant" as they stand. There is a "distance" between such pretheoretical beliefs and philosophical argument, and that distance is made up of Taylor's set of rational criteria of validation, which includes the need to consider things "objectively." Intuitions can thus only come to have a claim to epistemological validity once they have survived the test of rational validation criteria that they are required to meet.

Thus we have, from within the school of biocentric thinkers, radically divergent accounts of the nature, status, and utility of intuition. In the next section I go on to analyze this disagreement through an examination of the different ways these three authors decontest the notion of intuition. It will be seen that differences in the conceptualizations of the concept between these authors mean that for each, intuition is taken to perform different functions over differing ranges of objectives. This analysis will in turn allow us to arrive at more general conclusions about the limits to an acceptable and useful role for intuition in constructing environmental ethics.

Are Taylor, Goldsmith, and Naess discussing the same phenomenon when they use the term *intuition*? There is, it seems to me, a core intension to the concept that all three thinkers share ((1) below), but there are also elements of the concept that are idiosyncratic to each writer. The notion of intuition has in these works some or all of the following conceptual components, which help grant each conception its distinctive character:

1. A descriptive term for a belief about the world that is held on a nonrational and pretheoretical[46] basis, as a result of emotional attachment or "listening to one's inner voice," for example. This does not of course entail that there cannot *also* be rational reasons for holding such a belief.
2. The cognitive process by which beliefs of type 1 about the world come to be held.
3. A cognitive faculty capable of setting in motion a cognitive process of type 2 in order to gain beliefs of type 1.
4. Something like a "first guess" about an ethical or epistemological question or problem, based on an already inculcated set of beliefs and/or values. Such conjectures have yet to be tested against other beliefs, empirical evidence, standards of coherence and consistency, and so on. Such moral guesswork may alternatively be with regard to *prioritizing* conflicting moral principles rather than deriving them.

The important point about the first three distinctions is that it is possible to believe we can gain beliefs of type 1 without believing in any particular "process" of type 2 or faculty of type 3. One can also accept that there are beliefs of type 1 and/or a process of type 2 without believing that there is a particular faculty of type 3. The relevance of this will become clear as we proceed.

**Intuition and Systems of Thought**

The first element of intuition—(1) above—is a conceptual component common to all these authors' work on intuition—that is, all take the word *intuition* to refer to beliefs about the world that take such a form. Differences emerge as other conceptual components are added to flesh out the notion of intuition in the work of each thinker—although given these, there also some surprising similarities. Taylor, as noted, argues that intuitions have no bearing on the truth and falsity of claims made in moral discourse, and do not constitute grounds for accepting or rejecting a system of ethics. Because Naess says his work is primarily intuitions, we seem to have a methodological abyss between the two. Naess's intuitionism, however, is obviously meant to consist of more than guesswork and feelings about nature, and the question of how Naess wants to work his intuitionism up into a value system is relevant here. Naess explicitly says, "It is quite correct that outbreaks of feeling do not supply an adequate guide to a person's system of value. In environmental conflicts, for instance, expressions of love of nature are not enough. . . . What should count, are the norms and value priorities actively expressed in conflict."[47] What distinguishes these "values in action" from spontaneous feeling? According to Naess, "Spontaneous positive or negative reactions often do little more than express what a person likes or dislikes. Value standpoints are reflections in relation to such reactions: 'Do I like *that* I like it?' "[48]

So, we have come to the view with Naess that intuitions form the starting point of an environmental axiology, and that these intuitions have to be reflected on before they can be validated.[49] What factors should guide the reflective process? How can we know whether we should "like that we like it"? Do we merely move up one order of intuition, in that we have a "feeling" that it is acceptable to feel X? Naess seems to suggest that what is necessary is something like rational reflection on felt preferences. It is important, he says, "to clarify the relationship between spontaneous feelings, their expression through our vibrant voices, and statements of value or announcement of norms motivated by strong feelings but having a clear cognitive function."[50] This suggests that the feelings themselves do not have a cognitive function, a position that already seems a little closer to Taylor's. We have to "clarify" the relationship between

1. Feeling
2. Voice
3. Values/norms

How can we do this? Naess says that we do not want to eliminate feelings from discussion; they must instead be "clarified, and made explicit as the need arises."[51] Idiosyncratic elements of one's arguments have to be "sorted out" (presumably this means acknowledged as such, and put to one side) "if the debate is concerned with more or less general norms."[52]

This is all somewhat sketchy, but Naess seems to be suggesting something like the following. We bring our intuitions about nature into a debate with others. We then, as he recommends, "formulate strong, clear expressions of values and norms"[53] on the basis of having reflected about whether we are happy with the intuitions we have. As our values and norms are counterposed against the values and norms of others, we continue this reflective process and accept the need to remove value claims that seem based on idiosyncratic feelings. At the same time we need to remember that the only valid intuitions are those that are directly experiential, nonexperiential intuitions themselves being, perhaps, considered "idiosyncratic."

If our intuitions are to be mediated by rational reflection, this raises the question of Naess's understanding of rationality and the processes governed by it. Naess assesses rationality initially in instrumental terms, it is "relational: rational is rational only in order to reach human ultimate ends, whether in terms of happiness or perfection."[54] The *only* here seems to suggest that Naess sees rationality purely in instrumental terms, implying that it does not have a substantive role to play in setting the ultimate human goals it must serve. This seems, however, an improbable view, and a later passage on economics seems to suggest something else:

> In modern economics texts there is much talk about rationality and rational choice. . . . Rationality has to be measured in relation to basic norms. . . . Whenever we bring questions of rationality into economic life the ultimate norms of economy have to be considered. . . . Economic growth in the third world is still conceived largely in terms of non-normative economics, the "experts" being unaccustomed to reasoning from the maximally wide and deep perspective of a total view.[55]

Although the opening sentence suggests again merely the relational view of rationality, the latter part, in which policymakers are urged to "reason from a maximally wide and deep perspective" suggests that Naess wants us to reason about basic norms.[56] How does this relate to his idea that basic norms are intuitive?

For Naess, just as descriptive and normative statements are bound up together in gestalts, so emotive and reasoned arguments are also intimately related: "The activism of the ecology movement is often interpreted as *irrational*, as a 'mere' emotional reaction to the *rationality* of a modern Western society. It is ignored that reality as spontaneously experienced binds the emotional and the rational into indivisible wholes, the gestalts."[57] We are back with the importance of "spontaneous experience" as both the basic motivation of environmental norms, and the glue that holds together the gestalt of the rational and emotional.

So, in brief, for Naess intuitions provide the raw material for a set of attitudes toward nature, but insofar as they are "feelings" they are not in themselves enough, even if based on direct experience. These feelings have to undergo a process of second-order reflection and articulation in discussion with others before a legitimate system of ethical norms regarding the human-nature relationship can be said to be established. Feelings ground value claims; these relationships have to be explicitly stated and used to ground a system of ethics, which itself defines the ultimate goals by which secondary norms and behavior can be judged rational or irrational—thus are emotions and rationality bound together. This has to imply, if it is to mean anything, that initial intuitions about the world can be rejected or modified in order to become the kind of intuitions that "we like that we like." This in turn suggests two standards of intuition in Naess's theory, which we might call "basic" and "reflected." Basic intuitions provide the raw material to be fed into the process of reflection. Reflected intuitions emerge from this process, being those intuitions that we are happy to hold on to even after the process of rational reflection; they embody the gestalt of reason and emotion.[58]

I believe that a similar distinction can usefully be applied to Taylor's work, without an illegitimate amount of conceptual stretching taking place. Our "basic intuitions" are precisely those that we are warned to keep a critical distance from, being as they are socially constructed and therefore epistemologically unreliable. However, unless one cleaves to the possibility of a completely presuppositionless axiology, the outcome of the process of rational engagement and argument that Taylor insists is essential to both the construction and judgment of ethical systems seems, at least partially, to consist of the intuitions that have survived, or been transformed in, that process. Thus, rather more problematically than in the case of Naess, we could also consider (at least some of) the underlying principles of Taylor's ethical system of "Respect for Nature" as incorporating "reflected intuitions." For example, although Taylor insists that his theory is, as far as it is possible to be, empiricist on the basis of best scientific evidence, and rationalist in its construction and articulation, certain steps in the argumentation process seem to contain intuitionist elements. Moving from the "fact" that organisms pursue their own survival to the idea that we can consider them to have a "good of their own" provides an example. Then the step from them having a "good of their own" to the idea that this in itself is something worthy of "respect" relies, I would suggest, on implicit, reflected intuitive judgment on Taylor's part. To this extent, despite the endorsement of intuitionism on the one hand, and rejection on the other, some reconciliation of the actual employment of intuition in environmental theory building can be seen between Naess and Taylor.

Naess and Taylor, then, while sharing conceptual component 1 in their accounts of intuition, give different accounts of process 2. For Taylor intuitions are generated through a process of inculcation that takes place as we are socialized into a community's shared values and norms, whereas for Naess intuitions are generated through direct experience. This difference is of course significant, because the direct-experience process grants a prima facie validity to intuitions for Naess, whereas the socialization process is what denies intuitions similar prima facie validity for Taylor. Nonetheless, from these radically different starting points, both authors, as I have suggested above, come to positions with strong common elements when articulating the conditions that have to be satisfied for principles of environmental ethics to be valid. In both cases, intuitions are subject to a process of rational reflection before they qualify as valid principles, even though ultimately Naess believes we should balance reason and emotions, while Taylor believes that rationality and objectivity are the sole validating criteria.

Goldsmith has a different position, seeming to endorse conceptual components 1, 2, and 3 in his account of intuition. Thus intuitions are beliefs about the world that come to us through a nonrational process "that is itself mysterious and ineffable" but that arises through a capacity we have evolved, which is located in our "primitive brains." This might presumably also be the source of the "inner voice" to which Fox refers (see above). Such deep ecological proclamations of a capacity or faculty of intuition bring us back to an old epistemological debate, and raise the question of whether Goldsmith's evolutionary explanation can breathe life back into the corpse of epistemological intuitionism. The idea of intuition as a faculty or capacity for gaining access to truth—especially moral truth, as once proposed by Ross and Moore, lies demolished by the attacks from within analytical ethical theory from the 1950s by the likes of Toulmin and Nowell-Smith.[59] As a result of their attacks "an appeal to intuition is seen as a 'failure' in normal political discourse."[60] The question is why it should, to complete the quote have "a far more positive role to play in ecological thinking."

The argument against moral intuitionism comes down to this. In treating intuition as a faculty, intuitionists treat moral properties as equivalent to empirical properties—something that can be detected by the faculties that human beings possess. But treating values as empirical properties leads to absurd beliefs, such as that moral disagreement literally makes no sense. That is, if you know two people have the same sense faculties and share the same language, it is just nonsense for them to disagree over the color of a pillar-box (mailbox) observed under standard conditions. It is not, however, nonsense (whatever else it may be) for them to disagree about whether meekness is good. This difference between values and properties is crucial; it is a difference between contingent and necessary agreement.[61] It also leads to paradoxical results. Toulmin uses the following example: We have a man who all can agree acts well; he is noted for his high moral character, kindness, incorruptibility, modesty, thoughtfulness, and so on. When asked if he is conscious of observing any "nonnatural" property of goodness, he says he isn't. He says he does what he does because there are good reasons to do it, and that he isn't interested in any additional nonnatural properties of his actions. To be consistent, the objectivist (i.e., intuitionist) will have to say of this: "He may know *what things are good*, he may know *what it is to be good*, but he cannot know *what goodness is*."[62] But this man cannot be thought to be missing the one thing that really matters (knowledge of what goodness is), this would be ridiculous—all that we are interested in is that he comes by his actions for good reasons. If you are told that someone does not know what goodness is, you would not expect them to display the quality of consistently good actions. Conversely, "A philosopher who, out of fidelity to a theory, is driven into saying a thoroughly virtuous and upright man does not know what goodness is, is assuredly up the garden path."[63]

One would write similarly of an environmentalist who behaved in an exemplary "green" fashion but denied intuiting the intrinsic value of nature, basing her behavior instead on a version of humanism. Such attacks left intuitionism, which perceived moral truth as "like yellow, yet so unlike yellow" as nothing but intellectual ruins, to employ Bernard Williams's description.[64] Can the evolutionary argument employed by Goldsmith leave the building of deep ecology intuitionism intact, while all around has been demolished?

What work would the evolutionary explanation of a faculty of intuition have to perform in order to rescue conceptual component 3 from the anti-intuitionist critique? It would have to offer a credible account of why we could expect the detection of value to be an ability of the sort in favor of which there would be natural selection through time. If the evolutionary approach is to succeed, it has to show that there are good reasons to think that the ability to detect value is the kind of genetically endowed trait that would enhance survival chances sufficiently to have an impact on reproductive success. Over a long run of time, this advantage would not have to be enormous, merely sufficient.

It is worth noting that if such an argument can work for Goldsmith, it may also have something to offer in solution of the "epistemic crisis" that Holmes Rolston III claims that we in the West suffer. There has been considerable debate about Rolston's own epistemology,[65] but both his critics and his supporters grant that his approach is "value empiricist" and that he "rejects intuitionism." This, however, can only be true if we are prepared to grant Rolston a large assumption, which is that it is actually *possible* to be an "empiricist" with regard to value. That Rolston is both realist and empiricist with regard to our perceptions of nature itself is clear enough:

> Yes, but "nature" is a category we invent and put things we meet into, because there is a realm out there, labelled nature, into which things have been put before we arrive. Leaks or not, we do catch things in our buckets that come from some source out there. Nature is what is *not* created by the human mind. We can, through various constructs of the human mind, find out things that are not created in the human mind. Anyone who thinks that there is any knowledge of the material world believes that.[66]

This can reasonably read as paralleling part of Goldsmith's epistemology. Our senses grant us accurate epistemological access to the world, and although our inability to know that this is true directly leads to epistemological skepticism, we can know it indirectly. That is, if we grant the claims of both Goldsmith and Rolston that obtaining accurate sensory information about the world is going to be a better coping strategy than not, this strategy is also the most likely to be evolutionarily successful. We are creatures who have evolved, ergo, our sensory faculties do grant us accurate information about the world.[67]

Even if we grant Rolston all he claims about the human senses and epistemology—that is, that we really do "construct language around" our sensory perception in such a way that we give ourselves an accurate picture of nature, does it follow that we should therefore grant him the same concessions with regard to his account of natural value? He thinks we should:

> Every organism is a spontaneous, self-maintaining system, sustaining and reproducing itself, making a way through the world, checking against performance by means of responsive capacities with which to measure success. Its genetic set is a *normative set* in the sense that by such coding the organism distinguishes the what *is* and what *ought to be*. The organism is an axiological, though not a moral, system. . . . A life is defended for what it is in itself, without necessary further contributory reference. Every organism has a *good-of-its-kind*; it defends its own kind as a *good kind*. In this sense, the genome is a set of conservation molecules.[68]

This argument is redolent of the "autopoietic ethics" argument described by both Fox and Eckersley, and the "respect for nature" argument of Taylor. We can recognize that all organisms or "genetic sets" have goods of their own, that they are "ends in themselves" and therefore "value" themselves—thus we should value them too. Thus "value in nature, like value in human life, is something we can see and experience."[69] Thus all we require in order to appreciate natural value is knowledge of some basic biological facts.

But is it? Again, evolution will have favored, to say the least, organisms that were able to defend, repair, and maintain themselves, at least up to the point where they could reproduce. An inability to locate nutrition or repair damage would hardly be a good coping strategy for any organism. Thus does evolution select for certain properties and types of behavior. What is not clear to me from Rolston's account is why we make the epistemological step from claiming to know that "this organism locates nutrition and repairs damage to itself" to knowing that "this organism *values* itself"—never mind the following step to "therefore we should value it." But this is not the place to rehearse such old arguments about the is/ought distinction. Suffice it to say that if Rolston's empirical epistemology with respect to value does not work, the environmental ethicist might yet want to draw on the possibility of a faculty of intuition as suggested by Goldsmith.

An evolutionary account would have to offer reasons as to why we should have developed a faculty for detecting value, and thus why being the type of organism that can detect value would offer an evolutionary advantage. We should note, to begin with, that Goldsmith rejects the modern neo-Darwinist account of genetically driven competitive evolution. In accord with his endorsement of the Gaia hypothesis, he sees evolution as directive and biospherically controlled. The goal toward which evolution is directed is the maintenance of the equilibrium of the biosphere itself. Life processes evolve for a purpose, which is that they have a specific function in the biospheric system—that of contributing to its stability. Goldsmith is thus providing us with a functional explanation of the evolutionary process, in which function is taken as cause. Setting aside the problems with this account of the evolutionary process (how did the tendency toward dysfunctional biospheric behavior that Goldsmith thinks humans now engage in evolve?), we can ask whether, in either this or the more common neo-Darwinist account of evolution, we could expect an intuitive capacity for detecting the existence of value in the world to emerge.

Goldsmith holds that the seat of our instincts, emotions, *and values* lies in our primitive brain, with its intuitive capacity. These instincts and emotions grant access to "fundamental knowledge" about the "harmonious unity of nature" and the value of the biosphere, indeed for Goldsmith they seem to be our only epistemological tools. "Man" is not "designed" to act in a nonemotional way, and is thus incapable of objective knowledge and rational behavior.[70] This is as true of our scientific endeavors as any other. Thus ecology is a "faith," is "subjective," and "reflects the values of the biosphere." Ecology informs us about natural laws, and these natural laws should also form the basis of our moral laws, as they did for traditional societies. What is of value in this system is the correct functioning of the biosphere—the life-support mechanism for all life on earth. Traditional peoples "knew" this intuitively, but our overrationalized thinking processes have forgotten this and instead adopted the "modernist worldview," which holds that all that is of value is human made.

According to Goldsmith's own account of evolution, then, we would have developed an intuitive capacity to recognize value because the belief that the natural world is valuable and therefore worthy of moral respect is itself functional to the continued health of the biosphere, which *is* of value to us. Thus are our intuitive beliefs and reality brought together. The functional explanation offers reasons to believe that we would have evolved an intuitive capacity to recognize natural value. This of course only "works" as an explanation if we accept what is a highly problematic functional account of the evolutionary process. In a neo-Darwinian account, it is difficult to understand why we should develop this particular capacity. A capacity to consciously value anything could only emerge at an advanced stage of evolution, so notwithstanding Rolston's strictures about plants and animals valuing themselves, only human beings from a reasonably advanced stage of their evolution have had the capacity to *consciously* value. Even if we accept the realist argument with respect to natural properties in the world—the argument that our cognitive faculties will have evolved to give us accurate information—why should this also be true of value? We could even concede the probable evolution of a capacity to know what is *instrumentally* valuable to us—beings with the ability to consciously value who did not realize that food was instrumentally valuable would have an odd perception of the world. But *intrinsic* values? What attribute of intrinsic value could do the evolutionary work? Why shouldn't "intrinsic value" be something that we invent? Or that groups that emerge into positions of power invent for those who are subordinate to them?[71] There seems to be no reason why, from the standpoint of the generally accepted view of evolution, human beings should develop an intuitive capacity to detect the presence of intrinsic value. This leaves biocentric intuitionism as open to the attacks of the likes of Toulmin and Nowell-Smith as ethical intuitionism was, at least insofar as ecologists cleave to conceptual component 3 in their account of intuition. Thus unless one is also prepared to accept Goldsmith's functional account of evolution, I would suggest that the contention that our "primitive brain" has the capacity to directly intuit (intrinsic) value does not have any obvious means of support.

I should perhaps at this point say something in passing about conceptual component 4, which has not figured in our discussion of these biocentric authors but has been a central component of some important political works that have engaged with intuition. It features for example in much of Anglo-American analytical political theory, whereby putative theoretical principles are tested against common intuitions in the process of reaching something like a Rawlsian "reflective equilibrium." In his work on justice, Rawls thinks we have to begin with our intuitions, but shares Taylor's concern that our "everyday" conceptions of justice will be "strongly colored by custom and current expectations."[72] These initial judgments about justice may well conflict with (at least some of) a set of theoretically worked-out principles.

In this case we have a choice. We can either modify the account of the initial situation or we can revise our existing judgements. . . . By going back and forth, sometimes altering the conditions of the contractual circumstances, at others withdrawing our judgements and conforming them to principle, I assume that eventually we shall find a description of the initial situation that both expresses reasonable conditions and yields principles which match our considered judgements duly pruned and adjusted. This state of affairs I refer to as reflective equilibrium.[73]

Component 4 also features in Rawls's account as a way of attempting to weigh different and conflicting moral principles in a nonsystematic fashion—thus intuition*ism*, which he seeks to refute as an inadequate approach to the problem of justice. This "commonsense" intuitionism

takes the form of groups of rather specific precepts, each group applying to a particular problem of justice. There is a group of precepts which applies to the problem of fair wages, another to that of taxation, still another to punishment, and so on. In arriving at the notion of a fair wage, say, we are to balance somehow the claims of skill, responsibility, and the hazards of the job, as well as to make some allowance for need.[74]

Rawls rejects this use of intuition as inferior to a systematized theory of justice that can order principles in accordance with reflective equilibrium. Taylor does exactly the same with the development of priority principles in *Respect for Nature*.[75]

## Conclusion

The main purpose of this chapter has been to bring some analytical clarity to an explanation of the radically differing accounts of the status and utility of intuitions in a selection of biocentric literature. The method chosen to achieve this has been to break "intuition" down into its various conceptual components in the versions of it employed by biocentrists. I have suggested that component 1 is common to all accounts of intuition and takes a similar form in each case. With component 2 however, we see significant divergence. For deep ecologists such as Naess, Rothenberg, and Fox, the process by which valid intuitions are derived is from direct experience of nature; for Taylor, intuitions are socially and culturally imbibed, and it would make no sense to talk of "valid intuitions prior to any rational sorting process; for Goldsmith, intuitions are "fundamental knowledge" that is genetically inherited and stored in our primitive brains. Taylor's position rules out adoption of component 3, which is clearly an idea that he would anyway want to reject. Naess's position does not seem to imply component 3 either. To be sure, direct experience of nature has to be processed in some way before it is translated into intuitive belief, but this does not have to imply any particular capacity for gaining access to truth about the world. The reason direct experience is privileged is precisely because it *is* direct experience, not because it relates to a special cognitive capacity. Goldsmith does, however, include component 3 in his account of intuition; our primitive brains have a special capacity for revealing truths about the world to us, and that includes the perception of value. The Gaian account of evolution that Goldsmith offers in support of this can offer reasons for such a capacity to have developed. However, a standard neo-Darwinian theory of evolution can at most explain the evolution of a routine capacity for the detection of instrumental value, which is to say no more than a capacity to detect objects useful to our survival. With regard to intrinsic values, Goldsmith's theory, without is functional evolutionism, seems as vulnerable to the attacks of anti-intuitionists as the ethical intuitionism of the early twentieth century was.

Taylor also endorses component 4, seeing moral intuitions as no more than a "first guess" based on a set of socialized assumptions about the world, and equally no more than a first guess in terms of prioritizing principles. Although the Naess/Rothenberg positions are prepared to make stronger claims as to the "validity" of the appropriate types of intuition, they also endorse the contingent status of intuitions, and put forward the view that valid forms of argument will balance intuition with reason in the intuition-reason gestalt. Goldsmith's position is more purely intuitionist than any of these others, endorsing components 1, 2, and 3 but therefore rejecting 4 as an inadequate appreciation of what intuition is and is capable of. For Goldsmith, rational thought, or rather the attempt to think about the world in an objective and rational way, is what leads us astray and renders human thought and action heterotelic rather than homeotelic. Goldsmith, it seems, would have to reject any "reflective equilibrium," insisting instead that we should trust our deepest intuitions in the manner of traditional societies.

Where does this leave us? Hopefully with a clearer picture as to what biocentric authors are actually disagreeing about when they take such radically divergent views on the content, status, validity, and role of intuition. As to the larger normative question of whether there is a (or some) legitimate role(s) for intuition in environmental ethics, it seems to me that unless we believe in a completely presuppositionless political philosophy, intuition will always *be* there somewhere. The putative role for intuition set out by Goldsmith requires highly contestable accompanying epistemological and ontological baggage. There is, however, a debate of genuine interest between Naess's and Taylor's positions with respect to component 2, and this might make a useful starting point for research into this fundamental question.

## Notes

1. Holmes Rolston III, *Environmental Ethics: Duties to and Values in the Natural World* (Philadelphia: Temple University Press, 1988).

2. Paul Taylor, *Respect for Nature* (Princeton, NJ: Princeton University Press, 1986).

3. Avner de-Shalit, *Why Posterity Matters: Environmental Policies and Future Generations* (London: Routledge, 1995).

4. Arne Naess, *Ecology, Community and Lifestyle*, ed. and trans. David Rothenberg (Cambridge: Cambridge University Press, 1989); Warwick Fox, *Toward a Transpersonal Ecology* (Totnes: Green Books, 1995).

5. Robert Goodin, *Green Political Theory* (Cambridge: Polity Press, 1992).

6. John O'Neill, *Ecology, Policy, and Politics* (London: Routledge, 1993).

7. I assume a reason can be considered (morally) valid without being compelling, or both valid and compelling. It must also be possible that a reason can be compelling without being valid (such as when there is a threat of violence against a person unless they perform a particular action).

8. The various elements that might make up a conception of intuition are discussed more systematically at a later point in the chapter.

9. See for example Edward Goldsmith, *The Way: An Ecological Worldview* (Totnes: Green Books, Revised Edition 1996), chap. 8.

10. Taylor, *Respect for Nature*, 23.

11. As will become clear in the ensuing discussion.

12. Bill Devall and George Sessions, *Deep Ecology: Living as if Nature Mattered* (Salt Lake City: Peregrine Press, 1985), 75.

13. Devall and Sessions, *Deep Ecology*, 239.

14. Devall and Sessions, *Deep Ecology*, 69.

15. Warwick Fox, "Deep Ecology: A New Philosophy of Our Time?", *Ecologist* 14, nos. 5–6 (1984): 196.

16. Fox, "Deep Ecology," 198.

17. David Rothenberg, "Introduction," in *Ecology, Community and Lifestyle*, 3.

18. Naess, *Ecology, Community and Lifestyle*, 29.

19. Naess, *Ecology, Community and Lifestyle*, 2.

20. Rothenberg, *Ecology, Community and Lifestyle*, 7.

21. Rothenberg, *Ecology, Community and Lifestyle.*

22. Naess, *Ecology, Community and Lifestyle*, 63.

23. Naess, *Ecology, Community and Lifestyle*, 35.

24. Naess, "The World of Concrete Contents," *Inquiry* 28 (1986): 420; original italics.

25. Naess, "The World of Concrete Contents," 427.

26. Naess, "The World of Concrete Contents," 418.

27. Naess, "The World of Concrete Contents," 424.

28. Ibid.

29. Goldsmith, *The Way*, 87.

30. Ibid.

31. This is a reference to C. H. Waddington's *The Evolution of an Evolutionist* (Edinburgh: Edinburgh University Press, 1975).

32. Goldsmith, *The Way*, 88.

33. Goldsmith, *The Way*, 41.

34. Goldsmith, *The Way*, 45.

35. Goldsmith, *The Way*, 47.

36. It has been suggested to me that Goldsmith is not a good example of a biocentric ecologist for this chapter, due to the idiosyncrasy of his views, but as will be clear from his inclusion, I disagree. I believe that at the level of environmental philosophy Goldsmith's arguments have much in common with other biocentric writers, a view that Naess's endorsement of Goldsmith's major theoretical work supports (Arne Naess, "The Way," *Ecologist* 19, no. 5 (1989): 196–197).

37. Taylor, *Respect for Nature*, 22.

38. Taylor, *Respect for Nature*, 23.

39. Taylor, *Respect for Nature*, 23; emphasis in original.

40. Taylor, *Respect for Nature*, 24.

41. Taylor, *Respect for Nature*, 24.

42. Taylor, *Respect for Nature*, 24.

43. Taylor, *Respect for Nature*, 165.

44. Taylor, *Respect for Nature*, 162.

45. Taylor, *Respect for Nature*, 270n.

46. "Pretheoretical" in the sense that an intuition is never, as it appears, the "finished article," but always requires testing, confirming, elaborating, systematizing, or some such process. I do not mean that intuitions cannot *follow* from engagement with consciously theoretical work, as in "spontaneously" discovered solutions to theoretical problems. My thanks to John O'Neill for a discussion on this point.

47. Naess, *Ecology, Community, and Lifestyle*, 64.

48. Ibid.

49. That is, "experience-based" intuitions of the type discussed above are now seen to have only a prima facie claim to validity, although this is still a validity claim that nonexperiential intuitions lack.

50. Naess, *Ecology, Community, and Lifestyle*, 64.

51. Ibid.

52. Naess, *Ecology, Community, and Lifestyle*, 65.

53. Ibid.

54. Naess, *Ecology, Community, and Lifestyle*, 33.

55. Naess, *Ecology, Community, and Lifestyle*, 105.

56. See Henry S. Richardson, *Practical Reasoning about Final Ends* (Cambridge: Cambridge University Press, 1997) for the view that reason is no "slave to the passions," and that we can, indeed should, reason about our ultimate ends.

57. Naess, *Ecology, Community, and Lifestyle*, 63.

58. This can be articulated in the terms set out in the introduction to this book. Naess seems here to be suggesting a form of "public reflective equilibrium," but only partially. That is, if certain types of intuitions are disbarred because they are not based on "authentic" experience of nature, the set of beliefs and theories incorporated into the process of public reflection is delimited on a priori grounds. Such a set thus represents only a fraction of the full range of the beliefs and theories in society.

59. See W. D. Ross, *The Right and the Good* (Oxford: Clarendon Press, 1930); G. E. Moore, *Ethics* (Oxford: Oxford University Press, [1911] 1966); Stephen Toulmin, *An Examination of the Place of Reason in Ethics* (Cambridge: Cambridge University Press, 1958); P. H. Nowell-Smith, *Ethics* (Harmondsworth: Penguin, 1954).

60. Andrew Dobson, "Deep Ecology," *Cogito* (1989): 41–46, 42.

61. See Toulmin, *An Examination of the Place of Reason in Ethics*, 19–21.

62. Toulmin, *An Examination of the Place of Reason in Ethics*, 23.

63. Toulmin, *An Examination of the Place of Reason in Ethics*, 24.

64. Bernard Williams, *Ethics and the Limits of Philosophy* (Cambridge, MA: Harvard University Press, 1985).

65. See for example Bryan Norton, "Review of Holmes Rolston, III's *Conserving Natural Value*," *Environmental Ethics* 18 (1996): 209–214; Christopher J. Preston, "Epistemology and Intrinsic Values: Norton and Callicott's Critiques of Rolston," *Environmental Ethics* 20 (1998): 409–428.

66. Holmes Rolston III, "Nature for Real: Is Nature a Social Construct?", in *The Philosophy of the Environment*, ed. T. D. J. Chappell (Edinburgh: Edinburgh University Press, 1997), 42.

67. This argument rests of course on probability and plausibility, not deductive certainty. We could imagine a group of beings who received completely inaccurate sensory information about the world in which they lived, but for whom such information fortuitously increased their chances of survival.

68. Rolston, "Nature for Real," 61.

69. Rolston, "Nature for Real," 61.

70. Goldsmith, *The Way*, 83.

71. As in Bookchin's account of the emergence of hierarchy. See Murray Bookchin, *Ecology of Freedom*, rev. ed. (Montreal: Black Rose Books, 1991).

72. John Rawls, *A Theory of Justice* (Oxford: Oxford University Press, 1981), 35.

73. Rawls, *A Theory of Justice*, 20.

74. Rawls, *A Theory of Justice*, 35.

75. Taylor, *Respect for Nature*, chap. 4, sec. 2.

# 3

# The Justice of Environmental Justice: Reconciling Equity, Recognition, and Participation in a Political Movement

David Schlosberg

While the environmental justice movement has gathered much attention from academics, activists, and government officials alike, it seems odd that little has been written on what, exactly, is meant by the *justice* of environmental justice. For the most part, the concept has been used to illustrate the fact that low-income communities and communities of color face more environmental risks than more well-off or white communities; this is linked, of course, to the other injustices in economic and social conditions disempowered communities face. Environmental justice activists and academics call for more equitable distribution of environmental risks—or, more succinctly, for less risk overall, but especially in communities already unduly burdened. But there is more to the conception of environmental justice than just this distributional aspect, and, in fact, a focus solely on distribution is problematic. Environmental justice activists have also called for *recognition* of communities as unfairly affected, and insist on being seen and heard by both a mainstream environmental movement and a government that has, for the most part, ignored them. Further, the movement has insisted on changes in the way environmental policy is made, in order to bring in community *participation* in both the design and ongoing oversight of environmental risks.

So the concept of environmental *justice* in political practice deals with more than simply distribution. But, again oddly, there has been no thorough attempt to try to define exactly what the justice in environmental justice means. I attempt an initial foray into the issue, starting with an examination of the conceptions of justice as equity, recognition, and participation in the political theory literature. This illustrates the theoretical discussions of justice by contemporary political theorists such as Nancy Fraser and Iris Young. I then follow each of those notions through to an examination of the environmental justice movement. The argument here is that the movement embodies a number of different frameworks of justice, even if it is not always explicit about those differences. At various times, justice is defined as equitable distribution, recognition, and participation. The movement demonstrates, I believe, the possibility of employing these different notions of justice simultaneously in a comprehensive political project.

## Defining the Environmental Justice Movement(s)

One of the fastest-growing sectors of the environmental movement in the United States is the environmental justice movement—or, rather, the set of movements that make up a concern with environmental justice. The term *environmental justice* is used to cover two overlapping parts of the grassroots environmental movement: the antitoxics movement and the movement against environmental racism.[1]

The antitoxics movement got its start with Love Canal and the concomitant growth of awareness of the prevalence and dangers of toxics in communities. Dump sites and situations like Love Canal—contaminated communities with threats to human health—were the initial focus of the movement, and this focus continues. But the movement now covers a wide variety of issues relating to environmental threats to human health: not just old industrial waste sites (or new Superfund sites), but also municipal and hazardous waste dumps and incinerators, nuclear waste, industrial pollution in communities, pesticides, and dioxin exposure. A variety of networks tie the movement together, the largest being the Center for Health, Environment, and Justice (or CHEJ),[2] which began in 1982 as a response to the immense need for information communities began to request of the Love Canal Homeowners Association. The center now claims to have assisted over 8000 groups since its work began. But in addition to the CHEJ, there are a variety of networks focused on environmental justice, organized around issues such as, for example, oil refinery pollution and the effects of semiconductor manufacturing.[3]

The movement against environmental racism, which popularized the term *environmental justice*, focuses on environmental issues as they pertain to communities of color and the disproportionate risk those communities often face. Most academics and activists trace the beginning of this movement to a 1982 protest against the dumping of PCB-laden dirt in a new hazardous waste landfill in Warren County, North Carolina. Warren County was not only one of the poorest counties in North Carolina, but also had a population that was 65 percent African-American. This part of the environmental justice movement was empowered and emboldened by studies in the 1980s and early 1990s that showed not just connections between environmental risk and poverty, but specific connections between race and environmental hazards.[4]

When one discusses "environmental justice," the topic could be the antitoxics movement, the race-based environmental justice movement, or a combination of the two. There are certainly differences in these two parts of the movement, and many authors treat them separately. The antitoxics movement is discussed, for example, by Szasz as well as by Gould, Schnaiberg, and Weinberg.[5] Epstein argues that the differences between the parts are crucial, even though she still wants to regard them as one large "environmental justice/toxics movement."[6] I do the same in my own previous work on the movement.[7] I want to argue here that even given some of the differences in the greater grassroots environmental justice movement, there is a unity, of sorts, around the concept(s) of "justice." First, however, the theoretical terrain of the concept must be explored.

## Conceptions of Justice

### Justice as Distribution

In the literature of political theory, justice has been defined almost exclusively as a question of equity in the distribution of social goods. Rawls, for instance, calls justice "a standard whereby the distributive aspects of the basic structure of society are to be assessed." Justice, then, defines "the appropriate division of social advantages."[8] In his application of justice to the environmental arena, Brian Barry insists that justice only applies where distributive issues arise; other issues are merely questions of right and wrong.[9] Justice, in this reading, is the set of rules that govern our distributional relationship. Justice as distribution is centered on socioeconomic factors, rooted in the economic structure of society. This conception of distributional justice is typically used to critique the distribution in a given society, and to argue for social redistribution and more social equity. Of course, once one considers distribution as a means to justice, one needs to have a theory or set of principles regarding how, exactly, distribution (or redistribution) is to—justly—take placc. The whole point of Rawls's notion of "justice as fairness" is justice as just distribution—or, more properly, the rules that govern a just distribution. Similarly, Miller discusses, in a now classic text, three different possible principles of distribution: need, desert, and entitlement.[10]

While theories of just distribution tend, as does Rawls's, to focus on absolutely universal principles, Walzer began a move away from a concern with a universal theory of justice in favor of understanding the concept in historical and cultural place; this move has particular resonance in dealing with environmental justice. Still wed to the notion of distribution, Walzer attempts to introduce a language of difference. He argues "that the principles of justice are themselves pluralistic in form; that different social goods ought to be distributed for different reasons, in accordance with different procedures, by different agents; and that all these differences derive from different understandings of the social goods themselves—the inevitable product of historical and cultural particularism."[11] For Walzer, not only are different things valued differently by different people, but this means that the very criteria for distribution will differ according to how we value things. Social meanings of objects, procedures, and principles are historical and will change over time; hence Walzer introduces a notion of a "distributive sphere," where conceptions of justice are limited in place and time. Walzer's approach to the discussion of justice in a real, diverse, world is more complex and more grounded than Rawls's "veil of ignorance." Even so, Walzer remains tied to the concept, and language, of justice purely as a concept of distribution.

Without doubt, the discussions within the equity framework are vast, rich, and complex. And to the credit of environmental political theorists, the framework has been comprehensively examined with an eye toward *environmental* justice in works by Dobson as well as by Low and Gleeson.[12] Still, as thorough as these works are, I find them incomplete, especially because the distributive conception of justice itself has come under intense critical inquiry by political theorists.

**Justice as Recognition**

In the past decade there have been numerous challenges to the traditional way in which the concept of justice has been approached in the political theory literature. Iris Young has made the most direct and forceful challenge to a justice based solely on issues of distribution.[13] Injustice is not solely based on inequitable distribution, Young argues. In *Justice and the Politics of Difference*, she describes injustices based on a lack of recognition of identity and difference. Part of the problem of injustice, and part of the reason for unjust distribution, is a lack of recognition of group difference. Young begins with the argument that "where social group differences exist and some groups are privileged while others are oppressed, social justice requires explicitly acknowledging and attending to those group differences in order to undermine oppression."[14] In this, obviously, Young shifts the focus away from the more traditional territory of Rawls and other theorists of distributive justice, toward a focus on the postmaterial demands of new social movements around race, gender, and sexuality. For Young, distribution is not the only problem; a concept of justice needs to focus more generally on the elimination of institutionalized domination and oppression, particularly of those who represent "difference."

The basic thesis of the politics of recognition has been laid out by both Taylor and Honneth.[15] As Honneth argues, the key is a link between recognition from others and our own human dignity: "The language of everyday life is still invested with a knowledge—which we take for granted—that we owe our integrity, in a subliminal way, to the receipt of approval or recognition from other persons.[16] Taylor insists that in this sense, "due recognition is not just a courtesy we owe people. It is a vital human need."[17]

Taylor distinguishes between two kinds of recognition: (1) the equal dignity of all, and (2) the politics of difference, where everyone is recognized for their particular distinctiveness: "*Everyone* should be recognized for his or her unique identity. . . . With the politics of equal dignity, what is established is meant to be universally the same, an identical basket of rights and immunities; with the politics of difference, what we are asked to recognize is the unique identity of this individual or group, their distinctness from everyone else." This latter form of recognition causes Taylor some distress. The "demand for equal recognition extends beyond an acknowledgment of the equal value of all humans potentially, and comes to include the equal value of what they have made of this potential in fact. This creates a serious problem."[18] At this point, unfortunately, Taylor's discussion degenerates into a bit of neoconservative paranoia, criticizing what he calls at various points "incoherent," "radical," "subjectivist," "half-baked," "neo-Nietzschean" theories that support multiculturalism.[19] As a number of responses to Taylor have pointed out, he seems to want only some identities recognized. Recognition becomes especially difficult for him when it comes to the margins, innovation, newness, and any challenge to the universalizability of identity.[20]

Honneth's discussion is both a bit more complex and a bit more accepting of difference than Taylor's. There are, he argues, three different kinds of disrespect: the violation of the body (here Honneth refers to torture), the denial of rights, and the denigration of ways of life.[21] Recognition here is much broader than a simple tolerance; individuals must be fully free of physical threats, offered complete and equal political rights, and have their distinguishing cultural traditions free from various forms of disparagement.

As with Young, both Taylor and Honneth contend that a *lack* of recognition—demonstrated by various forms of insults, degradation, and devaluation at both the individual and cultural level—is an injustice not just because it constrains people or does them harm, but because it "impairs these persons in their positive understanding of self—an understanding acquired by intersubjective means."[22] Taylor asserts that

> "the thesis is that our identity is partly shaped by recognition or its absence, often by the *mis*recognition of others, and so a person or group of people can suffer real damage, real distortion, if the people or society around them mirror back to them a confining or demeaning or contemptible picture of themselves. Nonrecognition or misrecognition can inflict harm, can be a form of oppression, imprisoning someone in a false, distorted, and reduced mode of being."[23]

Lack of recognition, then, is a harm—an injustice—as much as a lack of adequate distribution of various goods is.

Fraser argues that Honneth's politics of recognition is problematic because it is tied singly to self-realization; she argues that he does not recognize the key structural and institutional manifestations of misrecognition.[24] But Honneth is keenly interested in the importance of self-esteem in the political realm, and the fact that such self-esteem comes from recognition by others—not just from individuals, but also from culture and the state—leads to a thorough critique of the effect of cultural and political institutions. Honneth's second notion of disrespect specifically "refers to those forms of personal disrespect to which an individual is subjected by being structurally excluded from the possession of certain rights within a society."[25] His third form of disrespect includes the cultural and institutional concurrence in the denial of self-esteem. Honneth argues that a focus on self-realization *and* the institutional limits to both self- and other-based recognition is at the core of existing social movement struggles. So the implications of Honneth's notion of recognition go far beyond a simple call for internal self-realization, as Fraser asserts; a structural and institutional critique is an absolutely necessary part of the call for recognition.

Obviously, numerous social movements have focused on responding to various forms of misrecognition; there is certainly a relationship between the everyday experience of disrespect and the emergence of social movements such as indigenous rights, civil rights, gay and lesbian rights, feminism, and the more general movement for multicultural acceptance. As Connolly argues, a form of resentment grows with misrecognition, disrespect, and disempowerment. This resentment is not just individual and existential, but becomes civil resentment as well. Social movements arise as responses to disrespect and misrecognition move from the individual and personal to the collective community. These movements are a "collective struggle for recognition."[26] One only has to recall the striking images of civil rights activists in the 1960s marching while holding a simple and poignant message: "I Am a Man." Certainly, the call there went beyond justice as distribution, into the realm of recognition—at both the individual and community level.

As Fraser, Honneth, and Young have all argued here, mis- or mal-recognition is a cultural and institutional form of injustice. This type of cultural injustice is "rooted in patterns of representation, interpretation, and communication."[27] In confronting the injustices of cultural domination, nonrecognition, and lack of respect, various movements focus on remedies based in cultural, symbolic, and, ultimately, institutional change.

**Justice as Procedure**

Material distribution and recognition are two absolutely key notions of justice in the contemporary political realm. But a third focus on justice as *process*, including demands for broader and more authentic public participation, is often seen as the tool to achieve both distributional equity and political recognition. For Honneth, one form of disrespect or misrecognition—the lack of rights—is directly linked to democratic participation. Citizens are subject to a form of personal disrespect when they are "structurally excluded from the possession of certain rights within a given society. . . . The experience of being denied rights is typically coupled with a loss of self-respect, of the ability to relate to oneself as a partner to interaction in possession of equal rights on a par with all other individuals."[28] There is a direct link, for Honneth, between a lack of respect and recognition and a decline in a person's membership and participation in the greater community, including their right to participate in the institutional order.

Young makes this connection clear as well. For Young, a concept of justice needs to focus more generally on the elimination of institutionalized domination and oppression. To accomplish this, justice must focus on the political *process* as a way to address a variety of injustices, including both the inequitable distribution of social goods and the inequitable distribution of social recognition. In dealing with issues of justice beyond the distributive, Young insists on addressing justice in the "rules and procedures according to which decisions are made."[29] She says that

> the idea of justice here shifts . . . to procedural issues of participation in deliberation and decisionmaking. For a norm to be just, everyone who follows it must in principle have an effective voice in its consideration and be able to agree to it without coercion. For a social condition to be just, it must enable all to meet their needs and exercise their freedom; thus justice requires that all be able to express their needs.

The central focus for Young, in addressing justice both as distribution and as the recognition of difference, is on decision-making structures, and she argues for "democratic decision-making procedures as an element and condition of social justice."[30]

Likewise, Carol Gould insists that taking differences seriously in public life requires "a radical increase in opportunities for participation in contexts of common activity. . . . For if individuals have an equal right to determine their own actions and further, if engaging in common activity is one of the necessary conditions for their self-development, then it follows that there is an equal right to participate in determining the course of such common activity."[31] Gould, like Young and numerous others who advocate a model of discursive or communicative democracy, insists that this participation needs to happen in a variety of social and cultural institutions, as well as in the more specific context of politics and government.[32] Discourse models and calls for more participatory democracy are thoroughly compatible with the varied notions of justice in both theory and practice; they certainly address cultural norms, social discourses, and the role of institutions of power in issues of both equity and recognition. In this sense, increased participation can address issues of distribution and cultural misrecognition. Arguments for justice as procedure, then, demonstrate how varied notions of justice can be incorporated into a single project.

**The Necessity of a Linked Approach to Justice**

Some on the traditional left have lamented the move toward justice as recognition, especially as it has been developed in the "identity politics" of social movements or the postmaterial critiques of the "cultural" left. Unfortunately, Fraser notes, theorists have also generally insisted on a dichotomy between distribution and recognition, by focusing on one or the other conception of justice. But this sort of interpretation misses the point of those like Fraser, Honneth, and Young, who insist on a thoroughly integrated understanding of justice.

The whole point of Fraser's forays into the examination of these various justice claims is to show that they are not antithetical. Fraser argues that this split in the academic left between "social" justice and "cultural" politics—justice as equity and justice as recognition—represents a false dichotomy. Fraser insists that "justice today requires *both* redistribution and recognition," emphasizing that "justice requires both, as neither is sufficient."[33] Communities, or collectivities, are, in fact, "bivalent"—they are often differentiated as a collective by both economic structure and the status order of society. In this case, neither a politics of redistribution nor one solely of recognition will suffice to remedy injustice. "In general, then," according to Fraser, "one should roundly reject the construction of redistribution and recognition as mutually exclusive alternatives. The goal should be, rather, to develop a two-pronged approach that can address the twofold need for both."[34]

Likewise, Honneth also notes a relationship between material equity and justice as recognition. He recognizes the more utilitarian struggle over the equitable distribution of goods, including cultural goods, as a motivator for collective action. This is contrasted with a model of social conflict that has the denial of social or legal recognition at its core. But Honneth does not want to replace the theoretical model for the former with one for the latter: "It is important to stress . . . that this second model of conflict, based on a theory of recognition, should not try to replace the first, utilitarian model but only extend it." Like Fraser, Honneth argues that social movements encompass both notions of justice.[35]

And, importantly, some theorists are beginning to note that the relationship between justice as equity and justice as recognition is played out in the procedural realm, because barriers in both areas can hinder the ability of individuals and communities to participate. In *Justice and the Politics of Difference*, one of Young's primary emphases is on institutions and the political process. While she argues that distributive justice does not go far enough because it does not include a recognition of differences in the social realm—differences that go beyond who has how much—Young goes on to examine the institutional features that lead to injustices in terms of *both* distribution and recognition. This leads directly to her insistence on participatory democratic structures to address existing injustices based in both distribution and recognition. In a similar vein, Fraser argues that when "patterns of disrespect and disesteem are institutionalized, for example, in law, social welfare, medicine, public education, and/or the social practices and group mores that structure everyday interaction, they impede parity of participation, just as surely as do distributive inequities."[36]

The point here is absolutely crucial: it is not just that political and cultural institutions create conditions that hamper equity and recognition, but that both distributive inequity and misrecognition hamper real participation in political and cultural institutions. Issues of justice are not just bivalent, but trivalent. In this case, improved participatory mechanisms can help meliorate both other forms of injustice, but those forms of injustice must be addressed in order to improve participation. For a "parity of participation," Fraser argues, we need both objective and intersubjective conditions to be met. Objective conditions include a distribution of resources to ensure participants' independence and voice. Subjective conditions require that "institutionalized cultural patterns of interpretation and evaluation express equal respect for all participants and ensure equal opportunity for achieving social esteem."[37] As many discursive and communicative democrats argue, moral respect and the recognition of the right of all to participate are key principles for improving and extending democratic action.

Justice, then, requires not just an understanding of unjust distribution and a lack of recognition, but, importantly, of the way the two are tied together in political and social processes.[38] These notions and experiences of injustice are not competing notions, nor are they contradictory or antithetical. Inequitable distribution, a lack of recognition, and limited participation all work to produce injustice, and claims for justice can—some would say *must*—be integrated into a comprehensive political project. My argument here is that the environmental justice movement represents just such a project.

## The Justice of Environmental Justice

It is quite interesting that even given the wealth of literature devoted to environmental justice and the environmental justice movement in the past decade, there is no single definition available that simply lays out a definition of the term. Rather, what one sees in the literature is a variety of framings of the issue, by a rich assortment of both academics and activists. Still, the argument here is that the movement represents understandings of justice as distribution, recognition, and procedure—with different notions appearing throughout the literature of the movement. In Robert Bullard's first edited collection on the movement, for example, there is no systematic attempt to define the broad term *environmental justice*. Still, mentions of equity (in the distribution of environmental ills), recognition (with a focus on cultural recognition), and participation (particularly authentic, as opposed to inauthentic or token, inclusion) are evident throughout the book. The same can be said for other collections on the topic.[39] Early on, activists in the movement took on the definition of *environment*, broadening it beyond the more mainstream organizations' notion of parks, wilderness, and lands "outside." A refusal of narrow definitions is at the heart of the movement; this is explicitly stated in the case of the term *environment*, but is no less evident in the notion of justice as well.

**Environmental Justice and Distribution**

Still, the most often cited, and most obvious, evidence of environmental injustice is in the realm of distribution—specifically the inequitable share of environmental ills that poor communities and communities of color live with. Here, the call for "environmental justice" focuses on how the distribution of environmental risks mirrors the inequity in socioeconomic and cultural status. As Dowie has noted, "While created equal, all Americans were not, as things turned out, being poisoned equally." Benjamin Chavis's argument outlining "environmental racism" makes the point bluntly: "People of color bear the brunt of the nation's pollution problem."[40] Studies that demonstrated such inequity, such as a 1983 United States General Accounting Office report and the 1987 study by the United Church of Christ (UCC), *Toxic Wastes and Race in the United States*, spurred and empowered the movement. The UCC study found race "the most significant among variables tested in association with the location of commercial hazardous waste facilities. This represented a consistent national pattern." Similar conclusions have been found in studies done with regard to hazardous waste disposal sites, various types of incinerators, polluted water, toxic releases from industry, lead poisoning, and other types of environmental dangers.[41] The antitoxics side of the environmental justice movement makes the same argument along class lines: that not just race, but poverty is a central indicator of the presence of environmental damage in a community. In addition, studies have shown that agencies such as the EPA enforce environmental laws in poor communities and communities of color less stringently than in wealthy white communities.[42] The bottom line is that the "unifying insight of environmental justice recognizes that neither the costs of pollution nor the benefits of environmental protection are evenly distributed throughout our society."[43]

**Environmental Justice and Recognition: People, Culture, and Communities**

While distributional equity seems to be the central definition of justice in the environmental justice movement, this type of equity certainly does not encompass all of the critiques or desires of the movement. A central concern of the environmental justice movement is a conception of justice as recognition. This interest is notable throughout the environmental justice movement's literature and political action. Laura Pulido argues that central to environmental justice struggles is an engagement of issues of cultural meaning, including, but not limited to, identity. As Bullard argues, the "focus of activists of color and their constituents reflects their life experiences of social, economic, and political disenfranchisement." Struggles for environmental justice "are embedded in the larger struggle against oppression and dehumanization that exists in the larger society." Krauss's study of women in the environmental justice movement demonstrates that for both white activists and activists of color, "the starting places for and subsequent development of their analyses of toxic waste protests are mediated by issues of class, race, and ethnicity." And Tesh and Williams argue that identity is crucial for the movement as well, especially in its insistence on the validity of the experiential, subjective knowledge of grassroots activists and communities.[44] The bottom line here is that environmental justice activists often see themselves as outside the cultural mainstream. As such, their identities are devalued. The movement, then, turns to recognition as a key component of the justice of environmental justice.

This question of recognition is discussed in the movement both at the personal level and at the level of community; misrecognition is experienced in both realms. The more personal issue comes up in numerous activist testimonials. Cora Tucker, an African-American activist, discussed her reaction at a town board meeting, when white women were addressed as "Mrs. So and So," while she was addressed simply as "Cora" by the all-white, all-male board: "I said, 'What did you call me?' He said, 'Cora,' And I said, 'The name is Mrs. Tucker.' and I had the floor until he said 'Mrs. Tucker.' . . . It's not that—I mean it's not like you gotta call me Mrs. Tucker, but it was the respect."[45] Lois Gibbs, of the CHEJ, tells a similar story of a public hearing in which representatives appeared not to be listening to her testimony. She stopped speaking, and when the hearing official finally noticed the silence and asked if she was through, she simply said she was just waiting until someone was listening. Gibbs then continued her testimony. During the campaign to halt a proposed incinerator in South Central Los Angeles, women's concerns were often dismissed as irrational, uniformed, and disruptive. As Hamilton argues, male city and corporate officials "used gender as the basis for discrediting women's concerns." In hearings regarding a proposal to build a hazardous waste incinerator in Kettleman City, California, observers noted the different body language county commissioners expressed when Mexican-American residents and representatives of ChemWaste were at the microphone—patronizing on the one hand, and respectful on the other.[46] Misrecognition and disrespect on the individual level is an everyday experience for these activists; again, authentic recognition is a key element of their demand for justice.

But the question of recognition obviously goes beyond individual experiences and needs; questions of community and cultural recognition permeate the movement as well. The issue is community and cultural survival in a system where recognition is denied and communities and cultures are thoroughly devalued. Pulido argues that one key difference between the members of mainstream environmental organizations and members of environmental justice organizations is that the latter "draw people who already exist as a social or spatial entity in some way," as workers, a class, or community—and these communities insist on recognition.[47] A demand for community recognition was obvious in the battle of the Mothers of East Los Angeles (MELA) against the construction of a toxic waste incinerator. As Pardo notes, the "Mexican American women living east of downtown Los Angeles exemplify the tendency of women to enter into environmental struggles in defense of their community." She offers a quote from MELA activist Juana Gutierrez : "As a mother and resident of East L.A., I shall continue fighting tirelessly, so we will be respected."[48] Certainly, activists make a direct connection between the defense of their communities and the demand for respect.

For many in the environmental justice movement, this defense of community is nothing less than a matter of cultural survival. This is certainly central to Native American and other indigenous communities and activists. Lance Hughes, director of Native Americans for a Clean Environment, makes clear the reason for his organization's focus on environmental issues: "We are not an environmental organization, and this is not an environmental issue. This is about our survival." Winona La Duke, a prominent Native American activist in the environmental justice movement, cites sovereignty issues and cultural survival as key reasons for her participation in the movement.[49] In one study, interviews with a variety of Native American activists show that they have "a genocidal analysis rooted in the Native American cultural identification, the experience of colonialism, and the imminent endangerment of their culture." For activists interviewed in another study of indigenous and Chicana women in the Southwest, threats "to the environment are interpreted as threats to their families and communities." They see "toxic contamination of their communities as systematic genocide."[50] As Peña argues, "to the extent that we construct our identities *in place*, whenever the biophysical conditions of a place are threatened, undermined, or radically transformed, we also see these changes as attacks on our identity and personal integrity." Communities often feel like endangered species, and environmental justice battles are battles for the preservation of the "homeland environment" and the local knowledges and senses of place that exist in those communities.[51]

So the environmental justice movement focuses on individual and community self-empowerment; the point is to gain recognition for oneself, for one's own community, and for the movement as a whole. The African-American feminist scholar Patricia Hill Collins notes that the phrase "coming to voice" is being increasingly used in both feminist and black feminist writing; this notion includes breaking silence, developing self-reflexive speech, and confronting or talking back to oppressors.[52] This coming to voice and self-empowerment has become a central part of the environmental justice movement. A mainstay of slogans of the movement is the notion that "we speak for ourselves." As Di Chiro argues, the question of agency inherent in "speaking for ourselves" is a key issue for activists in the movement. Bullard notes that "African-Americans and other people of color must be empowered through their own organizations and institutions if they are to effectively address the problem" of environmental injustice.[53] This holds not just for communities of color, but for all individuals and communities fighting the injustice of misrecognition. The objective of the Center for Health, Environment, and Justice has been to empower local citizens to do their own organizing and networking. Women's experiences in these organizations have often transformed how they perceive their own identities. Gottlieb argues that these transformative experiences and the focus on one's own community create a very powerful image of that community. As Austin and Schill observe, grassroots people have proven that they are capable of leading, speaking, and doing for themselves. Finally, the various actions of individuals and community groups bring something else tied to political recognition: a sense of efficacy in the political process.[54]

**Environmental Justice and the Centrality of Participation**

Without a doubt, the demand for political participation in decisions governing communities is central to the environmental justice movement. The construction of inclusive, participatory decision-making institutions—a "place at the table," or equal, informed, respectful participation—is at the center of environmental justice demands. Benjamin Chavis includes "the history of excluding people of color from the mainstream environmental groups, decisionmaking boards, commissions, and regulatory bodies" in his definition of the term *environmental racism*. In Freudenberg and Steinsapir's study of the movement, the first and major shared perspective across the grassroots is the "right of citizens to participate in making environmental decisions—emphasis on process as well as content of decision making." Gould, Schnaiberg, and Weinberg state that "from our perspective, these groups are attempting to exercise their rights as citizens. They seek to have some say in the local development of their communities, in order to ensure that the quality of their lives will be protected." Bretting and Prindeville found a strong belief in the rights of citizens to participate in making environmental decisions common among all indigenous and Chicana activists interviewed. Capek's environmental justice "frame" includes a demand for accurate information, respectful and unbiased hearing of claims, and democratic participation in deciding the future of contaminated communities.[55]

The focus is on fully realizing democratic participation in environmental and community decision making. Bullard argues that democracy is key: "What do grassroots leaders want? These leaders are demanding a shared role in the decision-making processes that affect their communities. They want participatory democracy to work for them." Likewise, Hamilton notes the expansion of the concept and practice of democracy to be more inclusive of community input. She argues that the focus of the movement is on ecological democracy, including new forms of citizen participation in governance: "Multiple decision making units (like neighborhood councils) should regulate development and ensure citizen input on economic and environmental decisions."[56]

The demand for this type of authentic, community-based participation comes out of the experience of disenfranchisement. Krauss has discussed how white women in the movement originally believe in the political system and quickly learn the lesson that the government does not necessarily work for them: "In the process of becoming activists, blue-collar women who believed in the system dramatically shift perspective in their understanding of political life. They recognize the failure of the system as a whole to act on their behalf and their own disenfranchisement from the policy-making process. They have to find ways to expose this and make the system democratic." Lois Gibbs, former organizer at Love Canal and now director of the Center for Health, Environment, and Justice, recounts the same story.[57] Krauss observes that women of color activists already understand this lack of response through their personal histories of exclusions based on race and class. The lack of participation in environmental decision making comes, in large part, from the limitations of race, class, and gender. These present a range of structural obstacles—including less access to political, legal, scientific, and other resources—to full participation in environmental decisions.

Basically, environmental justice groups argue that the injustices they suffer come from a lack of state oversight (often based in mis- or malrecognition). The demand to counter this is not just a call for recognition, but also a call for more thorough and participatory local input into, and control over, environmental decisions. The call is often for "a place at the table."[58] Groups do not want others—either mainstream environmental groups or government agencies—simply saying that they will take care of the community's interests; they wish to be consulted from the start, speak for themselves, work with a variety of other groups and agencies, and be offered a full partnership in the making of decisions. Hunold and Young examine this relationship between participation and recognition in the case of hazardous waste citing. Applying Young's theoretical framework, they argue that public deliberation is not only the most likely path to a distributively fair solution, but that the process itself respects the interests and autonomy of people.[59]

Robert Lake is one of the few others who have directly examined the conception of justice in the environmental justice movement; he complains that the movement "generally overemphasizes issues of distributive justice" and "adopts an unnecessarily truncated notion of procedural justice."[60] Lake argues, along the lines of Young, that one simply cannot have a thorough distributive justice without having justice in the *procedures* for producing that distribution. Lake suggests that the movement's focus on distributional equity not only takes away from procedural equity, but also misses the centrality of procedure in producing inequitable distribution. But he seems not to recognize the amount of attention actually given to the issue of procedural equity in the movement. While he argues that the concern with procedural equity in the environmental justice movement is both limited and truncated, I see this concern as central. There is much within both the movement's literature and its political action that demonstrates the very key focus on participatory process.

These types of concerns with procedural justice are evident not just in the analysis of the problem, but also in the demands of the environmental justice movement. At the First National People of Color Environmental Leadership Conference in 1991, calls for procedural equity were on par with demands for environmental and social equity. The Principles of Environmental Justice, adopted at the Summit, include demands that "public policy be based on mutual respect and justice for all peoples," "the right to participate as equal partners at every level of decision-making including needs assessment, planning, implementation, enforcement and evaluation," and "the fundamental right to political, economic, cultural and environmental self-determination for all peoples."[61] The Southwest Organizing Project (SWOP) has also developed a "Community Environmental Bill of Rights." It includes "the right to participate as equals in all negotiations and decisions affecting our lives, children, homes and jobs," and the "right of access without cost to information and assistance that will make our participation meaningful, and to have our needs and concerns be the major factor in all policy decisions." In another example, the National Environmental Justice Advisory Council (NEJAC) to the EPA includes a subcommittee on Public Participation, which has developed a Model Plan for Public Participation, to be used by federal agencies in designing a process for participation for communities affected by environmental policies under consideration.[62] Again, in a place where policy is developed for the movement as a whole, procedure is central. The Model Plan suggests that policymaking procedures must encourage active community participation, institutionalize public participation, recognize community knowledge, and utilize cross-cultural formats and exchanges to enable the participation of as much diversity as exists in a community. Obviously, through these principles and policy suggestions, a shared and respected role in the decision-making process is a key demand of the movement. This demand for procedural equity is not just in the *principles* of the movement, but in actions on the range of diverse issues the environmental justice movement addresses.[63]

The demand for more participation in community development of environmental policy represents another type of shift from the standard understanding of justice. Notions of social justice often rely on government intervention to implement or design a more just society. Most often, demands for distributional equity are made to government, which is where the remedy of such injustice is sown. But the environmental justice movement calls for government intervention only in part: to establish just laws on the distribution of environmental risk. But in its call for justice in terms of both recognition and procedure, the movement calls for more grassroots and community democratic participation. In other words, the movement recognizes that a central government can only go so far in establishing justice; for more extensive development of justice, democratic participation in the development of rules people live by in a community (including how industry lives in a community) is key as well.

## The Interplay of Equity, Recognition, and Participation in Environmental Justice

As in the discussion of justice on the theoretical level, these three notions of justice in the environmental justice movement—regarding equity, recognition, and participation—are, and must be, thoroughly integrated. Conceptions of justice, and, more important, the experiences of injustice, are confronted in numerous ways at once. Within the environmental justice movement, one simply cannot talk of one aspect of justice without it leading to another. Taylor, in Bullard's first edited collection on environmental justice, notes that the environmental justice movement "integrates both social and ecological concerns much more readily [than the traditional environmental movement] and pays particular attention to questions of distributive justice, community empowerment, and democratic accountability." And Bunyan Bryant, one of the academic founders of the movement, insists that a thorough definition of *environmental justice* (as distinguished from environmental racism and environmental equity) is broad, referring to cultural norms, people realizing their highest potential, personal empowerment, and democratic decision making.[64] Unfortunately, these statements are often lost in the wealth of literature available on environmental justice, and many still misread the conception of justice in the movement as purely equity based. The argument here, however, is that not only are three different conceptions of justice apparent in the movement, but the movement also recognizes that these notions of justice must be interrelated: one must have recognition in order to have real participation; one must have real participation in order to get real equity; further equity would make more participation possible, which would bring further recognition, and so on.

One common critique of the tripartite approach to justice is that everything flows from distribution, or that redistribution will satisfy all such broader demands. Indeed, some argue that environmental justice activists would be satisfied if the government simply improved the distribution of environmental bads. But given the broad focus of environmental justice, I do not see activists being satisfied without cultural recognition and political participation; more important, activists recognize that such government capitulation *will simply not happen* without a broader cultural recognition of the victims of environmental injustice and their inclusion in problem solving and policymaking. There are definite relationships between inequity, misrecognition, and lack of participation; the integration of these concerns in the achievement of environmental justice is at the heart of the movement, in both its critiques and its intended solutions.

First of all, one can look to a lack of recognition and validation of identity as a central factor in the distribution of environmental risks.

Hamilton, for example, notes that land-use decisions reflect class and racial bias: "Because they reflect the distribution of power in society, they cannot be expected to produce an equitable distribution of goods."[65] The simple point is that there is a crucial link between a lack of recognition and the inequitable distribution of environmental bads; it is a general lack of value of the poor and people of color that leads to this distributional inequity. We can use Honneth, on the theoretical level, to examine this link. One of the central notions of respect and recognition for Honneth is physical integrity: "The forms of practical maltreatment in which a person is forcibly deprived of any opportunity freely to dispose over his or her own body represent the most fundamental sort of personal degradation."[66] While Honneth refers to how acts such as physical injury, torture, and rape deny recognition, we can certainly add unwanted exposure to environmental risks as an example of seizing control of a person's body against their will. Exposure to risk is a type of physical abuse, especially given the direct health effects shown to be produced by, for example, exposure to lead in urban housing or to uranium-mine tailings on Native American reservations. Again, there is a direct relationship between a lack of recognition and environmental degradation. These events are not independent, nor should they be considered as such. Activists in the movement understand this linkage; hence their interest in, and insistence on, both environmental equity and cultural recognition.

In addition, one can certainly see a link between a lack of individual or cultural recognition and a lack of valid participation in the political process. Simply put, misrecognition due to racism and/or classism creates real structural obstacles to political participation. Activists and academics alike in the movement criticize the mainstream environmental movement for ignoring and devaluing the poor and people of color by devaluing the environment they live in.[67] The major groups in the United States, the argument goes, are much more interested in wilderness or the great outside than in urban environmental issues because they value one understanding of "environment" over another. This is a form of disrespect. For Lee, self-determination and participation in decision making about one's own environment is absolutely key to environmental justice; it brings with it an appreciation of diverse cultural perspectives and an honoring of cultural integrity. For Bullard, the reason for insistence on "speaking for ourselves" is the empowerment of disenfranchised people and their inclusion in a more fully democratized process. Again, activists in the movement understand the linkage between recognition and participation in the political process.[68]

Finally, the analysis comes full circle. The combination of misrecognition and a lack of participation creates a situation of inequity in the distribution of environmental dangers. Pulido argues that environmental justice movements are "simultaneously about both material concerns and systems of meaning" and how the various forms of injustice suffered are mutually constitutive. "The task is to identify the ways in which racism, cultural oppression and identity interact with economic forces to create unique forms of domination and exploitation"; such a concern with this linkage leads to a need to challenge "the various lines of domination that produce the environmental conflict or problem experienced by the oppressed group in the first place."[69] This means confronting material inequality, cultural misrecognition, and other power relations that deny meaningful participation. It is a focus on the political process, specifically on demands for public participation and community empowerment, which are seen as the tools to achieve both distributional equity and recognition. For the environmental justice movement, the demand for more public participation and procedural equity in the development, implementation, and oversight of environmental policy is the key to addressing issues of both distributional equity and recognition.

This dedication to a threefold understanding of justice was apparent—in both the process and the results—at the First National People of Color Environmental Leadership Summit, held in Washington, D.C., in 1991. Organizers paid particular attention to recognition and participation of the very different participants. One of the most important insights at the conference, as noted in an evaluation of the process by Isaiah Madison, Vernice Miller, and Charles Lee, was a vision of a truly democratic process in which people of color could speak for themselves and, just as important, listen to each other. As Dana Alston argued at the Summit, "Our vision of the environment is woven into an overall framework of social, racial, and economic justice."[70] Elsewhere, Alston noted that "all cultures coming to the table would be respected; there would be equity as far as participation and voice, across gender, race, ethnicity, and region."[71] This process was used to develop the Principles of Environmental Justice. As noted previously, the seventeen principles include three that directly address political democracy; in addition, three are specifically devoted to issues of cultural integrity and recognition.[72] Again, the point is that the movement embodies an integration of various conceptions of justice; in dealing with the inequity in the distribution of environmental risk, the movement's major gathering emphasized that recognition and participation were absolutely crucial in the struggle for equity—both in their internal process and in their external political demands. One could not have one element of justice without the others.

## Conclusion

The movement for environmental justice may not add anything to the theoretical literature of the study of justice, but its analyses, practices, and demands undoubtedly offer a real-world illustration of these theoretical concepts in political action. Certainly, and at the very least, it should be clear that environmental justice means much more than a lack of equity in the distribution of environmental ills. More broadly, what the environmental justice movement demonstrates is the possibility of addressing different conceptions of justice simultaneously, and bringing numerous notions of justice into a singular political project. In this respect, I would argue that the movement demonstrates that it is, indeed, possible to incorporate both material and postmaterial demands in a single and comprehensive political movement. As Pulido has noted, environmental justice offers "a positive example of how postmodern identity politics can be linked to concrete material struggle."[73] The project of environmental justice goes one further, however, combining elements of economic and quality-of-life issues with identity politics within the context of a struggle for political participation and real political power. The concept of environmental justice illustrates that the theoretical arguments about the nature of justice are more than academic exercises; the issues surrounding a struggle for justice on all fronts has been brought to life clearly and forcefully by a very active and passionate political movement. Hopefully, the movement will achieve the justice it seeks in political practice. And along the way, given the very real engagement with theoretical issues of justice in political practice, I see the movement helping to transform the study of justice in the academy as well.

## Notes

1. For overviews, see Bunyon Bryant and Paul Mohai, eds., *Race and the Incidence of Environmental Hazards: A Time for Discourse* (Boulder, CO: Westview Press, 1992); Bunyan Bryant, *Environmental Justice: Issues, Policies, and Solutions* (Covelo, CA: Island Press, 1996); Robert Bullard, ed., *Confronting Environmental Racism: Voices from the Grassroots* (Boston: South End Press, 1993); Luke W. Cole and Sheila R. Foster, *From the Ground up: Environmental Racism and the Rise of the Environmental Justice Movement* (New York: New York University Press, 2001); Daniel Faber, ed., *The Struggle for Ecological Democracy* (New York: Guilford, 1998); Richard Hofrichter, ed., *Toxic Struggles: The Theory and Practice of Environmental Justice* (Philadelphia: New Society, 1993).

2. Previously the Citizen's Clearinghouse for Hazardous Waste (CCHW).

3. The National Oil Refinery Action Network (NORAN, at http://www.cbecal.org/alerts/alerts_oil.htm), and the Campaign for Responsible Technology (CRT, at http://www.svtc.org).

4. United States General Accounting Office, *Siting of Hazardous Waste Landfills and Their Correlation with Racial and Economic Status of Surrounding Communities* (Washington, DC: Government Printing Office, 1983); United Church of Christ, *Toxic Wastes and Race in the United States: A National Report on the Racial and Socio-Economic Characteristics of Communities with Hazardous Waste Sites* (New York: United Church of Christ, 1987); Robert Bullard, *Dumping in Dixie: Race, Class, and Environmental Quality* (Boulder, CO: Westview Press, 1990); Bryant and Mohai, *Race and the Incidence of Environmental Hazards*.

5. Andrew Szasz, *EcoPopulism: Toxic Waste and the Movement for Environmental Justice* (Minneapolis: University of Minnesota Press, 1994); Kenneth Gould, Allan Schnaiberg, and Adam Weinberg, *Local Environmental Struggles: Citizen Activism in the Treadmill of Production* (Cambridge: Cambridge University Press, 1996).

6. Barbara Epstein, "The Environmental Justice/Toxics Movement: Politics of Race and Gender," *Capitalism, Nature, Socialism* 8, no. 3 (1997): 63–87.

7. David Schlosberg, "Challenging Pluralism: Environmental Justice and the Evolution of Pluralist Practice," in *The Ecological Community: Environmental Challenges for Philosophy, Politics, and Morality*, ed. Roger Gottlieb (London: Routledge, 1997); Schlosberg, "Networks and Mobile Arrangements: Organizational Innovation in the U.S. Environmental Justice Movement," *Environmental Politics* 6, no. 1 (1999): 122–148; Schlosberg, *Environmental Justice and the New Pluralism: The Challenge of Difference for Environmentalism* (Oxford: Oxford University Press, 1999).

8. John Rawls, *A Theory of Justice* (Oxford: Oxford University Press, 1971), 9–10.

9. Brian Barry, "Sustainable and Intergenerational Justice," in *Fairness and Futurity: Essays on Environmental Sustainability and Social Justice*, ed. Andrew Dobson (Oxford: Oxford University Press, 1999).

10. David Miller, *Social Justice* (Oxford: Clarendon Press, 1976).

11. Michael Walzer, *Spheres of Justice* (Oxford: Blackwell, 1983), 6.

12. Andrew Dobson, *Justice and the Environment: Conceptions of Environmental Sustainability and Dimensions of Social Justice* (Oxford: Oxford University Press, 1998); Nicholas Low and Brendan Gleeson, *Justice, Society and Nature: An Exploration of Political Ecology* (London: Routledge, 1998).

13. Iris Marion Young, *Justice and the Politics of Difference* (London: Routledge, 1990).

14. Young, *Justice and the Politics of Difference*, 3.

15. Charles Taylor, *Multiculturalism* (Princeton, NJ: Princeton University Press, 1994); Axel Honneth, "Integrity and Disrespect: Principles of Morality Based on the Theory of Recognition," *Political Theory* 20 (1995): 187–201; Honneth, *The Struggle for Recognition: The Moral Grammar of Social Conflicts* (Cambridge, MA: MIT Press, 1995).

16. Honneth, "Integrity and Disrespect," 187, 188.

17. Taylor, *Multiculturalism*, 26.

18. Taylor, *Multiculturalism*, 37–38, 42.

19. Taylor, *Multiculturalism*, 66, 70.

20. See, for example, Thomas Dumm, "Strangers and Liberals," *Political Theory* 22, no. 1 (1994): 167–176. Taylor is also critical of attempts to deconstruct identity, which often come together with calls for recognition. This is the case with subjugated and stereotyped identities, such as gays/lesbians or Native Americans.

21. Honneth, "Integrity and Disrespect," 190–191; Honneth, *The Struggle for Recognition*, 132–134. This tripartite distinction among forms of recognition Honneth reads out of Hegel and Mead. The reference to Hegel is interesting, because it demonstrates a concern with the importance of recognition in a much earlier era. For Hegel (G. W. F. Hegel, *Hegel's Philosophy of Right* (Oxford: Oxford University Press, [1821] 1967)), the state is a community of individualized subjectivities, bound together while being recognized as individual subjectivities. The dialectical overcoming of individuality comes with recognition from the state.

22. Honneth, "Integrity and Disrespect," 189.

23. Taylor, *Multiculturalism*, 25.

24. Nancy Fraser, "Social Justice in the Age of Identity Politics: Redistribution, Recognition, and Participation," in *The Tanner Lectures on Human Values* 19 (Salt Lake City: University of Utah Press, 1998), 24.

25. Honneth, *The Struggle for Recognition*, 133.

26. William Connolly, *Political Theory and Modernity*, 2nd ed. (Ithaca, NY: Cornell University Press, 1993), 171–173; Honneth, *The Struggle for Recognition*, 164.

27. Fraser, "Social Justice in the Age of Identity Politics," 7.

28. Honneth, "Integrity and Disrespect," 190.

29. Young, *Justice and the Politics of Difference*, 23.

30. Young, *Justice and the Politics of Difference*, 34, 23.

31. Carol Gould, "Diversity and Democracy: Representing Differences," in *Democracy and Difference*, ed. Seyla Benhabib (Princeton, NJ: Princeton University Press, 1996), 181.

32. See, for example, John Dryzek, *Deliberative Democracy and Beyond: Liberals, Critics, Contestations* (Oxford: Oxford University Press, 2000).

33. Nancy Fraser, *Justice Interruptus: Critical Reflections on the "Postsocialist" Condition* (New York: Routledge, 1997), 12; Fraser, "Social Justice in the Age of Identity Politics," 5.

34. Fraser, "Social Justice in the Age of Identity Politics," 23.

35. Honneth, *The Struggle for Recognition*, 165. Unlike Fraser, however, Honneth sees such an integrated notion of justice in past social movements as well (pp. 166–167). He reads a concern for recognition, along with material concerns, in the histories of class activism in England by both E. P. Thompson and Barrington Moore. These studies, Honneth argues, offer empirical support for the thesis that "social confrontations follow the pattern of a struggle for recognition" in addition to, or alongside, struggles for distributional equity (p. 168).

36. Fraser, "Social Justice in the Age of Identity Politics," 26.

37. Fraser, "Social Justice in the Age of Identity Politics," 30.

38. There are, however, some significant differences between Fraser and Young, especially given Young's desire to downplay distribution and Fraser's concern that inequitable distribution is at the heart of much of the oppression Young addresses. See Fraser's discussion of Young in chap. 8 of Fraser, *Justice Interruptus*.

39. Bryant, *Environmental Justice*; Bryant and Mohai, *Race and the Incidence of Environmental Hazards*; Bullard, *Confronting Environmental Racism*; *Environmental Injustices, Political Struggles: Race, Class, and the Environment*, ed. David Camacho (Durham, NC: Duke University Press, 1998); Faber, *The Struggle for Ecological Democracy*; Hofrichter, *Toxic Struggles*.

40. Mark Dowie, *Losing Ground: American Environmentalism at the Close of the Twentieth Century* (Cambridge, MA: MIT Press, 1995), 141; Benjamin Chavis, Jr., "Foreword," in *Confronting Environmental Racism*, ed. Bullard, 3.

41. United Church of Christ, *Toxic Wastes and Race in the United States*, 9. See also, for example, Bryant and Mohai, *Race and the Incidence of Environmental Hazards*; Bullard, *Confronting Environmental Racism*.

42. Marianne Lavelle and Marcia Coyle, "Unequal Protection," *National Law Journal* 14 (1992): A16.

43. Bob Edwards, "With Liberty and Environmental Justice for All: The Emergence and Challenge of Grassroots Environmentalism in the United States," in *Ecological Resistance Movements*, ed. Bron Taylor (Albany, NY: SUNY Press, 1995), 36. There are numerous arguments about the accuracy of these equity claims; some studies have attempted to show no racial or class bias. Differences in findings occur depending on the level of analysis (from state-level data down to census tract) and the nature of the environmental problem (toxic releases, incinerators, waste dumps, and so on). For a discussion of the criticisms of the inequity approach, and a response from one of the researchers in the United Church of Christ study, see Benjamin A. Goldman, "What Is the Future of Environmental Justice?", *Antipode* 28, no. 2 (1996): 122–141. For constructive overviews of the equity literature, see Andrew Szasz and Michael Meuser, "Environmental Inequalities: Literature Review and Proposals for New Directions in Research and Theory," *Current Sociology* 45, no. 3 (1997): 99–120; James Lester and David Allen, "Environmental Justice in the U.S.: Myths and Realities," paper presented at the annual meeting of the Western Political Science Association, Seattle, 1999; William M. Bowen, *Environmental Justice through Research-Based Decision-Making* (New York: Garland, 2001). While I personally am convinced by the data, the empirical disagreements are immaterial to the argument here regarding the overall conception of justice constructed by the movement.

44. Laura Pulido, *Environmentalism and Social Justice: Two Chicano Struggles in the Southwest* (Tucson: University of Arizona Press, 1996), 13; Bullard, *Confronting Environmental Racism*, 7–8; Celene Krauss, "Women of Color on the Front Line," in *Unequal Protection: Environmental Justice and Communities of Color*, ed. Robert Bullard (San Francisco: Sierra Club Books, 1994), 262; Sylvia Tesh and Bruce Williams, "Identity Politics, Disinterested Politics, and Environmental Justice," *Polity* 18, no. 3 (1996): 285–305.

45. Quoted in Krauss, "Women of Color on the Front Line," 267.

46. Cynthia Hamilton, "Concerned Citizens of South Central Los Angeles," in *Unequal Protection: Environmental Justice and Communities of Color*, ed. Robert Bullard (San Francisco: Sierra Club Books, 1994), 215; Magdelena Avila, personal communication, 1994.

47. Pulido, *Environmentalism and Social Justice*, 25.

48. Mary Pardo, "Mexican American Women Grassroots Community Activists: 'Mothers of East Los Angeles,'" *Frontiers* 11, no. 1 (1990): 6, 4.

49. Lance Hughes, quoted in Trebbe Johnson, "Native Intelligence," *Amicus Journal* 14, no. 4 (1993): 12; Giovanna Di Chiro, "Defining Environmental Justice: Women's Voices and Grassroots Politics," *Socialist Review* 22, no. 4 (1992): 117.

50. Krauss, "Women of Color on the Front Line," 267; John Bretting and Diane-Michele Prindeville, "Environmental Justice and the Role of Indigenous Women

Organizing Their Communities," in Camacho, *Environmental Injustices, Political Struggles*, 149.

51. Devon Peña, "Nos Encercaron: A Theoretical Exegesis on the Politics of Place in the Intermountain West," paper presented at the New West Conference, Flagstaff, AZ, 1999, p. 6; Peña, ed., *Chicano Culture, Ecology, Politics: Subversive Kin* (Tucson: University of Arizona Press, 1998).

52. Patricia Hill Collins, *Fighting Words: Black Women and the Search for Justice* (Minneapolis: University of Minnesota Press, 1998), 46.

53. Dana Alston, ed., *We Speak for Ourselves: Social Justice, Race, and Environment* (Washington, DC: Panos Institute, 1991); Di Chiro, "Defining Environmental Justice," 98; Bullard, *Confronting Environmental Racism*, 202.

54. Robert Gottlieb, *Forcing the Spring: The Transformation of the American Environmental Movement* (Washington, DC: Island Press, 1993), 210–211; Regina Austin and Michael Schill, "Black, Brown, Red, and Poisoned: Minority Grassroots Environmentalism and the Quest for Eco-Justice," *Kansas Journal of Law and Public Policy* 1, no. 1 (1991): 74; Camacho, *Environmental Injustices, Political Struggles*, 27. While these groups and community members feel a lack of respect and recognition coming from various corporations, agencies, and government entities, they do get recognition and validation from their own environmental justice networks and organizations. They finally get a respectful ear from these networks and organizations, and this recognition leads to solidarity. For more on the construction of solidarity and networks, and the part of recognition in this process, see Schlosberg, "Networks and Mobile Arrangements," or Schlosberg, *Environmental Justice and the New Pluralism.*

55. Chavis, "Foreword," 3; Nicholas Freudenberg and Carol Steinsapir, "Not in Our Backyards: The Grassroots Environmental Movement," in *American Environmentalism: The U.S. Environmental Movement, 1970–1990*, ed. Riley E. Dunlap and Angela G. Mertig (Philadelphia: Taylor & Francis, 1992), 31; Gould, Schnaiberg, and Weinberg, *Local Environmental Struggles*, 4; Bretting and Prindeville, "Environmental Justice and the Role of Indigenous Women Organizing Their Communities," 153; Sheila Capek, "The 'Environmental Justice' Frame: A Conceptual Discussion and an Application," *Social Problems* 40, no. 1 (1993): 8.

56. Robert Bullard, ed., *People of Color Environmental Groups 1994–95 Directory* (Atlanta: Environmental Justice Resource Center, 1994), xvii; Cynthia Hamilton, "Coping with Industrial Exploitation," in Bullard, *Confronting Environmental Racism*, 67.

57. Krauss, "Women of Color on the Front Line," 108; Lois Gibbs, *Love Canal: My Story* (Albany, NY: SUNY Press, 1982).

58. Carl Anthony et al. "A Place at the Table: A Sierra Roundtable on Race, Justice, and the Environment," *Sierra*, May-June 1993, pp. 51–58, 90–91.

59. Christian Hunold and Iris Marion Young, "Justice, Democracy, and Hazardous Siting," *Political Studies* 46, no. 1 (1998): 87.

60. Robert Lake, "Volunteers, Nimbys, and Environmental Justice: Dilemmas of Democratic Practice," *Antipode* 28, no. 2 (1996): 162.

61. *Proceedings: The First National People of Color Environmental Leadership Summit*, ed. Charles Lee (New York: United Church of Christ Commission for Racial Justice, 1992), xiii–xiv.

62. Southwest Organizing Project (SWOP), *Intel Inside New Mexico: A Case Study of Environmental and Economic Injustice* (Albuquerque: SWOP, 1995), 100; United States Environmental Protection Agency, *The Model Plan for Public Participation* (Washington, DC: EPA Office of Environmental Justice, 1996).

63. It is also crucial to note that the emphasis on procedural justice and participation is evident not just in the external demands made by the movement, but also in the internal processes many groups and networks set out for themselves. See the discussion in Schlosberg, *Environmental Justice and the New Pluralism*, chap. 6.

64. Dorceta Taylor, "Environmentalism and the Politics of Exclusion," in *Confronting Environmental Racism: Voices from the Grassroots*, ed. Robert Bullard (Boston: South End Press, 1993), 57; Bryant, *Environmental Justice*, 6. Still, it is odd that a number of academics who study various notions of justice in the movement continue to conceptualize environmental justice as purely equity based. Pulido, for example, deals extensively with cultural recognition and the forms of institutional power that deny that recognition to groups, but the title of her book—*Environmentalism and Economic Justice*—does not reflect this complexity. Certainly, Pulido understands the movement in more sophisticated terms than that. Others, however, seem not to recognize their own limited definition. In "Environmental Justice and the Role of Indigenous Women Organizing Their Communities," Bretting and Prindeville specifically study the threat of cultural destruction faced by Chicana and indigenous women activists, yet they define environmental justice as purely distributional in the introduction to their work.

65. Hamilton, "Coping with Industrial Exploitation," 69.

66. Honneth, *The Struggle for Recognition*, 132.

67. See, for example, Dorceta Taylor, "Can the Environmental Movement Attract and Maintain the Support of Minorities?", in *Race and the Incidence of Environmental Hazards: A Time for Discourse*, ed. Bunyan Bryant and Paul Mohai (Boulder, CO: Westview Press, 1992).

68. Charles Lee, "Beyond Toxic Wastes and Race," in *Confronting Environmental Racism: Voices from the Grassroots*, ed. Robert Bullard (Boston: South End Press, 1993), 39; Bullard, *Confronting Environmental Racism*, 13.

69. Pulido, *Environmentalism and Social Justice*, 13, 32, 192–193.

70. Isaiah Madison, Vernice Miller, and Charles Lee, "The Principles of Environmental Justice: Formation and Meaning," in *Proceedings: The First National People of Color Environmental Leadership Summit*, ed. Charles Lee (New York: United Church of Christ Commission for Racial Justice, 1992), 50; Dana Alston, "Moving beyond the Barriers," in Lee, *Proceedings*, 103.

71. Quoted in Di Chiro, "Defining Environmental Justice," 104.

72. More generally, it is important to note the influence and recognition of Native American perspectives throughout the Principles, particularly important given the more urban focus of the organizers of the Summit. Paul Ruffins's account of the Summit ("Defining a Movement and a Community," *Crossroads/Forward Motion* 11, no. 2 (1992): 11) includes a discussion of the effect of bringing together Native American and Hawaiian activists with more urban-based African-American activists. The atmosphere of the Summit enabled him to recognize the validity of indigenous views of nature. After years of bitter feeling about the white environmental community's focus on wilderness and animals rather than the urban environment, indigenous activists helped Ruffins to experience, for the first time, "the moral imperative of protecting animals and trees and land."

73. Pulido, *Environmentalism and Social Justice*, xvii.

# II

# Philosophical Tools for Environmental Practice

# 4

# Constitutional Environmental Rights: A Case for Political Analysis

Tim Hayward

The importance of providing for environmental protection at the highest political level is now widely recognized. Globally, more than seventy countries have constitutional environmental provisions of some kind, and in at least thirty cases these take the form of environmental rights.[1] No recently promulgated constitution has omitted reference to environmental principles, and many older constitutions are being amended to include them.

Within Europe, the constitutions of seven current European Union (EU) member states—including Germany, Greece, and the Netherlands—contain express environmental provisions, and so do those of all East European states. Express *rights* to environmental protection figure in the constitutions of Finland, Portugal, and Spain, as well as in those of the Czech Republic and Slovenia. At the "constitutional" level of the Union itself, environmental protection has been accorded the unique status of being a required component of the Community's other policies. To be sure, neither the European Convention nor the European Social Charter presently provide for a *right* to environmental quality, but it is reasonable to ask whether one or both of them should.[2]

Even within the United Kingdom, it is no longer inappropriate to pose the question whether citizens can or should be granted an environmental right.[3] While the absence of a codified constitution or formalized rights might once have seemed sufficient reason to disregard this sort of question, that reason is clearly ceasing to apply.[4] With the recent incorporation of the European Convention on Human Rights, and the possibility of further constitutional reforms, the question of what rights UK citizens want to have, and whether they should include environmental rights, is likely to become a more prominent item on the political agenda.

In short, the time is ripe to assess the merits of the case for substantive constitutional environmental rights. In what follows I present the case in favor of constitutional environmental rights and analyze its problematic aspects. After a preliminary clarification of the scope of the right under consideration, I go on to address four sorts of questions bearing, respectively, on the validity, necessity, practicability, and desirability of pursuing environmental protection by means of constitutional rights. In the course of doing so I will identify various issues that warrant further research in a UK and European context, and in the conclusion I indicate some associated tasks for political scientists and theorists.

## The Scope of the Right

The scope of the right under discussion is basically that proposed in the Brundtland report: "All human beings have the fundamental right to an environment adequate for their health and well-being."[5] A similarly worded right is now found in many of the national constitutions promulgated or revised since the report's publication. In choosing as the criterion of environmental protection the adequacy of the environment for human health and well-being, this right clearly does not capture all aspects of environmental concern. Its most obvious application would be with respect to pollution, waste disposal, and other sorts of toxic contamination, since the most immediate threats to health and well-being concern contamination of air, water, and food. However, depending on how health and, particularly, well-being are construed, many other issues could ultimately be brought under this rubric, including aspects of environmental concern that touch on the quality of life in aesthetic, cultural, and spiritual terms.[6]

Nevertheless, the right under consideration may be thought vulnerable to the more radical criticism that it is thoroughly "anthropocentric," because it considers the environment only under the aspect of its contribution to *human* health and well-being: no provision is explicitly sought for the nonhuman beings that coexist within our environment, and no mention at all is made of the environment "for its own sake." I will not offer a detailed response to this line of criticism, because I have already dealt with it at length elsewhere.[7] Here I think it suffices to make the following points. The first is that constitutional environmental rights (for humans) are not offered as a panacea for all of the problems arising from our interactions with nonhuman nature; they are proposed as just one, albeit significant, approach to dealing with them. This approach does not preclude others, and may indeed serve to support them and to enhance their potential for success. After all, even when environmental concern focuses on the good of nonhumans, its success depends on the political, economic, and legal resources available to the humans pressing the case. These, I believe, are on the whole more likely to be enhanced than hindered by certain entrenched rights. Furthermore, there is reason to believe that once a basic right is established, practical jurisprudence and wider social norms will develop progressively to support more ambitious aims.

One further point to clarify is the scope of the right's corresponding duties. Constitutional rights are typically held by individuals against the state. For some purposes, theorists distinguish between "positive" and "negative" obligations on states; furthermore, rights can also give rise to duties of individuals or other nonstate actors, thus having "horizontal" as well as "vertical" effect. For the purposes of this chapter, however, I will make the simplifying assumption that the duties at issue are primarily those of the state to implement and enforce laws that secure to the individuals the enjoyment of what is intended as the substance of the right.

## Environmental Protection as a Genuine Human Right

The first critical question to consider is whether environmental protection is the sort of thing that is appropriately made the substance of a human right. Taken as a general moral proposition, it would be hard to deny the claim that humans have a right to an adequate environment. Moreover, it is easy to appreciate the potential advantages for the environmental cause of making a linkage with human rights: the human rights discourse embodies just the sort of nonnegotiable values that seems required for environmental legislation; rights mark the seriousness, the "trumping" status, of environmental concern; they articulate this concern in an institutionalized discourse with some established mechanisms of enforcement. A less clear-cut question, though, is whether these potential advantages can be realized in practice.

At the international level, evidence that they can is not conclusive, but it is encouraging. A recognition in international forums that environmental law and human rights have many significant areas of overlap has been steadily growing over the past three decades. The link between human rights and environmental protection was first clearly established by Principle 1 of the Stockholm Declaration on the Human Environment of 1972, and given a further impetus with the Brundtland report of 1987, which presented the basic goals of environmentalism as an extension of the existing human rights discourse. More recently, the UN Commission on Human Rights commissioned a special report on human rights and the environment. The resulting *Draft Declaration of Principles on Human Rights and the Environment*,[8] suggests Popović, might serve "as a vehicle for development of a formal, binding international legal instrument that protects environmental human rights."[9] Furthermore, there have also been regional agreements providing environmental rights,[10] and, particularly significantly, an almost universally ratified treaty, the 1989 UN Convention on the Rights of the Child, provides, in Article 24, a right of the child to "the highest attainable standard of health," whose implementation requires "taking into consideration the dangers and risks of environmental pollution."

So, while it cannot yet be confidently asserted that international law recognizes a human right to an adequate environment, there have been significant moves in this direction. It would appear, therefore, that from the standpoint of jurisprudence there is no insurmountable obstacle to adding an environmental right to the existing catalog. For effective implementation, though, it is crucial that the right be given the force of a constitutional provision.

## Environmental Protection at the Constitutional Level: Is a New, Substantive, Right Necessary?

Constitutional provision for environmental protection can take a variety of forms, and the entrenchment of a substantive right is not the only potential option. On the one hand, it might not be necessary to entrench rights at all if other constitutional provisions, in the form, for instance, of policy principles, sufficed to accomplish everything sought by the right in question. On the other hand, it may not be necessary to entrench a new substantive environmental right if either existing substantive rights suffice to achieve the desired goals or if they can be achieved by purely procedural rights. Accordingly, since there are good general reasons for not inventing superfluous new rights, I will seek to assess the adequacy of these alternatives as they appear in the European context.

**Policy Principles and Directives**

The Treaty of Union provides that Community policy "shall be based on the precautionary principle and on the principles that preventive action should be taken, that environmental damage should as a priority be rectified at source and that the polluter should pay."[11] This is an unequivocal statement, with constitutional force, of principles of environmental protection. Nevertheless, the relevant legal provision, Article 130r, only lays down principles on which Community policy will be based; it does not impose any particular obligation to act in a particular way by member states: in the absence of determinate obligations there is a corresponding absence of rights. Moreover, even in principle, these policy principles seem unsuited to generating rights. The polluter-pays principle does not require the polluter to pay to an individual right bearer; rather, it provides an allocative mechanism to internalize negative externalities.[12] Environmental impact assessment largely entails procedural rights (see below). The precautionary principle may look like a more promising basis for generating rights, as Christopher Miller notes, but its implementation still depends on national courts recognizing it, and its interpretation is a matter of considerable controversy.[13]

If these principles do not generate rights, what about the directives? Miller emphasizes that directives provide goals rather than rights. Yet since directives may be capable of direct effect, and sometimes the conferral of individual rights is a condition for direct effect, some commentators believe that rights can be bestowed on individuals by quality standards that have a very general protective aim.[14] Nevertheless, it is not always the case that where a provision is capable of creating a right, that right can be effectuated, especially because this depends in practice on national courts' rules on establishing causation and liability, and these often amount to formidable obstacles in environmental cases.

**The Environmental Potential of Existing Substantive Rights**

Environmental issues have been brought to the attention of the European Convention institutions under a number of existing rights.

Article 2 of the European Convention provides a right to life. While traditionally understood as a protection against (arbitrary) deprivation of life by the state, some commentators believe this right might be invoked by individuals to obtain compensation in the event of death from environmental disaster, insofar as the state is responsible.[15] Yet it is debatable whether the right also involves positive obligations on the state to preserve or promote life expectancy, for instance, via less polluted water or air. Some legal theorists have argued that this right does have extensive implications for environmental protection: Stefan Weber[16] argues that it is the European legal provision most appropriate to protect the environment, invoking the suggestion that its rationale is the protection of life from all possible threats. Nevertheless, this is probably a minority view, and while the right to life may have some potential application in the environmental field, it has not yet been successfully invoked. Moreover, it would only seem to apply in relation to drastic and present harms, or at least to direct threats to life, and thus not cover other serious environmental concerns.

Many of the concerns not covered by the right to life might appear to be covered, though, by the right to health, provided in Article 11 of the European Social Charter Part I, which is a right of everyone to benefit from any measures enabling them to enjoy the highest possible standard of health attainable. The Committee of Experts has taken the view that this provision requires states to take measures to prevent air and water pollution, provide protection from radioactive substances, and facilitate noise abatement, food control, and environmental hygiene.[17] However, neither the wording nor the force of this right seems adequate to many aims of environmental protection. A similar point applies to Article 3 of the Social Charter, which provides a right to safe and healthy working conditions. This right does not seem to have had much impact at the national level.

Interestingly, a right that has been used to set potentially important precedents for environmental protection is provided by Article 8 of the European Convention, a right to respect for one's private and family life and one's home. A particularly significant precedent was the 1994 case of *Lopez-Ostra v. Spain.*[18] The applicant had suffered serious health problems from fumes from a tannery waste treatment plant, and her attempt to obtain compensation from the Spanish courts had been completely unsuccessful. The European Court of Human Rights held that there had been a breach of Article 8. This judgment may be claimed to have enhanced the legal protection of the environmental victim by opening the door to applying Article 8 to nearly all sources of pollution, not only to noise emissions as in previous cases. Nevertheless, this right is still ultimately tied to the concerns its words state—that is, private and family life—rather than the environmental well-being of individuals, whoever and wherever they are, in public or private spaces. So while the environmentally favorable interpretations of this right illustrate the possibilities for stronger environmental rights, it remains likely that such possibilities can only be fully realized through the instantiation of more specifically environmental rights.

Before saying more on this, however, it is important to acknowledge the role procedural rights can play, not only because these are significant in themselves but also because the prospects of success for substantive environmental rights are likely to depend on them. The issue will be to determine whether they are not, as a number of commentators and campaigners believe they are, sufficient as well as necessary.[19]

**Procedural Rights**

Procedural rights relevant to the civil, political, and legal possibilities for environmental protection include rights to information, rights of legal redress, and rights of participation. The potential of these rights has long been recognized, and it was given a significant additional impetus by "The Convention on Access to Information, Public Participation in Decision-Making and Access to Justice in Environmental Matters," generally known as the Aarhus Convention (1998). The Aarhus Convention, developed under the auspices of the UN Economic Commission for Europe (ECE), was signed, in June 1998, by 35 countries from this region, which covers the whole of Europe as well as parts of Central Asia, the United States, Canada, and Israel—although the North American countries opted out of the process.

At present a good deal of attention is focused, and quite appropriately, on the right to information. Article 10 of the European Convention guarantees the freedom to receive and impart information in general. This does not specifically mention *environmental* information. However, in 1990 the Council of the European Communities adopted the *Directive on the Freedom of Access to Information on the Environment*, whose objective is to ensure that information on the environment held by public authorities is made available to any person on request. Moreover, a new directive on environmental information, intended to meet the more stringent demands of the Aarhus Convention, is currently in preparation. While an adequately enforced right of access to environmental information is an indispensable prerequisite for effective and democratic measures to protect the environment, however, it is equally clearly not sufficient on its own, since there also has to be scope for putting the information to effective use.

Another necessary but insufficient condition is the loosening of the rules of standing in relation to rights to seek legal redress in the environmental arena, including "rights to object to ministerial and agency environmental decisions; and rights to bring action against departments, agencies, firms and individuals that fail to carry out their duties according to law."[20] There is already some movement in this direction within European states, including the UK.[21] But while such rights clearly have their use, they are a somewhat cumbersome resource: given the time, expertise, and, especially, costs involved in seeking civil remedies or judicial review, they tend to be exercised sparingly, and mostly by well-financed organizations.

Meanwhile, perhaps of greater practical significance are rights of participation in environmental policymaking, in decision-making processes regarding environmental developments, and in the determination of environmental standards. There is a legal basis at the EU level for consolidating such rights, and there have been encouraging developments at the member-state level.[22] Still, while democracy may be enhanced by procedural rights of participation in decision-making processes, effective opportunities to participate may be far from equal for all citizens. In fact, these rights are usually exercised most effectively by particular interest groups whose claim to represent the "public interest" is always open to legitimate contestation—by individual citizens as well as by disaffected members of the groups themselves, and, of course, by parties whose interest in the environment is primarily an instrumental economic one.

Thus procedural rights can also cut both ways, since parties who object to proposed action by the state to protect the environment also have a right to challenge such action. Moreover, historically, as well as currently, "environmentalist" claims of individuals have sometimes been supported by prevailing views of the "public interest" and sometimes pitted against it.[23] Such considerations can reasonably be argued to point to the need for a substantive environmental right that stands above any potential political and legal contestation.

Of course, giving the right that high status in relation to other social goals raises critical questions regarding both its practicability and its legitimacy, which are the topics for the next two sections respectively.

## The Question of Practicability

One problem is the notorious difficulty of getting clear and unequivocal interpretations of locutions like "decent" or "adequate" environment. However, this may not be a decisive objection since, as Nickel argues, "human rights typically set broad normative standards that can be interpreted and applied by appropriate legislative, judicial, or administrative bodies at the national level. . . . The proposed standard of '*adequate* for health and well-being' . . . provides a general, imprecise description of the level of protections against environmental risks that States should guarantee. Risk standards should be specified further at the national level through democratic legislative and regulatory processes, in light of current scientific knowledge and fiscal realities."[24] Thus the substantive meaning of the right may be possible to determine over time.

Nevertheless, issues of indeterminacy are not merely definitional. The nature of environmental problems is such that their causes are often difficult or impossible to identify with the degree of accuracy necessary to support legal action against specific alleged polluters; it is correspondingly difficult to assign specific duties to individuals or firms that are directly correlative with the right to an adequate environment. In this regard, it is important to note that the right under consideration is not necessarily intended to have the (horizontal) effect of directly creating liabilities of private parties. Rather, if the right is enforceable against the state (vertical effect), as it most often is in constitutions, and as I am envisaging, this difficulty need not arise at the point of claiming the right.

That is, if environmental quality has fallen below the threshold of protection that citizens have a right to have maintained, the responsible authority has a duty to take appropriate action (e.g., implementing new regulations or better enforcing existing ones). This approach will indeed involve working with some hypotheses about causation, but it would not need to identify causes with the same degree of precision as would be required in a civil or criminal suit against an alleged polluter. In fact, a strength of the rights approach is that it would underpin courts' powers and duties to deploy the precautionary principle and the principle of proportionality so that lack of evidence of harm is not automatically equated with evidence of no harm.

Of course, the uncertainty factor also means that courts may face formidable problems of knowledge, but courts routinely have to deal with testimony from experts in order to arrive at judgments, and if environmental cases really do prove too complex, a solution might be to establish a specialist environmental court.[25]

Another objection is that if the proposed constitutional right were made fully justiciable, it would open the "floodgates of litigation," placing unworkable demands on the courts.[26] However, it would be a mistake to take this as a knockdown argument,[27] and while accepting that it is in no one's interest for courts be deluged with unfulfillable claims, we should nevertheless appraise the different means for averting a possible flood. One is simply not to grant the right in the first place, but given the moral case for the right, there is no principled justification for that. Another is to restrict *standing* to press the right in courts. Again, however, this would contravene the moral principle that everyone should have equal access to the means for redeeming the right. A third possibility, representing a less drastic compromise of principle, but no loss of efficacy compared to the other two, is to specify what counts as a justiciable case in terms that would keep the potential volume of litigation proportional in relation to competing social values. In fact, most "social rights" at the constitutional or European level already are hedged in by appropriate qualifications of this sort, and if the restrictions and qualifications are reasonable, the right can still fulfill its purpose.[28]

Nevertheless, the right could still be unenforceable for other reasons. A particularly significant problem, according to some critics, is that courts just do not have the necessary powers to enforce it. This objection is sometimes expounded on the basis of a distinction between "negative" and "positive" rights. Whereas negative rights are taken to be relatively unproblematic to enforce because they essentially consist in freedoms from state interference and coercion, positive rights are more problematic because they require the state actively to provide certain substantive goods or services. Thus Cass Sunstein, for instance, argues that since courts cannot create government programs, do not have a systematic overview of government policy, and lack the tools of a bureaucracy, it is unrealistic to expect them to enforce positive rights, among which he counts the right to a clean environment.[29] Taking this view, the most that a constitution ought to do is provide policy guidelines, not enshrine actual rights. This claim is reinforced by the further consideration that environmental protection is just one social good that has to compete with others for a share of a nation's resources, and if a nation cannot afford the resources necessary for implementing an environmental right, constitutionalizing it might do more harm than good. For if some rights prove unenforceable—and Sunstein is particularly concerned about this in East European contexts—other civil and political rights might be undermined.[30] This worry, though, appears to lack empirical support, since, as Herman Schwartz observes, courts throughout Europe, East and West, where these "unenforceable" economic and social rights are constitutionally enshrined, have vigorously protected the other rights.[31]

Moreover, this view rests on assumptions about both the classification and the purpose of constitutional rights that can be questioned. On the one hand, the distinction between "positive" and "negative" rights is problematic both in principle and in practice.[32] While negative rights can require positive action—and cost money—to enforce, some so-called positive rights, and the right to an adequate environment would be a prime example, could very well be seen as a "right to be left alone." The effective enforcement of rights in general, however classified, and the allocation of resources to this end, are ultimately questions of political priorities. Thus, on the other hand, there is an issue about the *purpose* of constitutional rights. Commentators often stress that the basic point of constitutional rights is to provide campaigners and ordinary citizens with the judicial means to rectify environmental wrongs. Yet it is not the main purpose of constitutional rights to bring about a proliferation of legal suits, much less a generally litigious culture; rather, the goal is to secure individuals in the protection and enjoyment of the substance of the right.

It is therefore appropriate and important to consider other functions—legal and extralegal—of the right under consideration. Even within the realm of legal enforcement, some of the functions are not directly related to individual justiciability. One legal consequence of such a right would be the enactment of further environmental protection legislation, for while legislation can be inspired by provisions other than rights, rights provide a stronger stimulus due to their "trumping" force. Relatedly, it would guide judicial interpretation, serving "to stimulate a more environmental appreciative application and evolution of legal concepts by the judiciary,"[33] and would likely be enhanced by extralegal effects such as the positive feedback from society in what Eckersley refers to as an upward ratcheting effect on political expectations.[34] Stevenson also emphasizes the broader educational role of the right, particularly in fostering a publicly recognized environmental ethic.[35]

## A Democratic Deficit?

Having examined questions about whether environmental rights *could* be enforced, I turn to the question of the legitimacy of trying to enforce them. The point of constitutionalizing rights is to set them above the vicissitudes of everyday politics, but this is also effectively to raise them above the possibility of (routine) democratic revision. Thus a worry of Jeremy Waldron, for instance, concerning the entrenchment of *any* canonical list of rights, is that it transfers too much power from an elected legislature to the judiciary. Since courts do not simply enforce rights but also have to interpret them, Waldron argues that they "will inevitably become the main forum for the revision and adaptation of basic rights in the case of changing circumstances and social controversies."[36] Political theorists, he argues, should have grave misgivings about this prospect: "Our respect for . . . democratic rights is called seriously into question when proposals are made to shift decisions about the conception and revision of basic rights from the legislature to the courtroom."[37]

However, while a degree of caution is certainly appropriate, the genuine worry here can be overstated. For one thing, courts have a legitimate function in a democracy. Judicial enforcement of a written constitution means, to quote a venerable source, "that where the will of the legislature declared in its statutes, stands in opposition to that of the people declared in the constitution, the judges ought to be governed by the latter, rather than the former."[38] If democracy requires the rule of law, judicial powers cannot be seen as straightforwardly opposed to democratic principles. Indeed, as François Du Bois says, the "relationship between courts and legislatures is more subtle than the contrast between judicial and democratic institutions suggests."[39] Certainly, citizens' access to the institutions of justice is an important feature of any constitutional democracy, especially given that their effective ability to influence the legislature may be more limited than Waldron's position implies. Moreover, many environmental decisions are made not by the legislature but by the executive, and so, as Eckersley points out, "in so far as trade-offs must be made, it is better that they be made solemnly, reluctantly, as a matter of 'high principle' and last resort, and under the full glare of the press gallery and law reporters rather than earlier in the public decision-making process via the exercise of bureaucratic and/or ministerial discretion that is presently extremely difficult for members of the public to challenge."[40]

Another of Waldron's worries concerns the placing of binding constraints on future citizens, limiting their autonomy in policymaking through principles developed on the basis of historically superseded exigencies. However, although his argument purports to be directed against *any* canonical list of rights, it does not seem to apply to *all* kinds of entrenched rights. As noted, he refers approvingly to "democratic rights," and these cannot effectively be respected without guaranteeing at least the standard liberal civil and political rights into the future. However, this defense might be thought not to hold for other sorts of rights—those of the so-called "second" and "third" generations—or for environmental rights in particular. One way of maintaining a defense of the latter might be to assimilate environmental protection to the model of liberty protection. As Sax points out, if we focus on the idea that individuals have a basic "right" to choose, we recognize that our activities that impoverish the environment can constrain their objective possibilities for choosing. Yet since it is arguable that some development activities can also enhance those possibilities, this line of defense is problematic. Being too stringent with environmental and resource controls could compromise future people's effective capacities for productive development.[41] Indeed, this might apply to the current generation as well. In short, it may seem appropriate, especially in view of the high degree of uncertainty involved, to leave the striking of a balance between environmental risks and potential economic benefits to processes of democratic decision making. From this perspective it could be concluded that substantive environmental rights threaten to compromise democracy and freedom.

Nevertheless, it should be noted that the force of this objection depends in part on the availability of genuinely democratic environmental decision-making processes, and in a society that is free and democratic, a basic principle is that each individual (now, as in the future) ought to be able to choose which risks she is prepared to run. This provides a clear linkage between human rights and democracy: the rights issue is that no individual should be compelled to run a risk for the greater social good; the democracy issue is that since genuine choice is a key to self-government, "assuring that risks taken are the product of such genuine choice is fundamental to the legitimacy of environmental decisions."[42] To guarantee the possibility of genuine choice, various procedural rights (as noted above) are indispensable, but they are not sufficient. If democracy means the equal effective possibility of participating in decision-making processes and assenting to their outputs, it means that certain disadvantages have to be offset. It can and does frequently happen that risks chosen by majoritarian democratic processes fall particularly heavily on certain groups or individuals. While the latter may sometimes "assent" to the heightened risk, it is neither just nor truly democratic to allow a situation where poor people are compelled, out of material necessity, to accept environmental harm in return for some benefit.

It is therefore arguable that a basic environmental minimum is a precondition for democratic decision making. Hence, as Sax suggests, the majority might be said to owe to each individual a basic right not to be left to fall below some minimal level of substantive protection against hazard; thus a basic norm could be "that the least advantaged individual is insulated against imposition of risk below some minimal threshold within his or her own society."[43] He therefore believes that "a fundamental right to a substantive entitlement which designates minimum norms should be recognized."[44]

## Conclusion

Taken as a moral proposition, the claim that all human beings have the fundamental right to an environment adequate for their health and well-being is, I believe, unimpeachable. This fundamental right is also gaining ground as a norm of international law, is receiving explicit recognition in authoritative international documents, and has arguably already come some way to becoming an enforceable legal right in international treaties. Its progress in this direction is underpinned by the fact, which is also independently significant, that the right is explicitly provided in many national constitutions. Moreover, the appropriateness of at least some form of constitutional provision for environmental protection is even more widely accepted, and the case for procedural environmental rights—to information, participation, and redress—is all but unanswerable.

Where there is room for debate at the constitutional level, however, is over the question whether the moral right should necessarily be translated into a substantive right, rather than receive indirect protection by means of procedural rights and/or statements of policy principle. Whether a substantive right is necessary, practicable, or democratically desirable depends on a range of factors, the influence of each of which will vary according to different constitutional contexts. Focusing on the context of the United Kingdom, within the European, a number of the factors identified by legal theorists have a distinctly political dimension requiring analysis in its own terms.

Whether a constitutional environmental right is needed is not simply a question of whether there is a distinct area of environmental concern previously untouched by existing rights and environmental law. Other relevant questions are whether interpretations in practice are coherent with aims in principle, and whether implementation and enforcement are adequate; these questions invoke criteria—"coherence" and "adequacy"—that make implicit reference to basic social values that are politically chosen. At present, UK environmental law as a whole lacks a coherent set of principles and rights. Thus, in cases not clearly falling under a particular statute or common law precedent, judges are constrained to classify the environmental concern as a partisan cause, and although it is open them to treat the environmental concern as a public interest, this is a matter for their discretion. With constitutional recognition of a right to an adequate environment, there would be a greater incentive for legislators and judges to treat this not as a partisan cause but as a fundamentally important public interest.

Deficiencies in application could then be better addressed. In the United Kingdom, as in other EU member states, the "margin of appreciation" allowed in the implementation, enforcement, and interpretation of Community obligations relating to the environment means the latter are often inadequately complied with, especially when economic interests compete. A constitutional environmental right would both signal a commitment to a better balance between economic and environmental interests, and provide the basis for some means for achieving this in practice. For instance, the right could underpin changes in legal presumptions so as to mitigate the present obstacles to successful actions aimed at environmental protection. The burden of proof in relation to causation could be altered; stricter standards of liability introduced; rules of standing further liberalized; and, more generally, the efficacy of procedural rights in the interests of environmental protection could also be significantly bolstered.

In considering the practicability of the right, though, attention should not be restricted to issues involving rights claims in courts. The right would serve not only to enhance the quality of environmental legislation but also to support a view of the environment as a public interest within a more complete set of social values less skewed to narrow economic interests. This would help enhance citizens' understanding of rights and responsibilities in relation to the environment and allow greater democratic input into environmental decision making.

On the question of democratic legitimacy, a factor emphasized by numerous scholars is that the efficacy of constitutional provisions owes a good deal to institutions of judicial review. This raises the question whether strengthening these, and perhaps encouraging judicial activism, enhances or mitigates prospects for democratic decision making. While some commentators, and especially public interest lawyers, see a significant role for judicial activism,[45] others, and not only "conservatives," are critical of the scope this gives for blurring the separation of powers, and of the practical problem that judicial innovation can be quite selective. However, the issue is perhaps best addressed not as one of either judicial activism or passivism—since it is undesirable for judges either to exceed or to fail fully to exercise their constitutionally conferred powers—but rather as one of clarifying the values underpinning judicial interpretations. A constitutional environmental right could establish values that, without it, judges would quite properly be resistant to invoking.

Of course, constitutional environmental rights are not a panacea for all environmental problems. Indeed, new issues may be generated. For instance, enhancing environmental rights in rich industrialized countries could have the net effect of displacing environmental problems and further diminishing environmental rights in poor countries. So while an international regime of environmental rights would arguably be more robust if those rights are constitutionally recognized by state powers that are most influential in developing international law, the empirical question of whether enhancing environmental rights in rich countries would exacerbate existing environmental injustice requires research in international political economy and international environmental politics. Meanwhile, though, I would suggest that issues of balancing different people's rights (and also international obligations) should be addressed as that—as issues of balancing rights—rather than as possible reasons for not recognizing the rights in the first place. Moreover, at present, the most expansive constitutional environmental rights are found in constitutions of poorer and less influential states; as long as this remains the case, the view that these rights are "mere aspiration" will be encouraged.

Finally, to accord environmental protection the status of a fundamental *right* is not to give it absolute priority over other important social values, but it is to put it on more of an even footing with other—especially economic and property—rights. The most basic question is that of how the fundamental values and interests of a society are balanced against one another. Whatever view one ultimately takes of the proposal for a constitutional environmental right, a sufficient reason for serious consideration of it is that it brings that basic question into the open.

## Notes

This chapter forms part of a project funded by an ESRC research grant (R000222269), which is gratefully acknowledged. Earlier versions have been presented at the Oxford Centre for the Environment, Ethics and Society, Mansfield College, Oxford University, March 1998; the Political Studies Association Conference, Keele University, April 1998; the Social and Political Theory Seminar, University of Edinburgh, October 1998; and the Joint Conference of the International Society for Environmental Ethics and the Society for Applied Philosophy, Oxford, June 1999. I wish to thank participants at these events and at the ESRC-funded seminar "Constitutional Environmental Rights for Scotland?", in Edinburgh, November 1998, for many helpful comments and suggestions. Valuable advice on legal aspects of environmental rights was given by Alan Boyle, Christine Boch, Antonia Layard, and Leonor Moral.

1. See Tim Hayward, *Political Theory and Ecological Values* (Cambridge: Polity Press, 1998), 152–156, 178n2.

2. Related proposals have already been made, albeit unsuccessfully so far. See, for example, R. Desgagné, "Integrating Environmental Values into the European Convention on Human Rights," *American Journal of International Law* 89 (1995): 265n17.

3. Interestingly, the 1994 Labour Party publication *In Trust for Tomorrow: Report of the Labour Party Policy Commission on the Environment* proposed a Charter of Environmental Rights and an Environment Division of the High Court.

4. See Murray Hunt, *Using Human Rights Law in English Courts* (Oxford: Hart, 1998).

5. World Commission on Environment and Development, *Our Common Future* (Oxford: Oxford University Press, 1987), 348.

6. More expansive environmental rights have already been proclaimed in some constitutions. For instance, a number of South American constitutions refer to preservation of "ecological balance"—for example, Brazil's Art. 225; Costa Rica's Art. 50; Paraguay's Art. 7; so too does Portugal's Art. 66. Some also specify rights relating to biodiversity (e.g., Ecuador's Art. 44; Colombia's Art. 79). In the United States, reference is made in state constitutions to the preservation of the natural, scenic, historic, and aesthetic values of the environment (e.g., Pennsylvania's Art. 1 Sec. 27; Massachusetts's Amend. Art. XLIX).

7. Hayward, *Political Theory and Ecological Values*, 21–57.

8. Included in the report prepared by special rapporteur Fatma Ksentini, *Final Report of the UN Sub-Commission on Human Rights and the Environment*, 1994 (UN Doc. E/CN.4/Sub.2/1994/9).

9. N. Popović, "In Pursuit of Environmental Human Rights: Commentary on the Draft Declaration of Principles on Human Rights and the Environment," *Columbia Human Rights Law Review* 27, no. 3 (1996): 439.

10. In addition to the European agreements discussed below, see the 1981 African Charter on Human Rights and Peoples' Rights, Art. 24, and the 1988 Additional Protocol to the Inter American Convention on Human Rights, Art. 11.28.

11. Art. 130r2 of EC treaty as amended by Treaty of Union 1993.

12. See C. Miller, "Environmental Rights: European Fact or English Fiction?", *Journal of Law and Society* 22, no. 3 (1995): 374–397, esp. 388–389.

13. On whether the precautionary principle generates rights, see Miller, "Environmental Rights," 389; on controversies concerning the meaning of the precautionary principle, see, for example, D. Hughes, "Analysis of Duddridge Case?," *Journal of Environmental Law* 7, no. 2 (1995): 238–244.

14. *Direct effect* has a variety of meanings, according to C. Boch, "The Iroquois at the Kirchberg; or, some Naïve Remarks on the Status and Relevance of Direct Effect," *Jean Monnet Working Paper* no. 6, Harvard Law School, 1999. Most generally it refers to the fact that Community law is part of the national legal systems and can be invoked before national courts, but it can also refer to the capacity of a specific Community provision to be applied directly, as it stands; then again, it can refer to the capacity of a Community provision to confer rights on which individuals may rely. See also Miller, "Environmental Rights," 382f; D. Shelton, "Environmental Rights in the European Community," *Hastings International and Comparative Law Review* 16 (1993): 569ff; R. Caranta, "Governmental Liability after Francovich," *Cambridge Law Journal* 52 (1993): 272–297.

15. See Robin Churchill, "Environmental Rights in Existing Human Rights Treaties," in *Human Rights Approaches to Environmental Protection*, ed. Alan Boyle and Michael R. Anderson (Oxford: Clarendon Press, 1996), 90.

16. S. Weber, "Environmental Information and the European Convention on Human Rights," *Human Rights Law Journal* 12, no. 5 (1991): 177–185.

17. Churchill, "Environmental Rights in Existing Human Rights Treaties," 103.

18. For background to the case, see A. Rest, "Europe—Improved Environmental Protection through an Expanded Concept of Human Rights?", *Environmental Policy and Law* 27, no. 3 (1997): 213–216.

19. See, for example, G. Handl, "Human Rights and the Protection of the Environment: A Mildly Revisionist View," in *Human Rights, Sustainable Development and the Environment*, ed. E. D. Weiss et al. (Brazil: Instituto Interamericano de Derechos Humanos, 1992). Environmental campaigners also often tend toward this view, yet when pressing for procedural rights they may make their case on the basis of a substantive claim. See, for instance, "Charter on Environmental Rights and Obligations of Individuals, Groups and Organisations," attached as an appendix to David Rehling, "Legal Standing for Environmental Groups within the Administrative System—The Danish Experience and the Need for an International Charter on Environmental Rights," in *Participation and Litigation Rights of Environmental Associations in Europe*, ed. M. Führ and G. Roller (Frankfurt am Main: Peter Lang, 1991), 154.

20. Robyn Eckersley, "Greening Liberal Democracy: The Rights Discourse revisited," in *Democracy and Green Political Thought: Sustainability, Rights and Citizenship*, ed. Brian Doherty and Marius de Geus (London: Routledge, 1996), 230.

21. See M. Führ and G. Roller, eds., *Participation and Litigation Rights of Environmental Associations in Europe* (Frankfurt am Main: Peter Lang, 1991); Sven Deimann and Bernard Dyssli, eds., *Environmental Rights: Law, Litigation and Access to Justice* (London: Cameron May, 1995); S. Grosz, "Access to Environmental Justice in Public Law," in *Public Interest Perspectives in Environmental Law*, ed. David Robinson and John Dunkley (Chichester: Chancery Law Publishing, 1995).

22. See, for example, S. Douglas-Scott, "Environmental Rights in the European Union—Participatory Democracy or Democratic Deficit?", in *Human Rights Approaches to Environmental Protection*, ed. Alan Boyle and Michael R. Anderson (Oxford: Clarendon Press, 1996): 119–120; also Shelton, "Environmental Rights in the European Community," 575–577.

23. See, for example, E. Brubaker, *Property Rights in the Defence of Nature* (London: Earthscan, 1995).

24. James Nickel, "The Right to a Safe Environment," *Yole Journal of International Law* 18, 1 (1983): 285.

25. For a precedent, see P. Stein, "A Specialist Environmental Court: An Australian Experience," in *Public Interest Perspectives in Environmental Law*, ed. David Robinson and John Dunkley (New York: Wiley Chancery, 1995).

26. See, for example, J. B. Ruhl, "An Environmental Rights Amendment: Good Message, Bad Idea," *Natural Resources and Environment* 11, no. 3 (1997): 46–49, esp. 48.

27. A. J. Roman, "Locus Standi: A Cure in Search of a Disease?", in *Environmental Rights in Canada*, ed. John Swaigen (Toronto: Butterworths, 1981), notes that it is based on a conjecture that "ignores the reality that litigation is far too expensive, traumatic and inconvenient ever to become a popular pastime" (p. 17); see also Stein, "A Specialist Environmental Court."

28. This was illustrated in the case of *Lopez-Ostra* discussed above. Worth noting in the UK context is that the reasonable balancing of human rights against other social goods will require our courts to employ the concept, which is fundamental to the Convention, of proportionality. See Hunt, *Using Human Rights Law in English Courts*; R. Singh, *The Future of Human Rights in the United Kingdom: Essays on Law and Practice* (Oxford: Hart, 1997).

29. Cass Sunstein, "Against Positive Rights," *East European Constitutional Review* 2 (1993): 37.

30. Concerns about debasing the currency of human rights have also been voiced by M. Cranston, "Human Rights, Real and Supposed," in *Political Theory and the Rights of Man*, ed. David D. Raphael (Bloomington: Indiana University Press, 1967); Philip Alston, "Conjuring Up New Human Rights" *American Journal of International Law* 78 (1984): 607–621. Handl, "Human Rights and the Protection of the Environment."

31. H. Schwartz, "In Defense of Aiming High: Why Social and Economic Rights Belong in the New Post-Communist Constitutions of Europe," *East European Constitutional Review* 1 (1992): 25–29.

32. See, for example, Henry Shue, *Basic Rights: Subsistence, Affluence, and US Foreign Policy* (Princeton, NJ: Princeton University Press, 1980), esp. 35–53; also Singh, *The Future of Human Rights in the United Kingdom*, esp. 52–58.

33. C. P. Stevenson, "A New Perspective on Environmental Rights after the Charter," *Osgoode Hall Law Journal* 21, no. 3 (1983): 390–421, esp. 420.

34. Eckersley, "Greening Liberal Democracy," 220.

35. Stevenson, "A New Perspective on Environmental Rights after the Charter," 397.

36. Jeremy Waldron, "A Rights-Based Critique of Constitutional Rights," *Oxford Journal of Legal Studies* 13 (1993): 18–51, esp. 20.

37. Waldron, "A Rights-Based Critique of Constitutional Rights," 20.

38. James Madison, Alexander Hamilton, and John Jay, *The Federalist Papers* (Harmondsworth: Penguin, 1987), 439.

39. François Du Bois, "Social Justice and the Judicial Enforcement of Environmental Rights and Duties," in *Human Rights Approaches to Environmental Protection*, ed. Alan Boyle and Michael R. Anderson (Oxford: Clarendon Press, 1996), 158.

40. Eckersley, "Greening Liberal Democracy," 229.

41. See Tim Hayward, *Ecological Thought: An Introduction* (Cambridge: Polity Press, 1995), 130–136.

42. Joseph L. Sax, "The Search for Environmental Rights," *Journal of Land Use and Environmental Law* 6 (1998): 97.

43. Sax, "The Search for Environmental Rights," 101.

44. Sax, "The Search for Environmental Rights," 100.

45. See, for example, R. K. Gravelle, "Enforcing the Elusive: Environmental Rights in East European Constitutions," *Virginia Environmental Law Journal* 16, no. 4 (1997): 633–660; Du Bois, "Social Justice and the Judicial Enforcement of Environmental Rights and Duties"; Michael R. Anderson, "Individual Rights to Environmental Protection in India," and Martin Lau, "Islam and Judicial Activism," both in *Human Rights Approaches to Environmental Protection*, ed. Alan Boyle and Michael R. Anderson (Oxford: Clarendon Press, 1996).

# 5

# Trusteeship: A Practical Option for Realizing Our Obligations to Future Generations?

William B. Griffith

Only in the last three decades of the twentieth century did there begin to develop a fairly widespread awareness that the century's immense worldwide growth in human population, in the development of industry, in the use of energy and other natural resources, and in the generation of vast volumes of waste products, were having widespread detrimental effects on our physical environment.[1] Even more recent has been the (somewhat less widespread) recognition that at least some of our systematic activities and institutional practices were creating potentially irreversible, long-term deleterious effects, very likely to put at great risk the environment and natural resources available to future generations.[2] In this chapter, I assume that awareness of such large potential ill effects of our collective behavior will imply, in most contemporary ethical frameworks, that the present generation stands under a significant moral obligation to take such probable effects of its acts into account, to the extent possible. However, while philosophers argue strenuously as to how best to understand the grounds of such duties, I do not wish to bring these abstract philosophical issues into focus here. Rather I attempt to respond to the following stance implicit in the lack of attention to these philosophical arguments: it does not matter whether or not one agrees we have obligations to future generations, since no one has a good argument as to what such acceptance would commit us to doing differently. So the questions I wish to raise concern what such duties, if accepted, would require in the way of practice. That is, how can we best understand and articulate concretely the *content* of such duties as we may have to future generations, not abstractly but by proposing *institutional arrangements* by which we might take steps to satisfy them? My intuitive idea is that many of us might be willing to take such claimed obligations more seriously, if only we could get clearer on what sorts of specific actions were entailed by them, and how we ought to weigh these duties against our other present-day duties. I am especially concerned with how we might plausibly derive guidance as to how to begin acting collectively to carry them out.

Thus, my general approach is to try to connect philosophical discussions of the character of our ethical obligations to more practical developments worked out in a pair of related legal fields, to try to show how philosophical ideas might be given a form that should be helpful to those actively working in the environmental field to transform our institutions to deal with concrete problems. In taking this approach, I see this chapter as contributing directly to the overall aims of this book.

The way I intend to proceed is by linking three complex sets of ideas drawn from philosophy and law, in what I hope will be a mutually clarifying and supporting way. First, I review some philosophical discussions of the concepts of "trusteeship" and its analog, "stewardship," which propose them as superior ethical concepts or "lenses" through which the present generation should view its relationship to the earth's resources. Second, I review the very practical form in which the ideas above have been given institutional form in the well-developed law of private trusts, to test whether the philosophical ideas hold serious promise for effective institutionalization in the context of the larger societal problem, of protecting natural assets for future generations. I then turn to a rather narrow set of related ideas, developed more or less independently and without much public notice, in environmental and property law, the so-called Public Trust Doctrine. My aim is to show that this somewhat peculiar doctrine, which now operates chiefly to protect certain coastal tidelands against private appropriation, appears to be ripe for an as-yet little-explored possible extension, to a much larger body of environmentally precious natural resources in serious need of greater protection.

I will proceed as follows. In the second section, "Environmental Ethics and the Role of the Trusteeship Concept," I show how the concept of "trusteeship" has sometimes been presented as a way of resolving certain problems that currently bedevil most contemporary theories of environmental ethics, especially when one attempts to draw practical, political

will. In our context, one would be left with the necessity of interpreting the divine will—not a task most contemporary philosophers are well prepared to take on.

Philosophically, the attractions of the intuitive idea of trusteeship may be seen by recognizing here an example of a "second-order strategy" to improve decision making over a long term.[9] One such strategy is *precommitment*—that is, taking steps to make it less likely that one will make a "bad" decision at a later time by constraining the latter decision making in some way. Another such strategy for dealing with future decisions is *delegating* authority to make them, to someone we expect will be in a better position to do so when the time arrives. Thus, imposing on our present and future decision making the constraints of facing such decisions as trustees restricts our freedom by virtue of our fiduciary relationship, but also puts us in the position of requiring only that as agents we must make a "good faith" effort to do the best at any given moment for beneficiaries, without attempting to specify in advance exactly what those actions should be.

The philosophical suggestions above have largely remained at the level of intuitive guidance, and what is needed now is a clearer articulation of them. But notice that these suggestions have a couple of advantages helpful for our specific purposes. First, in making use of the analogy of a trust relationship, which is ordinarily initiated by a voluntary decision to protect certain properties on behalf of others, this position bypasses the problem of motivation and/or ground for such a decision (beneficence or duty), to focus on what is entailed by the relationship. Second, reliance on the instrument of trusteeship enables us to avoid excessive and difficult speculation as to what future generations (by which one normally means descendants considerably distant from us) will need and want, to concentrate on what we should do *today*, to act responsibly and prudently as a trustee should, to protect the assets under our control and pass them along to the immediate next generation at least undiminished, and if possible enhanced in value.

## The Law of Private Trusts

The intuitive idea that we might protect those who cannot look out for their own rights by charging another with the moral responsibility for managing affairs for them, has been most elaborately worked out institutionally in Anglo-American common law.[10] In what follows, I first set out what might be called the "standard" model of a trust, then give an account of the key institution by which the moral obligation of the trustee was enforced by society in an effective way by the courts of "equity." This in turn gave rise to several variants on the basic or paradigm concept of the trust.

## The "Standard" Model of Trusts

As historically understood, a "trust" in the technical sense was in the usual case created by an owner of certain assets or property ("settlor") conferring on another named person(s) (the "trustee(s)") legal authority over those designated assets (the "corpus"), ordinarily by an explicit act of disposition and instruction (the "trust instrument"), while simultaneously imposing trust or confidence in the named agent(s) to hold and manage those assets on behalf of, in the interests of, a designated beneficiary or beneficiaries.[11] This basic model is what most laypeople think of when they think about a trust, but the overall picture is considerably more complicated.

The conveyance by the trust instrument of property to the trustee made that agent a full owner of the trust assets as far as the common law was concerned. So in the early history of English common law, if the trustee were to abuse the confidence placed in him, say by selling to himself assets of the trust at a price favorable to him but not to the trust, this would have been within his legal authority to do. Thus, the cheated beneficiary who realized this malfeasance had no remedy at law. As things turned out, because of various rigidities in the system of common law, there seemed to arise pretty commonly cases in which legal rights and/or law-backed power were all on one side, and justice to an individual claimant seemed to fall on the other side.

Remarkably, there arose to fill this gap the institution of the appeal to the Chancellor of England, as the chief legal officer and (as a leading cleric) also the "conscience of the king," to "do justice." The first step would usually take the form of appealing to the conscience of the trustee, and if that were unavailing, of forcing the trustee, on pain of personal punishment, to yield up his ill-gotten gains and make whole the trust corpus.[12] Thus, what started as a purely moral obligation came to have the backing of an unusual enforcement power, one specifically charged with ensuring that justice was being done to an individual, rather than that the rights allocated by "the law" were being vindicated.

There are at least two important points to note in the above account. First, the personal appeal to the Chancellor, which fairly quickly came to mean an appeal to a separate court under the Chancellor, called a "court of equity," meant that a decision need not meet the rigid rules of the common law if different measures were needed to do justice in the individual case. (Philosophers will recognize the roots of this concept as coming from Aristotle's discussion in Book V of *Nichomachean Ethics*.) More specifically, the court of equity was empowered to protect those who might be at an unfair disadvantage in a court of law because of legal technicalities or inequalities of power. To that end, the courts of equity had available numerous remedies not available at common law, remedies that acted "in personem" (i.e., on the person of the defendant) and not merely "in rem" (i.e., on a thing, e.g., to transfer property), such as enjoining or forbidding a legally permissible act, or requiring performance of a contract where damages would be inadequate as a remedy. Upon the merger of "law and equity" in England and the United States, most courts were granted both legal and equitable powers, and choose which to exercise based on the nature of the issues, pleadings, and remedies sought in the case. But certain kinds of cases remained peculiarly "equitable" in nature, and enforcement of the kinds of fiduciary relationships involved in trusts were typical of these, so much so that legal systems that lacked something like equity courts were hampered in dealing with trust relationships.[13]

In any case, it was because of the power of courts of equity to create on their own motion specific trusts in property, that variations of the standard model arose in Anglo-American law, and helped make trust law the flexible instrument it came to be. That is, in the absence of an instrument voluntarily establishing a trust, a court could declare that a fiduciary relationship had been "implied" by things that had been said or done, or it could declare the "constructive existence" of a trust in certain assets, because a relationship of confidence had been relied on by some vulnerable person. The court could then force a person who had taken advantage of a vulnerable individual's trust to actually perform as a trustee would have, and yield up any gains from fraud, manipulation, or coercion to the victim's benefit.

**Trusteeship and the Environment Reexamined**

The preceding account helps illuminate the possibilities of borrowing from the private law of trusts to flesh out the philosophical concept of the present generation as a trustee for future generations. Several points merit further attention.

First, when the generational-trustee idea is initially advanced, some are immediately inclined to dismiss it, because it appears to fit poorly with the basic or standard model. Thus, one critically disposed could be expected to argue as follows. Granted that individual persons might make their own property and investment decisions with an eye to protecting the interests of future generations—say, sacrificing an increase in realized wealth for oneself and one's family by preserving an environmentally precious asset from development. This might well be a good thing, especially if some significant number of individuals undertook this kind of action. But how does this concept relate to the collective decisions we need to make as a society about the aggregate effects our actions are having on the environment and supply of natural resources? As to the latter, neither the federal nor state governments, nor any other collective body, *owns* most of the crucial environmental and natural assets that one might wish to preserve for the future. So these are not assets that anyone would be empowered to commit to a trust, without trampling on existing individual property rights. And in the absence of a benefactor's instrument establishing a trust, how could one know exactly which assets were to constitute the trust corpus? And who exactly are the beneficiaries conceived to be? Moreover, in the standard model, the trustee and the beneficiaries are distinct. But if the present generation were acting as trustee, it would have to include itself, along with future generations, as among the beneficiaries to whom it would be under a fiduciary obligation to act—a clear conflict of interest.

A deeper look at the literature of the law of trusts reveals, however, that the model of "standard" trusteeship on which these objections are predicated is actually only a particularly common form of what turns out to be a very flexible institutional arrangement. As one legal textbook says, "Each of these [standard] requirements must be tested against the inherent flexibility and history of the device," which permits of a variety of nonstandard deviations.[14] For example, charitable trusts are set up with unnamed, multiple beneficiaries, which may include the trustees from the present generation. Even outside of charitable trusts, a trustee having the status also of a beneficiary of a trust has not been found impermissible, it being only requisite that the trustee not be the sole beneficiary, and that she act with undivided loyalty to all the beneficiaries, not favoring some over others.

Of course, more significant is the objection that important parts of the body of natural resources about which environmentalists are concerned already fall within the scope of multiple individuals' property rights. These individual owners are on the whole not likely to concede willingly even an "equitable interest" held by future generations that would constrain their freedom of action. Further, many free-market supporters would find such an attenuation of private property rights anathema, as a matter of public law and policy. But it should be recalled from our discussion above that historically, one way a trust has come into being is through legal recognition of a "constructive trust." This is an after-the-fact recognition that a relationship of trust had already been in existence, and that an economic agent *should have acted as a trustee* but did not, with the result that the court would see that the outcome was the same as if a trust had been formally established. This is only to say that the law of trusts recognizes that in practice a trust may be established by legal authority retrospectively, and need not have been created by a voluntary initial benefaction. However, this is *not* to say that any court in our judicial system would be broadly recognized as having legitimate authority to effectuate such a constructive trust over broad environmental and natural resources.

All that has been shown so far is that the flexible concept of trusteeship as developed in the law of private trusts does not really support a number of prima facie objections to the idea of generational trusteeship. In fact, the latter idea appears consistent with the legal practices developed over the years in managing individual and charitable trusts. This does not show, of course, that the idea of such an extension is really justifiable, philosophically and practically. However, somewhat surprisingly, there is more support to be found in another, separate line of development in the law, one much less well known, to which we can turn to explore further this idea in its practical dimensions.

## The "Public Trust Doctrine" and Its Generalizability

### A Historical Doctrine Revived

While nearly all lawyers in common law systems are quite familiar with the concepts and legal doctrines pertaining to private and charitable trusts, and even many nonlawyers have a basic understanding of these institutional arrangements, what legal scholars call the "Public Trust Doctrine" is much less familiar to practicing attorneys and has even less public visibility. Nonetheless, it will be argued here, this doctrine has considerable potential for helping our society deal with its obligations to future generations, although as yet this potential has not been fully exploited.

In part, no doubt, this low level of visibility is because the doctrine, as usually understood, is one of quite narrow scope, and while within that narrow range it has largely unquestioned applicability, its public importance has not been perceived as very great. However, recent proposals for the extension of its reach and power have been troubled by a lack of clarity about its origins, and hence about the ultimate sources and reach of its authority.

The narrow range in which the doctrine appears authoritative is over "coastal areas, both ocean and freshwater, including the shores, submerged lands, tidewaters, and the coastal ocean."[15] As to these areas, the doctrine holds that they "are held in a trust by the citizens of the various states and municipalities to be used only for the benefit of the commonwealth. The doctrine . . . prohibits the alienation of these resources for private benefit."[16] (That is, the citizens' agent—the state—is not authorized to transfer title over trust lands, except when there would be some clear offsetting gain to the public welfare, and even then, only if the usual public rights of access and use were preserved.)

This traditional Public Trust Doctrine is presented in most accounts as founded on the notion of the several states succeeding, after the American Revolution, to the similarly constrained rights of the English crown in such coastal properties. Under British law, the sovereign (rather than the public) "owned" these lands, but not in a propriety way, to dispose of as wished, but rather held them "in trust" for the public, to use for navigation, fishing, and the like. (There is some reason to think this is inherited in slightly changed form from Roman law antecedents, designating similar natural assets as *res communes*, but the historical picture is far from clear.[17]) The application of the doctrine in the United States is presumed to be a consequence of the adoption of British common law by the new states of the union. Parallel rights then passed by parity to other states as they joined the union. Thus runs the usual summary account in a large range of law review articles.[18]

However, when this doctrine was nudged into greater prominence, usually credited to law professor Joseph Sax's highly influential 1970 article, as a suggested legal tool that could be used for much greater protection of environmental resources, the origins, status, and reach of this ancient doctrine quickly became quite contentious.[19]

**Expanding the Traditional Public Trust Doctrine**

Sax's original aim seems to have been to encourage courts to stand up against legislatures' willingness to give away valuable public resources to private firms for "development" purposes. In his survey, he was able to point to leading cases in both state and federal courts in which, as in a major dispute between the Illinois Central Railroad and the state of Illinois, judges had declared void transfers of assets by state legislatures when insufficiently respectful of the public's continuing interests.[20] Sax and his followers also wished to encourage courts to be aggressive in identifying new natural and environmental resources as assets held in trust for the public, and further, to expand the concept of "public uses" for which these were to be preserved.[21]

Not surprisingly, this bold thesis produced a counterreaction. Perhaps the leading line of criticism was that the Public Trust Doctrine, understood in this way, overly extended the power of the judiciary at the expense of the legislature. The doctrine was also attacked as superfluous after Congress and the states began to pass environmental statutes in the early 1970s. Moreover, the critics argued vigorously, the uncertain roots of the doctrine left great ambiguity about the justifiable scope of the doctrine. For example, was it best thought of as a part of trust law, or rather as a doctrine of "easements" in property law? On a closer look, there also appeared to be significant components drawn from the interaction of the civil law public-land doctrines of formerly French and especially Spanish possessions with the British-based common law (as Selvin has shown). And in any case, if it were truly a common law doctrine, as was usually asserted (see above), how could it be urged as authority to overturn legislative enactments, as though it were a constitutional provision? But if constitutional, where is it implicit in the language of the U.S. Constitution or in state constitutions?[22]

The struggle engendered by the revival of the Public Trust Doctrine won environmental lawyers some successes, but the criticisms just summarized showed that proposals to appeal to courts to expand its scope faced major limitations. One might have thought, for example, that a natural judicial extension of the scope of public trust assets would be to include the vast federally owned lands, comprising roughly half of the Western states. On investigation, however, this path appeared blocked by the judiciary whose support had been counted on—notably, by an earlier decision of the U.S. Supreme Court that had made clear that these lands were "owned" under Article IV of the Constitution, and not under the Public Trust Doctrine applicable to the states.[23] Moreover, in general, legal commentators found most courts (with the exception of those in California, Massachusetts, and perhaps Wisconsin) quite cautious in extending the Public Trust Doctrine as far as environmentalists urged, and critics predicted a public backlash against this antidemocratic "overreaching" by the courts.[24]

It should be noted that some legal scholars point to the doctrine's traditional flexibility, and see recent innovative uses of the doctrine by state courts as evidence that it will continue to have a role to play in struggles to protect the environment and natural resources. This seems particularly important in light of the requirement for state compensation for certain "takings-by-regulation" that the U.S. Supreme Court appears to have placed on environmental measures in a recent decision. Since the condition requiring compensation is that the constraints on use proscribed by the state were not already part of the "underlying" constraints on uses of property under state law, there will be an important question whether or not public trust constraints on uses will be found to be part of underlying state laws and reduce the requirement for expensive compensation.[25]

But given the slow pace at which such judicial intervention proceeds, it would appear that a new tack is needed if the potential of the basic idea of generational trusteeship is to begin to be realized.

## A Proposal for Expanded Use of the Public Trust Concept

Can we see how to rework this traditional doctrine, to better serve the aims of realizing our obligations to protect environmental and natural resources for future generations? In this section I hope to show how we might extract, from our discussions above of the trustee concept in environmental ethics, the law of private trusts, and the traditional Public Trust Doctrine, an expanded reach and scope for the concept of generational trusteeship of important natural and environmental resources.

### The Basic Ideas for Further Expansion of Trusteeship

In the writings of the chief proponents of a large expansion of the doctrine, there have been a variety of suggestions as to what might be included. For example, trust protection has been proposed for the public lands, as mentioned before; for air and water; for fish and wildlife; for mineral deposits; for soil quantity and productivity; and for whole ecosystems.[26] The motivation of such proposals is clear: to give some standing to the interests of those who would oppose economic exploitation of these resources without regard for their importance to noneconomic uses or to the environment, broadly understood.

The opposition to such proposals is also quite clear. To summarize, critics attack the proposed extension of trusteeship to such a broader base of assets on the grounds mentioned before: private owners already have extensive property rights in some of these assets (e.g., soil and mineral deposits); many of the assets mentioned are not "owned" by the state (e.g., air, water, ecosystems); even among those that are owned (e.g., public lands), the state's rights have not previously been clearly understood to be held in trust, and hence are already interlocked with private property interests permitted to make some use of these resources (lumber companies, ranchers). In the face of such complications, what court could be expected to find precedents that would justify expansion of the traditional Public Trust Doctrine in the ways proposed? And even if courts were to adopt such expansive authority, wouldn't its assertion without legislative action imperil trust in and overall effectiveness of the judiciary?

In response, several general points may be offered, drawing on the ideas developed in preceding sections, to show how such objections might be overcome.

First, while it is correct to assert that a trust ordinarily is established legally only by someone who already owns the assets to be placed in trust, and many of the most important environmental assets are not "owned" by any government entity, it is important to recall that "constructive trusts" have been recognized without such instruments by courts under the law of private trusts. That is, without any voluntary transfer of property rights to a trustee, where a court responsible for protecting the vulnerable found that an appropriate relationship of trust existed in the matter of certain assets, it could and would require that the properties be treated "as though held in trust."

Second, while it seems clear that the possibilities for vigorous action by the courts in extending the reach of trusteeship are limited, there is reason to think that *legislatures* might be urged to seize on the ideas involved and declare that "constructive trusts" in such assets already exist by virtue of the implicit "fiduciary" relationships (discussed in the philosophical literature) that any society has with both preceding generations from which it has "inherited," and especially with future generations "vulnerable" in the face of its decisions. The national or state legislatures could then write this trust status into statutes or even propose them as constitutional amendments. Such steps would negate the criticism that the historical Public Trust Doctrine makes expansion of trusteeship inherently "antidemocratic," in its dependency on courts' interpretation of ancient, esoteric doctrines.

Third, it should be noted that one limitation on environmentalists, in current judicial struggles over protection from development of environmentally precious assets, is that the parties in these disputes are often perceived as equally self-serving "special interests," among whom "the Greens" (i.e., environmental groups) just represent an affluent elite with strong preferences for leisure in beautiful surroundings over the economic gains from development.[27] But recasting the disputes over permissible uses of environmental assets as involving an effort by courts to protect the interests of *future* as well as diverse present beneficiaries, would greatly illuminate and clarify what is at stake.

All this is just so much theoretical possibility, one might object. How could such ideas be made to serve practical purposes of those working to preserve the environment? Perhaps the debate could be advanced by discussing a particular example of concrete institutional steps that might be taken, to expand and enforce trusteeship over important natural assets. So to that we now turn.

### A Specific Potential Application of an Expanded Trusteeship Concept: The Federal "Public Lands"

Although (as mentioned above) the U.S. Supreme Court has made clear that it does not find persuasive arguments to the effect that the federal government must understand its ownership of the great public lands of the Western United States as held under the Public Trust Doctrine, the Court has held that there is a kind of public trust involved. This allows the federal government to act as more than a simple proprietor, for example, in using its authority to protect the public lands against the depredations of neighboring activities in ways a mere owner could not.[28] So there is a strong presumption that the Court would have no objection to a statutory declaration by Congress to the effect that the public lands would henceforth be held in public trust for the enjoyment of both present and future generations.

What would be the effect of such a legislative action? By itself, with no further legislative steps, probably nothing much different from current practices would result. But drawing on the experience of our society's legal system with trusteeship in both private and public law, the Congress might go further and take important implementing steps.[29] The following would appear to be most crucial.

1. Congress would need to alter the current statutory instructions to the various federal managing agencies, which generally tell them to aim for "multiple-use-sustained-yield."[30] This guidance would need to be reformulated in a way that stressed the importance of considering, not only current users' demands, but the potential "uses" by future generations. While these cannot be known with precision, whatever they will be they should not be precluded by exhausting or severely damaging the resources. Further, new guidance should insist that "sustained yield" of

public lands, forests, and so on ought to mean more than "sustained" over the next few years, the normal agency's horizon, but rather over the long term.

2. Congress would also need to establish, perhaps in part by pulling together elements from existing agencies, a "futures-information-gathering" resource (e.g., an agency, commission, or council). The assigned task would be to assemble the results of the best of the current research on the impact of current practices on the diminution of the life-quality possibilities of future generations—for example, studies of global warming, of the impact of depletion of oil reserves, of the storage of nuclear waste, and so on. Such an agency could also be charged with sponsoring additional research, as needed to fill in the gaps or answer questions left unresolved by current research efforts. It would be especially important to monitor, support, and assist in the development of new, more responsible macroeconomic goals, such as "sustainable development," to replace the currently nearly universally accepted goal of maximal economic growth.[31] Ultimately, its responsibility would be to serve as a source of information, analysis, and advice to the existing agencies to be charged with protecting natural and environmental assets for future generations, so as to render more feasible their task of deciding what uses of the present generation should be permitted, and which should be denied as inconsistent with the nature of the assets as held in trust for both present and future generations.

3. Perhaps most important, Congress would also need to set up an oversight and appeal process, modeled on that exercised by courts of equity in overseeing private trusts and protecting beneficiaries not able to fight for their own interests. (The well-known, depressing record of federal agencies in mismanaging such trust assets as the Indian Trust Lands shows that such enforcement mechanisms are as much needed here as anywhere.) Very likely such reviewing courts would require the support of a specially designated "public defender" agency of attorneys, charged to represent in court the interests of otherwise-silent future generations against the pressures sure to be exerted on trustees by interested parties in the present generation.

The suggested model of the courts of equity as overseers is meant to be taken seriously here. One reason such courts were able to function as well as they did in supervising trustees was that, as noted above, equity doctrine stressed "doing justice to the parties" rather than following as closely as possible legal rules and precedents as common law courts did. Further, as noted, courts of equity developed especially flexible remedies, unknown to the common law, to achieve the just results they sought. It seems clear that courts attempting the immensely difficult task of overseeing, and ruling on challenges to, management by society's trustees of resources for the benefit of both present and future generations would badly need the same kind of flexibility. For how likely would it be that precise rules and standards could be written into law or regulations to control the actions of trustees? Wouldn't it be, as in the case of private trusts, a question of attempting to give legal backing to what is essentially a deep moral obligation, one that requires constant interpretation and reinterpretation by the agent? A reviewing court that also had access to the information and advice of the "futures" agency would no doubt be able to control the most egregious misfeasance of trustees, but in the end, the society would have to hope the court would be able to appeal to the "conscience" of the trustees charged with protecting those who cannot look out for themselves.

## Some Tentative Conclusions

The discussions of these complex ideas in the sections above have been necessarily somewhat sketchy and incomplete. My hope is that they will provide the basis for further scholarly exploration of these ideas, so as to make them clear, concrete, and flexible enough that they might provide a theoretical/conceptual framework for environmental practitioners to organize their efforts around. I would particularly like to stress the point that what environmental philosophers, lawyers, and economists probably cannot expect to do effectively while remaining within the bounds of their own disciplines, might be achieved by moving across disciplinary lines. I believe there is convincing evidence—for example, in the uneven development of environmental ethics and policy—that we need to make better use of both highly theoretical thinking and very practical work, if we are to advance in such a new and staggeringly complex field.

Much scholarly work surely needs to be done, by both philosophers and legal scholars, to better understand what it might mean for the present generation to serve as trustee of natural and environmental resources for future generations. For example, the historical origins, development, and original justifications of the Public Trust Doctrine need much more careful study, because the authority and exact articulation of legal ideas is so rooted in their provenance. Similarly, I think we have only begun to take seriously figuring out what it might mean to try to control economic growth, and substitute more defensible developmental social goals, ones that focus more on fairness in distribution of the benefits and costs of economic activity, and less on maximizing economic output, treating all resources as fungible and the earth's wastesinks as infinite.

In other directions of research, if the above arguments have shown anything of importance about the practical usefulness of the idea of "trusteeship," then I hope philosophers would return to the discussion of the basic character and foundations of the moral obligations we appear to have to future generations, in particular taking up the important issues concerning the motivation to accept and act on such duties.

No doubt there are some who would say that my proposal to rely primarily on democratic processes, via the elected legislature and executive, rather than on the courts, is hopelessly naive politically, and hence likely to be of no interest to practical environmental activists. I confess that this is possibly true, but I would offer the following caution to those who would discourage efforts to lobby along these lines.

Think back to the political and social context of the 1960s, just before the beginning of the surge of environmental legislation that has continued unabated for the last thirty years. In particular, recall the inspired language of the preamble of the National Environmental Policy Act of 1969, one of the real breakthroughs of our time, which asserts that it is "the continuing responsibility of the Federal Government" to use all practical means, consistent with other considerations of national policy, to improve and coordinate federal plans, functions, programs, and resources "to the end that the nation may . . . fulfill the responsibility of each generation as a trustee of the environment for succeeding generations." Although this grand declaration has not been taken as seriously as it should have been, who would have predicted in advance its emergence in national legislation from the turmoil and conflict of the 1960s? Who then would venture to deny the possibility that an ambitious chief executive or a majority party in Congress, committed to the long-term protection of the environment and willing to risk some political capital in an effort to explain its importance to the general public, could push this sense of responsibility further, in some of the directions outlined above? Such determined skepticism may prove correct, but it is hard to see it as the only view that recent environmental history supports.

## Notes

An earlier version of this chapter was presented at the conference on Moral and Political Reasoning in Environmental Practice, June 27–92, 1999, at Mansfield College, Oxford University. The research reported here represents the belated continuation of work begun in 1994–95 but left dormant due to the press of other obligations. The original motivation was a U.S. Department of Energy contract with the National Academy of Public Administration and the Battelle Institute to study the federal government's obligations to future generations with respect mainly to the disposal of nuclear waste. I owe thanks to Bayard Catron, Research Director, who asked me to join that project, in conjunction with which I prepared a background paper on this general topic: "Protecting and Providing for Future Generations: The Present Generation as Trustee," in *Deciding for the Future: Balancing Risks and Benefits Fairly across Generations: Issue Papers*, B. Catron, ed. (Washington, DC: National Academy of Public Administration/ Battelle Institute, 1994). I also am indebted to a young attorney, John Hartung, then a doctoral student in public administration, for assistance with the legal research involved.

1. J. R. McNeill offers an excellent overview of what he calls this "gigantic uncontrolled experiment on the earth," of heedlessly altering ecosystems with great intensity and speed, and of social reactions to it, in *Something New under the Sun: An Environmental History of the Twentieth Century World* (New York: Norton, 2000).

2. Bryan Norton has called the recent discovery of such potential long-term effects a "third wave" of environmental problem recognition, in *Towards Unity among Environmentalists* (Oxford: Oxford University Press, 1991).

3. Hans Jonas, "Technology and Responsibility: The Ethics of an Endangered Future," a 1972 address, reprinted in *Responsibilities to Future Generations*, ed. Ernest Partridge (Buffalo, NY: Prometheus Books, 1981), pp. 26, 28. Brian Barry eloquently expresses similar views ("We scan the classics in vain for guidance on this question") in "Justice between Generations," in *Law, Morality and Society*, ed. P. M. S. Hacker and Joseph Raz (Oxford: Oxford University Press, 1977), reprinted in Barry, *Democracy, Power, and Justice* (Oxford: Oxford University Press, 1989), 494. For a good discussion of the omitted problem of motivating regard for future generations, see Norman Care, "Future Generations, Public Policy, and the Motivation Problem," *Environmental Ethics* 4 (1982): 195–213.

4. A leading defender of this viewpoint is Paul Taylor, *Respect for Nature* (Princeton, NJ: Princeton University Press, 1986); other important figures arguing along similar lines include Holmes Rolston III, *Environmental Ethics: Duties to and Values in the Natural World* (Philadelphia: Temple University Press, 1988).

5. For an exposition of this view, see Joel Feinberg, "The Rights of Animals and Unborn Generations," in *Philosophy and Environmental Crisis*, ed. William Blackstone (Athens: University of Georgia Press, 1974). This view has by no means been universally welcomed by philosophers; see, for example, Ruth

Macklin, "Can Future Generations Correctly Be Said to Have Rights?", in *Responsibilities to Future Generations: Environmental Ethics*, ed. Ernest Partridge (Buffalo, NY: Prometheus Books, 1981), 151: "The question . . . must be answered in the negative."

6. See Donald Scherer's editor's "Introduction" to *Upstream/Downstream* (Philadelphia: Temple University Press, 1990), in which he argues for a similar skepticism about deriving policy prescriptions from "intrinsic value" theories, and stresses the advantages of appealing to well-explored moral and legal concepts.

7. Among prominent proponents of the "trustee" perspective are Edith Brown Weiss, "Our Rights and Obligations to Future Generations for the Environment," *American Journal of International Law* 84 (1990): 198, and Peter Brown, "Climate Change and the Planetary Trust," *Energy Policy*, March 1992, pp. 208–222. Generally supportive of the idea are Annette Baier, "The Rights of Past and Future Persons," in *Responsibilities to Future Generations: Environmental Ethics*, ed. Ernest Partridge (Buffalo, NY: Prometheus Books, 1981), 171–183, and Robert Goodin, *Protecting the Vulnerable* (Chicago: University of Chicago Press, 1985). For a clear, coherent account of the "bundle of property rights" as generally understood from a legal standpoint, see A. M. Honore, "Ownership," in *Oxford Essays in Jurisprudence*, ed. Anthony Guest (Oxford: Oxford University Press, 1968).

8. On the roots of this religious concept, see Robin Attfield, *The Ethics of Environmental Concern*, 2nd ed. (Athens: University of Georgia Press, 1991), who is responding in part to an earlier discussion by John Passmore, *Man's Responsibility for Nature* (New York: Scribner, 1974).

9. Much in this paragraph rests on the illuminating discussion in Cass Sunstein and Edna Ullmann-Margalit, "Second-Order Decisions," *Ethics* 110 (1999): 5–31, as well as on the earlier work of Jon Elster, *Ulysses and the Sirens*, rev. ed. (Cambridge: Cambridge University Press, 1984).

10. Although the notion of a trust is rooted in Roman law, it appears that the full-fledged development of the law of trusts was confined until fairly recently to the common law countries. See Maurizio Lupoi, *Trusts: A Comparative Study*, trans. S. Dix (Cambridge: Cambridge University Press, 2001), 8–10.

11. For one standard account, see George Bogert, *Trusts*, 6th ed. (St. Paul, MN: West, 1987).

12. A very readable account of the development of equity jurisprudence is found in Peter Charles Hoffer, *The Law's Conscience: Equitable Constitutionalism in America* (Chapel Hill: University of North Carolina Press, 1990).

13. See S. Subrin, "How Equity Conquered Common Law," *University of Pennsylvania Law Review* 135 (1987): 918–920, Lupoi, *Trusts*.

14. Thomas L. Shaffer and Carol Ann Mooney, *The Planning and Drafting of Wills and Trusts*, 3rd ed. (Westbury, NY: Foundation Press, 1991), 99–100.

15. Jack H. Archer et al., *The Public Trust Doctrine and the Management of America's Coasts* (Amherst: University of Massachusetts Press, 1994), 1.

16. Molly Selvin, *This Tender and Delicate Business: The Public Trust Doctrine in American Law and Economic Policy 1789–1920* (New York: Garland, 1987), 1.

17. Selvin, *This Tender and Delicate Business*, 17–26. There is no unanimity among interpreters as to whether ownership of trust assets rests in the public, with the state acting as its agent, as Selvin indicates in the quotation, or whether, as stated below, the state is viewed as owning the assets but as trustee.

18. For examples, see two law journal symposia devoted to the Public Trust Doctrine: *U.C. Davis Law Review* 14 (1980), and *Environmental Law* 19, no. 3 (1989).

19. Joseph L. Sax, "The Public Trust Doctrine in Natural Resource Law," *Michigan Law Review* 68 (1970): 473–492. This is widely viewed as one of the most influential law journal articles ever published, having stimulated an extraordinary flood of follow-up studies. For various views of Sax's influence, see the symposium in *Ecology Law Quarterly* 25 (1998), especially the valuable critique by C. Rose, "Joseph Sax and the Idea of the Public Trust," 351–362.

20. *Illinois Central R.R. v. Illinois*, 146 U.S. 387 (1892).

21. Sax, "The Public Trust Doctrine in Natural Resource Law," passim, and the huge number of follow-up articles, many of which are cited in the Symposium mentioned in note 19 above.

22. For representative criticisms of the doctrine, see for example R. Lazarus, "Changing Conceptions of Property and Sovereignty in Natural Resources: Questioning the Public Trust Doctrine," *Iowa Law Review* 71 (1986); J. Huffman, "A Fish out of Water: The Public Trust Doctrine in a Constitutional Democracy," *Environmental Law Review* 19 (1989). The best discussions known to me of its roots and status are in C. Wilkinson, "The Headwaters of the Public Trust," in *Environmental Law* 19 (1989) and Z. Plater, R. Abrams, and W. Goldfarb, *Environmental Law and Policy*, 2nd ed. (St. Paul, MN: West, 1998), chap. 22.

23. See *Alabama v. Texas*, 347 U.S. 272 (1954).

24. See C. J. Lewis, "The Timid Approach of the Federal Courts to the Public Trust Doctrine," *Public Land and Resources Law Review* 19 (1998), for an example of the first view; for the second, see G. R. Scott, "The Expanding Public Trust Doctrine: A Warning to Environmentalists and Policy Makers," *Fordham Law Review* 10 (1998): 1.

25. See *Lucas v. So. Carolina Coastal Council*, 505 U.S. 1003 (1992). Hope Babcock argues for the doctrine's potentially increased usefulness in the aftermath of the *Lucas* decision, in "Has the U.S. Supreme Court Finally Drained the Swamp of Takings Jurisprudence?", *Harvard Environmental Law Review* 19 (1994). For other enthusiastic arguments for the doctrine's general usefulness, see Plater, Abrams, and Goldfarb, *Environmental Law and Policy*, chap. 22—for example, their discussion of the "Mono Lake" case, *Nat. Audobon Society v. Superior Court of Alpine County* (Supreme Ct. of Calif.), 33 Cal. 3d 419 (1983).

26. The most extensive proposal I have found is in Rick Applegate, *Public Trusts: A New Approach to Environmental Protection* (Washington, DC: Exploratory Project for Economic Alternatives, 1976). Other less ambitious but similar suggestions are found in (among others) G. Meyers, "Variations on a Theme: Expanding the Public Trust Doctrine to Include Protection of Wildlife," *Environmental Law* 19 (1989), and A. Rieser, "Ecological Preservation as a Public Property Right: An Emerging Doctrine in Search of a Theory," *Harvard Environmental Law Review* 15 (1991).

27. See Mark Dowie, *Losing Ground: American Environmentalism at the Close of the Twentieth Century* (Cambridge, MA: MIT Press, 1995).

28. This case law is too complex to summarize here, but for details, see the discussion in Plater, Abrams, and Goldfarb, *Environmental Law and Policy*, chap. 24: "The Public Lands."

29. This section borrows from two of my papers that are not yet published: "Public Lands, Property Rights, and Intergenerational Justice," presented at the Seventh Annual Meeting of the Society for Philosophy in the Contemporary World, Fort Estes, CO, July 2000, and "The Reach of Property Rights When Environmental Effects Are Not Local," forthcoming in *Values in an Age of Globalization*: *Selected Proceedings of the 28th Conference on Value Inquiry, April 2000*, ed. K. Dobson.

30. Such instructions are formulated in *Multiple Use, Sustained Yield Act of 1960*, 16 U.S.C.A. Sec. 528 et seq., and reiterated in various other statutes such as *Federal Land Policy and Management Act of 1976*, 43 U.S.C.A. Sec. 701-1782. The latter act even mentions the "long-term uses of future generations" (S.1702), but this has been generally lost sight of.

31. This concept emerged into prominence with the publication of *Our Common Future*, by the World Commission on Environment and Development (Oxford: Oxford University Press, 1987). Excellent discussions from a philosophical standpoint are found in Andrew Dobson, ed., *Fairness and Futurity: Essays on Environmental Sustainability and Social Justice* (Oxford: Oxford University Press, 1999). Economic studies are plentiful—for example, see Sylvie Faucheaux, David Pearce, and John Proops, eds., *Models of Sustainable Development* (Cheltenham: Elgar, 1996).

# 6

# Ecological Utilization Space: Operationalizing Sustainability

Finn Arler

Ever since "sustainability" was introduced as the main concept in the modern debate on environmental protection, it has been criticized for its vagueness. It needs to be made more "operational," the critics argue; otherwise it will end up as yet another fancy catchword with an indeterminate meaning. It is necessary to give it a more specific content in order to be able to use it in the decision-making process. We have to know exactly which restrictions it puts on our ambitions. This must be a case for neutral and value-free experts who can tell us in detail how far we can go before we overstep the line between sustainable and unsustainable ways of living. When this is settled, we can carry on with all the things we prefer to do within the specified limits.

In this chapter I will focus on one, or rather, as we will see later, two proposals on how to operationalize sustainability through the introduction of the so-called ecological utilization space. First, however, I will present a short version of a fairly well-known story about modern life. This story will set the stage for the demand for operationalization. It is an important story to tell, because it constitutes part of the basis on which many decisions are built, especially in relation to environmental questions.

## Modern Life and Personal Free Scope

In modern societies we all, or at least many of us, find it important to preserve an appropriate degree of personal freedom or autonomy. We want to have a certain scope within which we can act in accordance with our own personal conception of the good life. We want to be able to dismiss forms of interference that we have not asked for and that we do not find helpful and inspiring in our personal lives. Nobody else should be able to determine our personal priorities and life choices. This is our own business. We may all be seeking happiness and the good life, but we despise paternalistic individuals who intrude in our affairs and attempt to dictate how we should live our lives. We want to do things our own way.

Consequently, we are all keen on establishing and preserving individual rights of noninterference. Neither the central state nor our fellow citizens should be allowed to intrude in our lives against our will, nor should they have the right to require us to live in accordance with values and ideals that we do not share, or that we have not accepted voluntarily. We simply want them to stay off our backs. This is a two-way street, of course. As long as other people are not interfering with our lives, we will not interfere with theirs. They are allowed to have a free sphere, too, similar to our own, where they can act in accordance with their personal values and ideals. If we want to, we can combine our spheres of activity, form friendships and voluntary associations, and live a common life together with other people. We do not want to be forced into anything, however. We want to share our lives with the people we love and care for, or share values and ideals with, and nobody should be able to order us to do anything else. Nor will we order other people to live their lives in ways they would never choose for themselves.

In this sense, we see each other as equals, as human beings whose individual freedom, autonomy, or self-determination cannot be considered as just one more particular value among many others. It must be seen as a transcendental value, the impersonal prerequisite for all personal values. It is therefore overriding in comparison to all the specific values related to the particular conceptions of the good life. The protection of individual freedom always comes first. Nothing can overrule it, because no specific earthbound values can move upward into the transcendental realm, where the basic rules of the game are settled. Protection of the right, the moral law of noninterference, is always a prerequisite to the enhancement of any kind of good.

If no reasons can be given for deviations from equality in the transcendental realm of rights, individuals of future generations must be given the very same status as those in the present generation. It does not seem possible to state any impersonal reasons for granting special privileges to individuals currently alive. Everybody must be treated as equals, no matter where or when they live. Like current individuals, future individuals ought to have their own free scope to act within in accordance with their personal ideals and values.

This is a very general outline of the story of modern life. It is not the full story, but it is a story with enough truth in it that it has set the agenda for many discussions about environmental regulation and environmental politics (as well as for many other kinds of discussions, of course). Many people are eager to find a way of dealing with environmental issues that does not force them to take a stand on questions of value (apart from the transcendental value of personal freedom), because value questions are seen as personal, difficult, and basically subjective matters. In this kind of setting "operationalization" is bound to become a critical term. Modern bureaucrats and state officials are looking for neutral and impersonal methods that make it possible to operationalize a concept like "sustainability," leaving all questions of value to be dealt with in the personal realm. They want to find a solution that is impersonal and neutral with respect to all the conflicting personal values and conceptions of the good.

In the English-speaking world, the effort to operationalize has mainly been interpreted in economic terms. I will return to this later. In various European countries, however, including my own country, Denmark, a new concept was introduced in the mid-1990s as another means of solving the problem of operationalization, the problem of remaining neutral with regard to values and conceptions of the good life: the idea of "ecological utilization space" (or "ecological scope," or "environmental space"). It is a bit difficult to trace the origin of the idea, but the German Wuppertal Institute has been one of its main promoters. The concept is closely related to various other older concepts like, for instance, "carrying capacity," which had been used by forestry economists for decades before it was adopted in various United Nations reports at least as far back as the early 1980s, the Dutch concept of "ecocapacity,"[1] and, beyond that, the Canadian idea of measuring "the ecological footprint."[2] Other sources could be acknowledged as well. In the discussion below, however, I will mainly be dealing with the definition of the concept used by the Danish Ministry of Environment and Energy, a definition fairly close to the one used in a report from the

Wuppertal Institute, sponsored by the European section of Friends of the Earth.

## Ecological Utilization Space

Let us take a look, then, at these definitions. In the Wuppertal report the concept is introduced in the following way: "The amounts of energy, water, land, non-renewable resources and forests which can be used without reducing the possibilities of future generations is called the ecological utilization space." The report continues a little later: "Principles of equality and social justice are incorporated into the concept 'ecological utilization space per person' by distributing the permissible use of resources equally among everybody."[3] Both neutrality and equality are thus maintained in the definition. The distribution is based on simple equality, and concepts like "possibilities" and "uses" are employed in a way that leaves no trace of values.

Every fourth year the Danish Ministry of Environment and Energy publishes a broad exposition or statement of the results from the preceding years and of the plans for the years to follow. In the 1995 edition the concept of ecological utilization space was given a fairly prominent place. The definition of the concept was quite close to the one in the Wuppertal report: "The ecological utilization space [*økologisk råderum*] is defined—from a global point of view—as the amount of natural resources (air, water, land, minerals, energy sources, nature areas, plants, animals, and so on) that can be used per year without preventing future generations from having access to the same amount and quality. Every human being shall have a right to his or her part of the ecological utilization space." The statement then continues with the following sentence, which gives the concept a somewhat different meaning, and which I will return to later: "Everybody should have a chance to reach the level of material welfare that the ecological utilization space and the technological capacity allow."[4] In the first part of the definition we find the same insistence on equality as in the Wuppertal report, and even though the loaded term *quality* is applied, the standards against which the quality is to be evaluated are not made clear, so the neutrality demand cannot be said to have been violated.

One can, accordingly, find two basic points behind both approaches to the concept of ecological utilization space. The first is epistemological. It is assumed that politically neutral experts using value-free natural science methods can specify the ecological utilization space. In the Danish governmental exposition it is explicitly seen as a virtue of the concept that it defines "a possible way of operationalizing the environmental demands of sustainable development."[5] We need natural science to tell us exactly how far we can go before we begin to act unsustainably. It is like walking on the edge of a cliff on a foggy day: one wrong step and we fall, and only scientists can see where the edge really is through their instruments. Society is sustainable as long as it stays within the proper limits. If it moves beyond these limits, future generations will end up on a lower plateau with fewer resources where they have less free scope than present generations. This way it seems possible to avoid value questions. The only presumed value is the transcendental value of autonomy. Everybody can behave as they please, in accordance with their own personal values, as long as they stay within their own ecological space. This space is defined by natural scientists without reference to any particular conception of the good.

The second assumption is ethical. It states that the distribution of natural resources ought to be based on a principle of (simple) equality, because value-free, deontological ethics cannot discriminate between people. In the Wuppertal report this is stressed several times. It is a separate goal to secure "just and equal access to the resources for all human beings,"[6] now and in the future. The report refers explicitly to Kant's categorical imperative as the philosophical basis of the claim. The Danish exposition agrees on this point: each and every human being should have an equal right to his or her share of the ecological utilization space. We must leave the "same amount and quality" to future generations—a distant echo of the Lockean proviso to leave "enough and as good" for others, but not quite the same, as we shall see shortly. This is all very much in line with the story of modern life. If there is no common conception of the good, and if there is only a limited amount of goods or resources, it makes sense to say that we should supply each person, now and in the future, with an equal right to utilize the same amount of each and every kind of resource as everybody else.

## Can the Limits of Sustainability Be Determined by Value-Free Science?

Let us take a closer look at the two assumptions. We begin with the epistemological assumption that natural science by itself, and in a value-free way, can determine the limits we have to stay within in order to remain sustainable—that is, in order to leave similar resources for future generations to use. The best way to test the assumption is to see how well it works in the specific analysis of different kinds of resources.

### Nonrenewable Resources

There are various kinds of resources. A basic distinction is the one between renewable and nonrenewable resources. In the first case the limit is set by the flow per unit of time; in the second case the limit is set by the total stock. The distinction is somewhat blurred, but we do not have to worry about that here.[7] Let us just stick to the well-known distinction, and begin with the nonrenewable resources. Can natural science tell us how much we can use in order to leave the "same amount and quality" for future generations? It does seem that we face a serious difficulty right from the start. If these resources are nonrenewable, and if human beings are going to stay on earth for a long time, how can we be entitled to use anything at all? As long as we do not know how much longer humans will exist (and even natural science cannot answer that question), we cannot set the limit. If we assume that there will be humans alive thousands and thousands of years from now, from the stated premises we will have to conclude that we cannot use anything at all—at least not relatively limited resources like fossil fuels. (We did not need much natural science to reach that conclusion.)

A bad start, indeed, but let us change the premises a little. Instead of talking about the "same amount and quality," let us make use of the Lockean proviso and say that there should be "enough and as good" left. In this case we have an extra opportunity: if we (now or within the foreseeable future) can fully replace a limited resource with another one, we are allowed to use all of the resource up. In this case, "enough" means enough until a substitute is found that is just as good. For instance, if there are other energy sources that are just as good as fossil fuels in every important respect, we are entitled to use the fossil fuels without disregarding future generations (at least as long as we are only talking about fossil fuels as a resource question).

Can we now leave it to the natural sciences to operationalize the concept of sustainability? Before we accept the offer, we should consider a couple of disturbing problems. The first problem is the word *good*, which is included in the Lockean proviso. If we are to avoid the discussion of values, what are we going to do with *good*? This is obviously one of those four-letter words we are told not to use. When we dropped the phrase "same amount and quality," the reason was that it imposed too many limitations, and we replaced it with "enough and as good," because this allowed us to make various substitutions. In each case, however, we will have to ask whether the substitute is good enough. This is a major problem.

Let us look at alternatives for fossil fuels, for instance. Which other energy sources would count as good enough? Is nuclear power good enough, even though we will be leaving radioactive waste, which will be potentially dangerous to human beings and other living creatures for thousands of years? Is hydraulic power good enough, even if we have to flood some beautiful or historically significant valleys and block the salmon pathways? Are windmills good enough, even though some people find that they disturb the scenery? Is natural science really capable of settling such issues on its own? Of course, it is not. There are obviously value questions involved of the kind, which were transported to the free scopes of individuals in the story of modern life.

A similar problem turns up when we try to estimate the total amount of nonrenewable resources. Figure 6.1 depicts the so-called McKelvey box, used by the U.S. Geological Survey (and many other geological institutions around the world) to classify resources. The disturbing issue is not only that there is much uncertainty about the total amount of a certain nonrenewable resource (the $x$-axis problem), but even more that the estimate depends on factors foreign to natural science (the $y$-axis problem). The question of how much can be used how quickly without disregarding future generations cannot be answered without answering a couple of other questions, which are difficult to deal with via the methods of natural science.

The estimates depend on two factors. First, there is a technological factor. When technology improves, more resources will be discovered

<table>
<tr><td colspan="2" rowspan="3"></td><td colspan="3">Identified resources</td><td colspan="2" rowspan="2">Undiscovered resources</td></tr>
<tr><td colspan="2">Demonstrated</td><td rowspan="2">Inferred</td></tr>
<tr><td>Measured</td><td>Indicated</td><td>Hypothetical</td><td>Speculative</td></tr>
<tr><td colspan="2">Economic</td><td colspan="3">RESERVES</td><td colspan="2"></td></tr>
<tr><td rowspan="2">Sub-economic</td><td>Para-marginal</td><td colspan="5" rowspan="2">RESOURCES</td></tr>
<tr><td>Sub-marginal</td></tr>
</table>

**Figure 6.1**
The McKelvey box, classifying mineral resources (U.S. Geological Survey/Bureau of Mines)

and more will become accessible. Estimates are made from a specific point in history, however, and it is impossible to predict future technological improvements. Second, there is an economic factor, or, to be more precise, a priority factor: some potential resources will never be used because they are too costly to extract and utilize. "Too costly" is an expression of estimates, however, which lies beyond the scope of natural science. These estimates have to be imported from types of discourse that lie beyond ordinary natural science. In the McKelvey box, the estimates are imported from economics. We will see later that this choice of conceptual framework can hardly be called accidental. Still, the basic point is that we are faced with weightings including value judgments, which cannot be arrived at using the ordinary methods of natural science.

### Renewable Resources

Let us now turn to the renewable resources and see if natural science is able to do any better here. In contrast to the nonrenewable resources, it is not the total stock of resources that sets the limit, but the potential flow per unit of time. We cannot utilize more resources than those flowing through the system. We may thus be able to establish some general demands that should not be violated. First, the exploitation rate should never be higher than the regeneration rate. Forests should not be utilized beyond their net production, fish should not be caught more quickly than the shoals can regenerate, and so on. Second, emissions should never be so great that the ecosystems receiving the emissions cannot neutralize them. $CO_2$ emissions should never exceed the sinks' absorption capacity; the release of nitrogen from agriculture should never exceed the denitrification rate, and so on. In economic terms, the capital stock should not be reduced, wherefore the rate of exploitation should never exceed the rate of regeneration, which can be compared with the rate of interest.

In some cases, however, we can move beyond the rate of regeneration for some time. We may, for instance, for a period of time catch more fish than the rate of reproduction seems to allow, as long as the shoal does not disappear altogether. For a number of years there will be fewer fish to catch—in other words, a smaller flow of resources—but the transgression of the regeneration rate is possible for a while. In other cases, unfortunate side effects may occur, but this does not make transgressions impossible. Eutrophication of lakes and watercourses results in unfortunate changes, but there is nothing absolute about this process. We can live with lower water quality, even if the salmon cannot. There are a few cases, of course, where we simply cannot move even temporarily beyond the limit. For instance, there is only so much solar energy coming in through the atmosphere, and we cannot possibly transgress the upper limit. We cannot live without the benefits solar energy brings us, and we cannot replace solar energy with anything else either. However, there are only a few similar cases (if any), and these cases do not seem to be among the most urgent ones.

Can natural science set limits here without making value judgments? The answer cannot be anything but negative. We will be facing exactly the same kinds of difficulties as in the case of nonrenewable resources. Some resources can be replaced, whereas tampering with others may lead to unfortunate consequences, but these unfortunate consequences are seldom so damaging that we cannot live with them. There are definitely some upper limits (we could not keep on living, for instance, if we were the only species left), but there are lots of reasons to react long before we even get close to such limits. Or, if the word *reasons* is not considered acceptable in relation to value, let us just say that there are enough value preferences around to motivate reactions long before we get even close to these absolute limits.

In figure 6.2, I have tried to present an overview of the different kinds of resources discussed above. In the entrances to the left I have separated nonrenewable resources from the two kinds of renewable resources. In the entrances at the top I have separated cases where the transgression of limits is impossible, no matter which values one subscribes to, from cases where replacement is possible with comparable alternatives, and cases where transgressions are unfortunate because of some foreseeable consequences. Some of the examples I have included in the figure could be placed in more than one room (there is, for instance, some upper limit on how much biodiversity can be reduced

| | | **Transgression of limits absolutely impossible (value-free judgment)** | **Transgression possible through replacement by comparable alternatives** | **Transgression unfortunate (value assumptions)** | |
|---|---|---|---|---|---|
| | | | | **Global/ international** | **Local** |
| **Nonrenewable resources —total amount (stock) decisive** | | e.g., arable land areas | e.g., fossil fuels, various metals and minerals | e.g., biodiversity (irreversible loss of species) | e.g., various local resources (irreversible losses of local nature qualities) |
| **Renewable resources** | **Flow or capacity per unit of time decisive —problem even in case of temporary transgression** | e.g., solar energy | e.g., various annual crops | e.g., protecting ozone layer (if CFC-gasses causes irreversible degradation) | e.g., watery ecosystems (where emissions of nitrogen causes eutrofication) |
| | **Flow or capacity per unit of time decisive —only a problem if transgression is permanent** | e.g., resilience and regeneration capacity in ecosystems, which are vital to survival of human beings | e.g., wood (exploitation of forests greater than net production) | e.g., atmospheric stability (greenhouse gas emissions larger than sinks) | e.g., clean water (emissions of pollutants larger than natural self-purifying capacity) |

**Figure 6.2**
The possibility of transgression of limits in relation to the different kinds of resources

without putting the survival of our own species at risk), but this is only a minor problem. What I hope to show is that only a few resources have to be included in the column where the transgression of limits is absolutely impossible.

My conclusion is that it is not possible to "operationalize" the concept of sustainability by the use of natural science methods alone. There are too many value judgments involved. We may not be able to conclude in general that there is no way at all to operationalize the concept, but it seems obvious that it cannot be done by the ordinary methods of natural science.[8] There may be other possibilities, and I will deal with one of them below. Before I do that, though, let us take a look at the second assumption behind the idea of determining an ecological utilization space: the ethical assumption that the utilization space should be distributed on an equal basis.

## Simple Equality as Distributive Criterion?

Simple equality is the criterion we use whenever we have no good reasons for making distinctions. This is the case with resources as well as with persons. If persons are considered simply to be persons as such, autonomous beings living a life in accordance with their own personal values, and nothing else, and if we are not allowed to judge which kind of life is most valuable, everybody must be treated as equals. As long as there are no common values or standards, we cannot discriminate between people. Similarly, if value judgments are not allowed to enter into the distributive scheme, all resources, together with potential resources, must be treated on an equal footing. If we cannot make judgments about the good (life) and its ingredients, it does seem quite reasonable to say that each and every kind of good or resource must be left for future generations in exactly the same amount as today. As soon as we say that it is more important to preserve one kind of resource over another, we make value claims of a kind that, according to the story of modern life, only belongs in the private sphere. Therefore, each person ought to have (access to) exactly the same amount as everybody else of every single kind of resource, which he or she is allowed to use in accordance with his or her personal values or private preferences.

Let us begin with the second premise, the equality of resources, which seems to be the easiest one to deal with. We do not have to think very far about consequences before we realize that it is quite absurd to claim that we should leave each and every kind of thing in exactly the same state and amount in order to be able to let future generations have (access to) exactly the same potential resource for their activities, not knowing which kinds of tastes the future will bring. First, it is simply impossible to fully comply with the claim. Natural changes inevitably occur that we cannot control, and we cannot even function ourselves if we are forced to keep everything in exactly the same state forever. We have no alternative but to give priority to the resources we consider most important, and to leave others in a state of constant change.

Second, when trying to establish priorities, we have to recognize that some resources, or potential resources, are simply too worthless to preserve, or even less desirable than that. Theoretically we could imagine a world, of course, where people would get a kick out of getting, say, malaria, or where it would be a status symbol to have chemical waste barrels piled up in the backyard. These scenarios are not likely, however, and I must confess that, like Avner de-Shalit,[9] I find it difficult to see myself committed to leaving those hypothetical individuals the necessary resources for their peculiar, unpredictable, and probably short-lived kind of lifestyle. There may be fewer tons of a specific kind of toxic bacteria in the world than there are of gold, but my hope and my guess is that future generations will be grateful to us for leaving them the gold, not the bacteria.

To avoid these absurd consequences without leaving the story of modern life with its demand for operationalization, we have to find a neutral way of dealing with valuations of resources. The solution usually chosen by people eager to operationalize sustainability is to let money become the common standard,[10] thus letting economics get in through the backdoor. I will return to some of the consequences of this approach in the next section.

Until then, let us return to the first of the two premises, the equal right of persons to the same amount of each and every resource, which they can use in accordance with their private conceptions of a good life. Is this a reasonable claim? First of all, it does not seem very sensible to give everybody exactly the same kinds of things, if they have different needs and wants. I do not know what to do with my amount of plutonium, for instance, and I would rather not have it at all. I am not even sure that I would be able to do much with my shares of iron or mercury—apart from selling them, of course. I do have a garden, so I would like to have at least some of the earthworms, but if I were living in an apartment, I think I would feel better without them.

Second, in a dynamic society any initial distribution will soon be changed when people begin to use and exchange their resources in different ways. Some are hard-working people constantly trying to improve their lot and acquire more goods, whereas others prefer to have fewer material goods if this gives them more free time. Some have special talents that make it possible to get something extra out of their share, whereas others are less talented. Most people would probably exchange a large part of their resources with others, and these exchanges would almost automatically make some people's shares more valuable than those of others. It would, in principle, be possible to redistribute the resources, say, once a year, but it does not seem fair to those who have put great effort into maximizing their share just to transfer it to the lazy ones, nor does it seem reasonable to nullify all exchanges of resources once a year (they would then have to be repeated again after the redistribution). Moreover, these recurring redistributions would obviously contradict the idea behind the story of modern life, by restricting the long-term opportunities within individuals' private spheres.

To avoid such absurdities it seems necessary to make some changes in the equality claim. There are at least two possible strategies, if we want to keep on having the possibility of operationalizing sustainability. The first is to talk about communities instead of individuals. In this case, it is not future individuals who are entitled to the same amount and quality of resources as current individuals. Instead, it is future communities that are entitled to the same amount and quality as current communities. One of the advantages of talking about communities—nations, for instance—instead of individuals is that there will be a broader spectrum of preferences and needs, so that it is more likely that there will be some preferences and needs corresponding to each kind of resource. This way we also avoid futile discussions about how many people are going to exist in the future and who they will be. If it is communities we are talking about, it does not matter exactly how many people they include, nor whether the individuals will be different when different decisions are made.[11] Furthermore, the choice of distributive criteria would be more open within each single generation: the simple equality criterion could be supplied with criteria like merit, needs, abilities, luck, and so on.[12]

One of the disadvantages, however, when seen from the standpoint of the story of modern life is that there is no guarantee that each individual will receive a share that can be considered equal, or at least equitable, as compared with those of others. In fact, there is an obvious problem of measuring equality: how are we to compare the relative value of different resources, if everybody has a unique set of preferences? This problem could be solved, though, if we changed the original claim and said that, instead of an equal share, the average future individual would only be entitled to an equal opportunity to have an average share. Another disadvantage is that the inevitable valuations—that is, the determination of what can be considered "enough and as good"—will have to be made on the communal level, if we choose to focus on the community only. But this would run counter to the story of modern life, which only allows private valuations.

Both these problems can be solved most easily, though, if we accept a second possible change in the equality claim: letting economics in through the backdoor again, and saying that all individuals should not have an equal right to the same amount of all the particular resources, but only an equal opportunity to obtain the same amount of generalized resources—that is, money, which they can exchange for whatever kinds of goods and services they prefer. This way of changing the claim is very much in line with the second part of the definition of ecological utilization space in the statement from the Danish government: "Everybody should have a chance to reach the level of material welfare that the [now: common] ecological utilization space and the technological capacity allow."[13] *Material welfare* can thus be interpreted as just another term for generalized resources, or money to spend. This can also be seen to be in line with the Wuppertal report, which recommends that the price system should be changed in a way that makes it possible to reflect the "true value" of environmental resources.[14] In a system where everything is valued at its "true price," there does not seem to be any need to preserve and distribute each and every resource separately.

In this case, the claim is that every individual now and in the future should have an equal opportunity to reach the same average level of material welfare—that is, to acquire the same amount of money (or economic value)—as the average person in the current generation. This does not imply anything like the utopian demand that every single person in the future should be granted the right to have and to keep the same amount of money as the average person of today, no matter what his or her priorities are, but only that average future individuals should have the same opportunities as current average individuals, leaving it up to them whether they are interested in taking advantage of these opportunities or not.

This is evidently a flexible clause that can be interpreted in quite a few ways, but one fairly obvious interpretation would be to combine both of the two previously described ways of changing the equality claim, and thus continue talking about communities as well as of economics. We are then left with the claim that the community as a whole should continue to be as well off economically as today (or maybe that the average future individual should have at least the same opportunity as current average individuals to obtain average portions of the common pie).[15] This way, however, the demand to operationalize the concept of sustainability turns into exactly the kind of claim that neoclassical welfare economists prefer. In the next section, I will consider where this will take us.

## Letting Economics in

Some of the objections I have offered to the concept of ecological utilization space have also been stated in slightly different ways by various Danish (as well as many other) economists.[16] They draw the conclusion that this concept has little validity or utility, and that the notion of economic sustainability gives us a much better foundation for the operationalization of sustainability. This position seems to be playing an increasingly prominent role in the Danish debate, more so than I would have believed possible just a few years ago.[17] Thus, it seems necessary to take a closer look at the assumptions we will have to accept (or shall I say: buy), if we let neoclassical welfare economics in through the backdoor. As will be clear in the following paragraphs, these assumptions fit extremely well with the story of modern life.

Basically, as in the story of modern life, all goods are considered to be goods simply because they are useful for satisfying preferences. All values can be reduced to, or at least can be dealt with as if they are, expressions of private preferences or emotional stances of varying intensity.[18] People's preferences and emotional perceptions are revealed by the choices they make. Some people have a strong preference for coffee or the survival of the blue whale, others do not, and matters of preference, taste, and emotion cannot be discussed rationally but only recorded as matters of fact. Everybody seeks to obtain the greatest possible satisfaction of preferences, in accordance with a unique set of private emotions related to various conceptions of the good life.

Common decisions must accordingly be based on the assumption that the private consumer is sovereign in his or her choice of goods. Communities are best understood as collections of private individuals, and they do not have any independent goals apart from maximizing the satisfaction of preferences. If common decisions are to be rational, they must therefore be conceived of as optimal aggregations of private choices. Whichever social welfare function one prefers, it has to be based on individualistic assumptions: economists "generally assume consumer sovereignty. That is, each individual's utility . . . is determined by that person's own judgments, not the judgments of society more generally."[19] *Sustainability* therefore means leaving future generations "the option or the capacity to be as well off as we are," or "to leave behind a generalized capacity to create well-being"[20] (which future generations can make use of or not). We cannot be specific about exactly which goods to leave behind, because we do not know the tastes of future individuals, and we are not allowed to interfere with them—"it is none of our business."

The total quantity of goods is limited, however, and it is necessary to make choices and trade-offs. Everything has a price, and nothing has an infinite price. Nothing is so important, so useful in satisfying preferences, that it cannot be exchanged or replaced with something else: "Goods and services can be substituted for one another. If you don't eat one species of fish, you can eat another species of fish. Resources are, to use a favorite word of economists, *fungible* in a certain sense. They can take the place of each other. That is extremely important because it suggests that we do not owe to the future any particular thing. There is no specific object that the goal of sustainability, the obligation of sustainability, requires us to leave untouched."[21] All goods and resources can be replaced and will be replaced as soon as the price is right. A consensus exists among most economists, we are told, that lack of one resource in most—if not all—cases can be fully compensated for by the presence of other resources.[22] All kinds of natural and cultural "capital" must be dealt with on an equal basis.

The measuring rod for the value of all kinds of resources is economic value or money.[23] Natural and cultural capital are simply various kinds of exchangeable capital, and they can all be measured by the same denominator: the market price. This way comparability across resources and preferences is made possible. If the market mechanism is working freely without external interference, the market price is an expression of the average intensity of present preferences toward it, and therefore of the good's expected utility value, or preference satisfaction value. It is a completely neutral measure that operates in real life. It is not the economists who value the environment; it is the sovereign consumers themselves. In this sense the methods of economists appears to be completely free of value judgments: they only "observe that individuals have preferences . . . and that those preferences are held with varying degrees of intensity."[24]

The economic value of preferred goods that for some reason are not or cannot be exchanged on the market (so that the related preferences cannot be represented) has to be fixed methodologically—using "Willingness To Pay" or "Willingness To Accept" surveys or by examining the various indirect ways in which the goods can be said to be valued economically by private consumers (prevention costs, replacement costs, wage differences, property price effects, travel expenses, and so on).[25] This way the direct-use values, indirect-use values, option values, and existence values not registered directly on the market can be brought within the horizon of the private consumers—for example, by taxing goods and services that have a negative impact on these externalities. The right taxes (the so-called Pigovian optimum taxes) are those that reflect the true external costs best: "The purpose of economic evaluation is to reveal the true costs of using up scarce environmental resources. . . . Valuation is essential if the scale of the tax or the strength of the regulation is to be determined."[26] However, to obtain accurate, not just virtual or indirectly estimated prices, ways of getting all goods into the market should be considered insofar as possible—for example, through privatizing goods that are still common property, or by issuing marketable pollution or utilization permits, marketable preservation bonds on threatened species and ecosystems, and so on.

Although various welfare functions can be described, most welfare economists would argue that common decisions should be oriented toward obtaining the greatest amount of utility, or preference satisfaction, or welfare as measured in economic terms. Or the decisions should comply with the Pareto or Kaldor/Hicks optimality principles—that is, all the projects should be promoted that can make somebody better off economically without making anybody else worse off, or, if some people do suffer a loss, it should, in principle, be possible to offer them economically appropriate compensation that does not eat up all of the extra benefit. Or they should comply with the principle behind Coase's theorem: in all situations with potentially conflicting interests, the one solution should be sought (through negotiation or otherwise) that gives all affected parties the greatest advantage and, in principle, nobody any disadvantages. If all goods and services were privatized, and all benefits and costs therefore had specified prices, the market would deal with this problem automatically. Consequently, there would be no significant differences between the various criteria.

There is yet another important feature connected to this way of operationalizing sustainability: future goods have to be discounted in accordance with the present market rate of interest. Otherwise suboptimal decisions will be made, giving inappropriate priority to projects that bring fewer economic benefits than the more profitable ones. Without discounting, environmental investments will replace other and more profitable investments to an unreasonable degree: "The criterion for optimal social and economic development is that the marginal total benefits from the different investments should be equal regardless of what the investments are aiming at. In other words, the social discount rate should be equal for all investments. . . . Discounting is necessary in order to compare costs and benefits at different time periods. Attempts to avoid discounting or to apply a different discount rate for climate measures [or other environmental investments] than for other investments will inevitably result in an inefficient policy."[27]

This is not simply a matter of pure time preference, putting higher value on current than on future preference satisfactions. This would run counter to the claim that everybody, now and in the future, should be considered as equals. The basic point is that future generations will be better off as well, if current generations invest in projects that yield the highest returns. If all goods were truly priced and discounted appropriately, *sustainability* would be just another word for "economic optimality." The main reason the average rate of interest is positive, in spite of the fact that nonrenewable resources are being used up, is that technology is improving, becoming more efficient, and giving access to new resources, so that future people can also be expected to become richer than the current generations. The combination of resources will be different, but the aggregated amount, measured in common equivalents, will be larger. It would therefore be irrational and inequitable for current generations to make sacrifices for the sake of future generations.

**Operationalization Reconsidered**

Can economy do the job of operationalizing sustainability in a neutral way? I have to admit that I have never met an economist who did not have any reservations at all about the use of economic calculation in relation to problems like, say, the increasing greenhouse effect—that is, problems stretching out not just a few decades, but centuries into the future. All the economists I have met tend to believe that these problems lie at the border of, or way beyond, the capacity of economic science, though they usually believe that economic analysis can enlighten certain aspects of the problem.

To illustrate the absurdities a straightforward economic analysis can end up with in long-range problematics, the Danish economist Alex Dubgaard has calculated how much the flooding of Denmark caused by an increased greenhouse effect 500 years from now would cost, when ordinary calculation methods are used (current valuations, current rate of interest, and so on). His result is that a fair compensation to future generations would be to make an investment on the order of $8 or the value of a fried chicken meal with potato chips—$8 for a country that people have been willing to die for.[28] One must be very fond of the economic paradigm if one cannot see any absurdity in this.

Consider another case where a resource fundamental to human survival—freshwater, for instance—is likely to disappear 500 years from now if a certain project is accepted. If current water prices, based on current preferences for (and availability of) water, are used in the calculations, the project would probably not have to be extremely profitable in order to pass the sustainability test. It does not seem to be too much of a consolation that when people begin to die from the water shortages, they will be extremely rich. The problems in this case are related to a problem inherent in the economic method: the lack of information about future values, preferences, and prices. A "true" calculation has to include all valuations, now and in the future. This is impossible, of course, but the use of current valuations will not even give us an approximately reliable account, if the circumstances are going to change as radically as the economists themselves expect when they discount the future. If future generations are really going to be as rich as the use of the current rate of interest suggests, one can easily imagine that their values and preferences will be quite different from those of current people. Judging from current trends, it seems quite likely, for example, that they will value environmental goods and unspoiled nature areas much more highly than is the case today.[29] The inclusion of such guesses about future preferences in the economic calculus would probably undermine its claim to scientific validity, however, and these guesses would definitely undermine its claim to be a necessary decision-making instrument. There seems little need for its calculations if all the basic valuations on which the calculations are based are totally hypothetical.

For these and similar reasons, most economists are hesitant when asked to make economic cost-benefit calculations that include valuations of goods and circumstances that will exist more than a couple of decades from now, especially when a large array of goods and activities needs to be included. It may be argued that the picture I am presenting is one of a strawperson or a scapegoat. Few economists appear ready to walk the plank and say that sustainability can be fully operationalized by economic methods. Most of them agree that there are at least some critical resources that lie beyond their own sphere of application—resources not exchangeable in the same way as the less critical kinds of goods.

Still, the conviction is widespread that it is necessary or appropriate to try to find some "objective" and "operational" measures and indicators of sustainability in relation to the critical resources—that is, measures and indicators independent of any conception of the good life. In cases where economists cannot meet the challenge, scientists step in instead. While economists continue to make calculations concerning exchangeable and replaceable resources, scientists look for "operational indicators" related to critical resources like the resilience or robustness of ecosystems, or the capacity of resistance in the human body. This way economists and natural scientists seem to be able to divide up the various kinds of resources between themselves—the exchangeable resources go to the economists, the critical ones to the natural scientists—and the definition of *sustainability* can continue to be neutral and avoid the pitfall of value judgments about the components of the good life.[30]

This line of argument is still not fully convincing, however. There seems to be an important omission. Take the flooding of Denmark again. This is not a problem of survival in a biological sense. The citizens can move somewhere else. As a country, Denmark is not a critical resource in the sense describe. Is it an exchangeable resource, then? Can the disappearance of a country really be fully compensated for in terms of money? Can countries be bought and sold as exchangeable resources? If the answer is negative, it seems that we have to take a third kind of resources into account. These resources are neither easily exchangeable nor critical in terms of survival. Let me call them "unique resources."[31]

Unique resources are resources so important to us in one way or another that their disappearance would cause a profound sense of loss and serious damage to our sense of who we are. They cannot easily be replaced by something else, nor bought and sold in an ordinary bargain, because they are loaded with meanings that are more or less crucial, not for our survival as biological creatures, but for our identity. These resources make up the cultural and natural heritage that is fundamental to the way we understand ourselves and that we are proud to pass on to our descendents. In a physical sense we can live without them, but we may be losing some basic part of ourselves once we begin to sell out.

Many economists would say that if the price is high enough, the unique resources will turn into exchangeable resources. There is some truth in this, of course, as long as there are enough comparable goods left or similar goods to buy instead. In Rome, for instance, the citizens are forced to set priorities in preserving the relics of the past. The heritage of the ancient world is so extensive that not all of it can receive the care it deserves. People may even have to sacrifice part of their heritage in order to be able to afford the good life themselves. The past can sometimes be a too heavy burden to maintain, as Nietzsche pointed out in the second part of his socalled "untimely reflections".[32] If it hampers life, it may be necessary to release the burden.

However, the more sacrifices we are forced to make, the more disturbing our reactions are likely to be. We begin to realize that there are values we cannot sell without suffering an identity crisis, where we no longer seem to know who we really are and what is truly important to us. Goods and values exist that cannot be conceived of as mere means to the satisfaction of casual whims. Without these goods and values, we would lose the bedrock of our lives. If we have any sense of identity at all, there are things so important to us that we are willing to make major sacrifices for them, or, to put it more aptly, there are things we are willing to devote a significant part of our lives to, without first calculating whether this will bring us more money or satisfactions.[33]

In figure 6.3, I have tried to show the differences between the three kinds of resources. Exchangeable (or easily replaceable) resources are related to values that are "soft" in the sense that they are easier to do without than the hard-core values, to which the two other kinds of resources are related. Critical resources are basic in relation to hard-core values like physical health and survival, whereas the various kinds of unique resources are important in terms of identity. Critical resources like (sufficiently) clean water and air will be needed in all kinds of societies, while the unique resources tend to be more specifically related to a particular culture or tradition. This does not mean that they can only be understood and appreciated by a very local culture. The temples of the Acropolis, for instance, do have a specific significance to the people of Greece, but this does not prevent others from acknowledging their significance or from considering them an important part of the common heritage of Europe and of humankind as well.

There are areas of overlap between the three kinds of resources, of course. It is not completely obvious, for example, where and when clean water can be considered an exchangeable resource, and when it turns critical. For some people the critical line appears much earlier than for

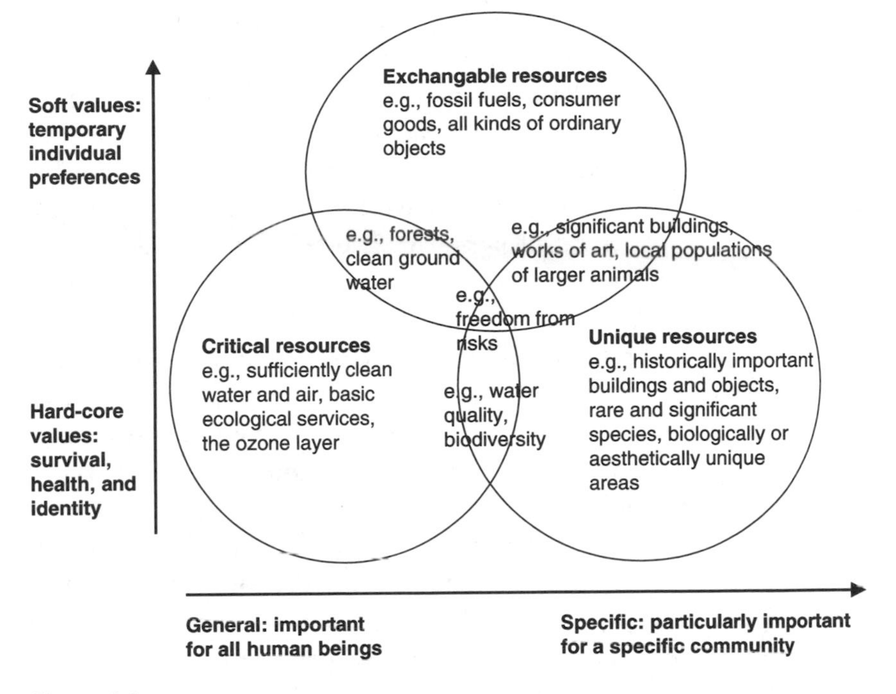


Figure 6.3
The three kinds of resources: exchangeable, critical, and unique

others. Similarly, the borderline between exchangeable and unique resources does not lie out there in the open, but can only be drawn by each community through political deliberation. Some communities are likely to let private individuals decide in most cases which part of a heritage to preserve, and only consider a few things worthy of attention from the entire community, whereas other communities tend to classify more things as unique resources that deserve to be preserved or taken care of by the community as a whole.

If we agree that there is a category of resources that are neither easily exchangeable nor crucial for our physical survival, two important things follow. First, we will have to move away from our interpretation of the sustainability demand as basically a negative claim, and adopt a more positive perspective. Sustainability no longer simply means that we have to restrain ourselves in order to leave an adequate legacy for our descendants. I will not deny, of course, that problems remain, which can best be described as conflicts of interest between current and future generations, and where principles of distributive justice and equity in a fairly narrow sense are relevant. The introduction of unique resources into the sustainability debate does mean, however, that we have to move away from the kind of zero-sum game where a sustainability demand inevitably results in taking away some of the goods of current generations in order to give them to the following ones. Instead, sustainability is just as much a question of preserving and enhancing the goods and values that we ourselves are most devoted to. This includes taking a position on behalf of future generations, preserving or enhancing certain kinds of goods and resources before others, but isn't this actually a sign of true respect—trying to leave our descendants those parts of our cultural and natural heritage that we find most valuable in the sense that they are crucial to our understanding of who we are and of what we see as the basic values of life?[34]

Second, it will no longer be possible to operationalize sustainability through the use of more or less mechanical methods using value-free measures and indicators. The identification of unique resources can never be devoid of values, nor can disinterested parties determine them externally. These resources can only be recognized from within: you have to know (or discover) what you care most about in order to be able to identify it. I will not rule out the possibility of operationalizing certain features, once we have determined a specific set of basic goals, nor would I ever dream of denying that natural science as well as economics can make significant contributions to the goal-defining process. What I do deny is that the whole process can be left to one or another group of experts in operationalizing agencies, by the use of one or another predefined set of operational rules.

It is interesting to notice that the identification of unique resources often takes place parallel with, but usually quite independent from, the discussion of sustainability. This is what happens, for instance, in the Danish government's statements on environmental politics from 1995 and 1999. In chapter after chapter all the specific goals are set up for water and air quality, preservation of the cultural heritage in the city as well as in the landscape, protection of endangered species and habitats, and so on. Only in a few cases are we talking about truly critical resources, where survival is at stake, and the reason why goals are defined in the first place by the community's political representatives is that the relevant features are not conceived of as resources that can be exchanged on the market. The specified goals are first of all goals concerning unique resources that the community has agreed on through a deliberative process. Still, the claim to operationalize sustainability is put forward almost as if these chapters were nonexistent.[35]

## The Story of Modern Life Reconsidered

If it is true that such unique resources exist, we will have to make revisions in the story of modern life. Individuals can no longer be described exclusively in terms of their personal preferences, nor can we see the role of government agencies simply as that of ensuring that individuals are able to satisfy these preferences. We have to find a place for identity as a more stable element in life than the ever-changing preferences.

I find it difficult not to closely associate the questions of identity with the presence of a community. When we grow up as individuals we always do it as part of a community. Our identity, the way we understand ourselves, the conceptional resources we use when we try to come to terms with the various ingredients of the good life—all of these things originate from a communal source. No matter how one chooses to define it, no community is without factions, of course, and as we grow older we all end up with a unique combination of values and preferences. But still, we cannot understand ourselves without the conceptual and behavioral resources we have inherited from the community. We share our most basic traditions with those around us.

Does this mean that we have to skip the whole story of modern life, and let go of the theme of individualism? Not at all, but we have to realize that the story's conception of individuals is much too one-sided. It cannot tell the whole truth if it ignores the communal side and only informs us about the private dimension. It becomes one-sided, too, when it focuses only on a limited range of individual rights. To the extent that it is concerned only with the individual, it can only consider the rights to a private sphere of activity. It is unable to see individuals as citizens with the right to participate in deliberations about common affairs. Thus the concept of autonomy becomes one-sided as well.[36] The only place autonomy is sought is in the private sphere, whereas the political sphere is either deserted or occupied and taken over by "experts" trying to specify the demands of sustainability through predefined methods. No allowance is made for shared values, and definitely not for common decisions about the good life. These are very strong limitations placed on our personal autonomy, and they are extremely difficult to reconcile with claims of neutrality.

Even though the way we see our identity is intimately connected to our community and cultural traditions, it can never be conceived as one seamless web. The individual can never disappear and become one with its community. This is why the story of modern life is still worth telling, provided that it takes all sides into account. The individual needs protection, not only in order to be in control of his or her own private life, but also in order to be able to contribute to the communal life. If we had to get to the bottom line, this is probably what sustainability in a modern society would be about at its very deepest level: the protection and continuation of a democratic process where individuals can participate in communal affairs without having to give up much personal autonomy. However, their lives will certainly be richer if they can also inherit not only the necessary critical resources and an equitable amount of exchangeable resources, but also the unique resources deemed most precious by their ancestors. At the end of the day, future generations may not place the same priority on the resources we consider most valuable, but our legacy to them may still give them a stronger foundation for the creation of their own version of the good life.

## Notes

1. R. Weterings and J. B. Opshoor, *The Ecocapacity as a Challenge to Technological Development*, Publikatie RMNO no 74A (Rijswijk: Advisory Council for Research on Nature and Environment, 1992).

2. Mathis Wackernagel, *How Big Is Our Ecological Footprint?* Draft (Vancouver: The UBC Task Force on Healthy and Sustainable Communities 1993); William E. Rees and Mathis Wackernagel, *Our Ecological Footprint: Reducing Human Impact on the Earth* (Philadelphia: New Society Publishers, 1996).

3. *Mod et bæredygtigt Europa* [Toward a sustainable Europe], report from the Wuppertal Institute (Copenhagen: NOAH/Friends of the Earth Denmark, 1996), 11.

4. *Natur- og miljøpolitisk redegørelse 1995* [Statement on nature and environmental policy 1995] (Copenhagen: Danish Ministry of Environment and Energy,

1995), 27. In the English summary of this statement (*Denmark's Nature and Environment Policy 1995: Summary Report* (Copenhagen: Danish Ministry of Environment and Energy, 1995)), the presentation of the ecological utilization space (or ecological scope) is much more vague than in the full statement.

5. *Natur- og miljøpolitisk redegørelse 1995*, 57.

6. *Mod et bæredygtigt Europa*, 9.

7. First, some of the nonrenewable resources can be reused over and over again. There is a limit, though, in that there will always be a certain loss of the resource every time it is recycled. It may not disappear altogether, but it will be dispersed in the environment in tiny amounts, which cannot be brought back into the recycling process—at least not without using a disproportionately large amount of energy. Second, some resources considered renewable may actually be used up. Clean water, for instance, is a renewable resource as long as there is enough of it, and as long as it stays clean. If it is used too quickly, there may be nothing left, and if it is heavily polluted, it may no longer be useful—at least not for a considerable amount of time. Similarly, when a desert is spreading, the renewable resources that used to be there—the trees and shrubs, for example—may not be able to regenerate. Or, to take one last example, an endangered species is renewable; if it becomes extinct, it is not.

8. Similar conclusion were reached in a report sponsored by the Danish Ministry of Environment and Energy (*Økologisk råderum—en sammenfatning* [Ecological Utilization Space—Project Summary], Miljøprojekt no. 433 (Copenhagen: Danish Ministry of Environment and Energy, 1998)), as well as in the Netherlands Scientific Council for Government Policy's evaluation of the use of the notion of "environmental utilization space" in Dutch environmental policy (*Sustained Risks: A Lasting Phenomenon* Report to the Government no. 44 (The Hague: Netherlands Scientific Council for Government Policy, 1995)). (I thank Robin Attfield for bringing the Dutch report to my attention.)

9. Avner de-Shalit, *Why Posterity Matters* (London: Routledge, 1995).

10. This is not just the case in the sustainability debate, of course. For instance, the first thing Ronald Dworkin does in his discussion of equality of resources is to let the inhabitants of his thought experiment—the survivors of a shipwreck, washed up on a desert island—invent local money and set prices on all goods at an auction (Ronald Dworkin, "What Is Equality? Part 2, Equality of Resources," *Philosophy and Public Affairs* 10, no. 4 (1981): 283–345). Instead of money, one could use "manna," like Bruce Ackerman, or some other kind of apparently neutral utility chip.

11. Thus we avoid Derek Parfit's Non-Identity Problem, as well as the Repugnant and other Counter-intuitive Conclusions, and so on; see Derek Parfit, *Reasons and Persons*, 3rd ed. (Oxford: Clarendon Press, 1987).

12. See also Andrew Dobson, *Justice and the Environment: Conceptions of Environmental Sustainability and Dimensions of Social Justice* (Oxford: Oxford University Press, 1998), 130.

13. *Natur- og miljøpolitisk redegørelse 1995*, 27.

14. *Mod et bæredygtigt Europa*, 10.

15. This interpretation does lend itself to counterintuitive conclusions like increasing the average income opportunities by decreasing the number of people, or increasing the total income opportunities by increasing the number of people. I will leave these problems aside here, however. (See also John Broome, *Counting the Costs of Global Warming* (Cambridge: White Horse Press, 1992), chap. 4.)

16. Thus, for instance, J. Munksgaard and A. Larsen, "Miljømæssigt råderum—et vildskud? [Environmental utilization space—An aberration?]," *Samfundsøkonomen* 1999, 1.

17. In the latest governmental statement on Danish environmental policy, *Natur- og miljøpolitisk redegørelse 1999*, the 1995 definition of the ecological utilization space is quoted in a chapter on sustainability, but this time alongside an exposition of the concept of economical sustainability (or weak sustainability) taken over from a exposition made by the Danish Economic Council (an independent agency staffed by economic experts and sponsored by the government): "Bæredygtighed, Balance mellem generationer" [Sustainability: Balance between Generations], in *Dansk økonomi* (Copenhagen: Det økonomiske Råd, 1998), 171–256. The statement does not make clear which of the two concepts the government was planning to use as its primary guideline (*Natur- og miljøpolitisk redegørelse 1999* (Copenhagen: Danish Ministry of Environment and Energy, August 1999), chap. 36, 503ff). In a governmental proposal outlining a strategy for sustainable development, made in the summer of 2001 by the social democratic/social liberal government and intended to be presented at the Rio + 10 World Summit in Johannesburg 2002 (*Udvikling med omtanke—fælles ansvar* [Development with Care—Common Responsibility] (Copenhagen: Miljøstyrelsen, June 2001)), all references to the concept of "ecological utilization space" (or "environmental space") were removed, whereas the economists' concept of "true savings" (measured in terms of economic value) was given a more prominent position as a main indicator of sustainability. The use of economic concepts and measures is even more prominent in the new liberalist/conservative government's revision of the former government's proposal (*Fælles fremtid—udvikling i balance* [Common Future—Balanced Development] (Copenhagen: Miljøstyrelsen, April 2001)).

18. For a description of the growing formalization and subjectivization of the concepts of "utility" and "well-being" in modern economic debate, see John O'Neill, *The Market: Ethics, Knowledge and Politics* (London: Routledge, 1998), esp. chap. 3.

19. Kenneth J. Arrow et al., "Intertemporal Equity, Discounting, and Economic Efficiency," in *Climate Change 1995: Economic and Social Dimensions of Climate Change*, ed. James P. Bruce, Hoesung Lee, and Erik F. Haites (Cambridge: Cambridge University Press, 1995), 142n18 (see also p. 138).

20. Robert M. Solow, "Sustainability: An Economist's Perspective," in *Economics of the Environment: Selected Readings*, ed. Robert Dorfman and Nancy S. Dorfman (New York: Norton, 1993), 181, 182. A similar definition can be

found in David Pearce, *Environmental Values and the Natural World* (London: Earthscan, 1993), 48, 55f, as well as in many other similar books and articles.

21. Solow, "Sustainability," 181.

22. Arrow et al., "Intertemporal Equity, Discounting, and Economic Efficiency," 140, 141n6.

23. Pearce, *Environmental Values and the Natural World*, 50. Some economists do say that this is only true up to some point, where trade-offs between cultural and natural capital cannot be continued, because some fundamental natural services would be irreversibly damaged (for example, David Pearce and R. Kerry Turner, *Economics of Natural Resources and the Environment* (New York: Harvester Wheatsheaf, 1990), 24f, 56f). Others would argue that there is no limit that cannot be seen from the market if everything is priced properly.

24. Pearce, *Environmental Values and the Natural World*, ix.

25. A couple of good overviews of these valuation techniques can be found in Pearce, *Environmental Values and the Natural World*, appendix II, and in David Pearce and Dominic Moran, *The Economic Value of Biodiversity* (London: Earthscan, 1994), chap. 5.

26. Pearce, *Environmental Values and the Natural World*, 5.

27. M. Munasinghe et al., "Applicability of Techniques of Cost-Benefit Analysis to Climate Change," in *Climate Change 1995: Economic and Social Dimensions of Climate Change*, ed. James P. Bruce, Hoesung Lee, and Erik F. Haites (Cambridge: Cambridge University Press, 1995) 166. The question of discounting is a matter of much controversy, of course, and many economists do argue that the discount rate should be reduced to zero, whenever the time horizon lies beyond a couple of decades. A good overview of the arguments for and against the use of positive discount rates can be found in John Broome, *Counting the Costs of Global Warming* (Cambridge: White Horse Press, 1992), chap. 3.

28. Alex Dubgaard, "Bæredygtighed og forsigtighedsprincippet" [Sustainability and the principle of precaution], in *Fremtidens pris. Talmagi i miljøpolitikken* [The price of the future: Number magic in environmental policy], ed. Henning Schroll et al. (Copenhagen: Mellemfolkeligt Samvirke/Det Økologiske Råd, 1999), 291. Similar illustrative examples of the absurdities of economic calculations when driven outside a fairly narrow scope can be found, for instance, in C. W. Clark's famous article about the economic sense in killing all blue whales in the ocean and transferring the profits to growth industries (C. W. Clark, "Profit Maximization and the Extinction of Animal Species," *Journal of Political Economy* 81 (1973): 950–961), and in Peter Wenz's calculation that discounting at a 5 percent rate would make one life today worth more than 16 billion lives in less than 500 years (Peter Wenz, *Environmental Justice* (Albany, NY: SUNY Press, 1988), 230).

29. Alan Holland has referred to this as a "Cambridge change"; although nothing has happened to an area, its exchange value may change radically, or, conversely, although the aggregated exchange value of unspoiled nature areas is not diminishing, the areas in themselves may be changing in the most radical

way (Alan Holland, "Sustainability: Should We Start from Here?", in *Fairness and Futurity*, ed. Andrew Dobson (Oxford: Oxford University Press, 1999), 57f).

30. See Herman Daly and John Cobb's point that the identification of "carrying capacity," which determines the maximal and optimal volume of economic activity, measured in physical units, is a job for the biophysical sciences, whereas the job of the economists is to suggest taxes and other kinds of regulations of the market in order to keep the activity at its "optimal volume" (Herman E. Daly and John B. Cobb, Jr., *For the Common Good* (Boston: Beacon Press, 1989), chap. 7). A similar point is made in the Danish Ministry's previously mentioned exposition (*Natur- og miljøpolitisk redegørelse 1999*, 515).

31. Holland has made a similar point: "There are human-made features for which there are no natural substitutes, and probably some for which there are no other human-made substitutes either" (Holland, "Sustainability," 53).

32. Friedrich Nietzsche, "Unzeitgemässe Betrachtungen, Zweites Stück, Vom Nutzen und Nachtheil der Historie für das Leben" (1874), in *Werke I*, ed. K. Schlechta (Frankfurt am Main: Ullstein 1976).

33. See also Michael Sandel's point against "purely preferential choice": "In consulting my preferences, I have not only to weigh their intensity but also to assess their suitability to the person I (already) am. I ask, as I deliberate, not only what I really want but who I really am, and this last question takes me beyond an attention to my desires alone to reflect on my identity itself. . . . While the notion of constitutive attachments may at first seem an obstacle to agency—the self, now encumbered, is no longer strictly prior—some relative fixity of character appears essential to prevent the lapse into arbitrariness which the deontological self is unable to avoid" (Michael J. Sandel, *Liberalism and the Limits of Justice* (Cambridge: Cambridge University Press, 1982), 180). I have discussed this point further in "Levn, levninger og brugt natur. Om forpligtelsen over for eftertiden" [Relics, remnants, and used nature: On the responsibility toward posterity], in *Naturminder—levnenes betydninger i tid og rum* [Nature memorials—The significance of relics in time and space], eds J. Guldberg and M. Ranum (Odense: Odense University Press, 1997).

34. See Avner de-Shalit's point that in order to do justice to future generations "we need not know the preferences of future people at all, but, rather, can decide what to leave to future generations on the basis of our own values" (Avner de-Shalit, *Why Posterity Matters: Environmental Politics and Future Generations* (London: Routledge, 1995), 130). A similar point has been put forward by Mark Sagoff: "We cannot avoid paternalism with respect to future generations. . . . We want them to have what is worthy of happiness. We want to be able to respect them and to merit their good opinion. How may we do this except by identifying what is best in our world and trying to preserve it? How may we do this except by determining, as well as we can, what is worth saving, and then by assuming that this is what they will want?" (Mark Sagoff, *The Economy of the Earth: Philosophy, Law, and the Environment* (Cambridge: Cambridge University Press, 1988), 63ff).

35. Even though the Danish governments' proposals for a strategy for sustainable development put much focus on the economically defined concept of "true savings" (see note 17), they also stress the need to involve other measures. The following sentence appears on page 3 in the former as well as in the new government's proposals: "We shall avoid critical effects on environment, nature and health, and we shall protect and preserve special and unique natural values, which cannot be regenerated if they disappear." Formulations like these indicate that the strategy is best understood as a compromise between one position, represented by the Danish Economic Council with its focus on "true savings" in economical terms (see note 17), and another position, until recently represented by the counterpart, the so-called Nature Council (Naturrådet), which is more oriented toward the critical and unique resources (Naturrådet, *Dansk naturpolitik. Visioner og anbefalinger* [Danish Nature Politics: Visions and Recommendations], (Copenhagen: Naturrådet, 2000)). However, the new liberalist/conservative government has recently abolished the Nature Council and set up a new (economic) environmental evaluation agency with the "skeptical environmentalist" Bjørn Lomborg as director, so the balance has changed now in favor of an economic position with a more positive view on the replacement of resources.

36. See also the discussion of autonomy and the necessary combination of liberal, republican, and procedural rights in Jürgen Habermas, *Faktizität und Geltung* (Frankfurt am Main: Suhrkamp, 1992).

# 7

# The Environmental Ethics Case for Crop Biotechnology: Putting Science Back into Environmental Practice

Paul B. Thompson

What should environmental philosophers think about agricultural biotechnology? Work by Dale Jamieson and Laura Westra suggests that their opinion should be rather negative. On the contrary, I will argue that environmental philosophers should be more welcoming of techniques for developing new crop varieties using recently developed techniques for the mapping and transfer of genes. I can summarize my position as follows. Plants modified by rDNA gene transfer do not pose significantly greater environmental risks than plants modified by breeding and selection techniques practiced over the last century. By its very nature, crop agriculture requires an endless succession of environmental compromises. Environmental philosophers should be actively engaged in improving our understanding of these compromises, but there are no reasons to think that transgenic plants pose a threat to the environment that would warrant exceptional treatment. In particular, wholesale opposition to crop biotechnology on environmental ethics grounds is unwarranted.

Given that scientists have proposed a number of environmentally beneficial applications of crop biotechnology, and that the ethical arguments that oppose it on environmental grounds fail, this provides the basis for a cautiously optimistic environmental ethics case in favor of crop biotechnology. Environmentally informed laypersons (including environmental philosophers) should not be trying to stop agricultural biotechnology, nor should they presume that transgenic crops will have adverse consequences for the environment. Scientists should evaluate specific proposals for crop biotechnology through informal scientific networks, and through existing (and evolving) regulatory mechanisms. The rest of us should work with agricultural researchers to bring about the most ecologically adventitious agriculture possible, relying jointly on the regulatory system and the professional ethic of plant scientists to preclude irresponsibly risky actions. In addition to making the case for crop biotechnology on environmental ethics grounds, this chapter deploys a pragmatic-democratic conception of environmental philosophy in sketching how informed laypersons (the "rest of us" noted above) can most effectively participate in building networks for ecologically sound and sustainable agriculture.

I do not argue that crop biotechnology should be welcomed unquestioningly. Elsewhere I have argued that individuals and ethnic communities have the right to decide whether they want to grow or consume transgenic organisms, and that this right entails a currently unmet moral obligation to provide exit (that is, the ability to opt out of the industrial food system) and informed consent. I have also argued that consolidation within agriculture and the emergence of a global system of intellectual property has disturbing implications for economic justice.[1] These conclusions were not based on what I would call environmental grounds, though they might provide the basis for one to oppose the emergence of a multinational complex of companies and research organizations promoting agricultural biotechnology. I have also argued that it is reasonable to interpret a broad array of social, personal, and even religious concerns under the heading of risk.[2] Hence it is logically possible to articulate concern about the emerging biotechnology complex in terms of risk, and it is then a small step to express this concern as a matter of environmental justice. Following this line of reasoning, it might be permissible for an environmentalist (or an environmental philosopher) to express opposition to agricultural biotechnology "on environmental grounds" within the context of politics and policy debate. However, there is an important distinction between this kind of strategic use of the phrase "environmental grounds," and the belief that transgenic crops really do pose such significant hazards for human and ecological health that all applications of crop biotechnology should simply be abandoned.

The environmental ethics case for transgenic crops developed in this chapter focuses specifically on environmental impacts as traditionally construed. These include effects on uncultivated plant and animal communities, the ecological sustainability of farming, and the general integrity and stability of both wild and human-dominated ecosystems, understood at scales that vary from the local to the global and from the genome to population levels. It could include issues in environmental justice, such as situations where poor or marginalized groups are placed at differential degrees of risk from environmental hazards. In fact, the focus of the chapter will be to argue that biotechnology does not involve environmental hazards that differ from those associated with other forms of agriculture. It follows that issues of environmental (as well as social) justice should be addressed through a comprehensive philosophy of agriculture, rather than through a conceptual approach that construes transgenic crops as involving a singular type of threat. As such, issues of social *or* environmental justice, as well as issues of food safety, animal welfare, or consumer choice, fall outside the scope of the discussion.

Even this narrowed focus is far too broad for a single chapter. A major part of the dispute over the environmental risks of crop biotechnology cannot be resolved without empirical confirmation of contested claims. Philosophy can play only a limited role in such a resolution, hence my discussion will focus on the role of philosophical assumptions that frame the debate over ecological risks from transgenic crops. Accordingly, I will review four philosophical arguments that find fault with transgenic crops on environmental grounds. The first is a *preservationist argument* that asserts the impurity of genetic manipulation. The second is a *teleological argument* that interprets transgenic organisms as an affront to the order implicit within nature. Third is a *precautionary argument* that applies the precautionary principle in a general indictment of genetic engineering. Finally, there are *risk analogies* that involve the judgment that crop biotechnology is similar to something else that is justifiably viewed in a prejudicial way. I will argue that all of these arguments fail to make their case against genetically engineered crops.

After disposing of the philosophical arguments against crop biotechnology, what can be said in its defense? In the interests of giving readers what the title advertises, the chapter includes a brief discussion of possible environmental benefits from crop biotechnology, followed by a brief and illustrative (rather than exhaustive) discussion of biological approaches to the evaluation of environmental risks. My intent in discussing benefits and risks is *not* to suggest that classical risk-benefit analysis is the right way to evaluate crop biotechnology, but rather to link this chapter to the larger issues and approaches that provide the thematic unity for the entire book. Environmental philosophers and environmental activists might continue to oppose biotechnology after reading this chapter, but if so, they will hopefully do so with a richer understanding of agriculture's place in environmental philosophy, as well as of the actual potential of recombinant techniques. A pragmatic and democratically committed philosophy of agriculture can become the basis for new institutional alliances among philosophers, environmentalists, farmers, and agricultural scientists. Hope for an environmentally sound agriculture should be placed in such networking and community building. Energy spent opposing biotechnology is wasted if it is not grounded on such an approach.

A few general comments are in order before turning to the arguments. First, given the need to work within reasonable space limits, the chapter presumes some familiarity with basic biological concepts in biotechnology. Fincham and Ravetz offer a concise introduction to these concepts at a level appropriate for college undergraduates.[3] Second, with a few significant exceptions, philosophers have not been prominent critics of genetic engineering. Criticism voiced by philosophers tends to emphasize economic or procedural issues. Third, the popular press is replete with allegations that genetically engineered crops are environmentally risky. Yet even when scientists are quoted, only the barest sketch of a rationale for these allegations is ever found. Fourth, even the most critical of scientifically based arguments against crop biotechnology do not provide arguments for a unilateral halt to the genetic engineering of crops. On the basis of these summary remarks, I point out a significant gap between what one (with trepidation) might call the "popular perception" of environmental risk from biotechnology and that of knowledgeable critics. It is reasonable to surmise that there are religious and philosophical presumptions implicit within this so-called popular perception. I have found it necessary to undertake a fair amount of reconstruction in recounting the four arguments mentioned above. Finally, there is the subsequent possibility that this chapter attacks strawpeople. Nevertheless, these presumptions deserve to be articulated and opened for debate. Others are welcome to attempt not only a more respectable statement of these arguments, but to frame alternatives.

## The Preservationist Argument

The view that I am calling the preservationist argument is actually more of a simple presumptive judgment than an argument. Worldwide, environmental ethics is committed to the preservation of wild species, natural areas, and pristine ecosystems. Crop biotechnology seems inimical to this goal on two grounds. First, human manipulation of genes at the cellular level is paradigmatically "unnatural," given the conception of nature that environmental philosophers have wished to promote. Some of the most celebrated environmental philosophy justifies preservation by drawing a distinction between the utility that nature has for humans and its inherent or intrinsic value. Living things, in this view, have a "good of their own," and it is certainly plausible to ask whether the use of recombinant transformation techniques shows proper respect for intrinsic value. Although this kind of proposal is, perhaps, more readily applied to proposals for the use of genetic transformation of animals,[4] it has also been introduced as an approach for thinking about the morality of transforming plants.[5]

This first way of framing a preservationist argument moves into some pretty familiar territory in environmental ethics, and the claim that genetic engineering of animals violates a deontological norm of respect has been subjected to some pretty predictable critique.[6] Since the editors of this book have voiced a commitment to avoid such well-trodden ground, I will not rehearse how the response would go for its application to plants. Beyond this point, it is not at all clear how one uses such a line of argument to conclude that the slicing, dicing, and chemical or radioactive violence done to plants in the course of plant breeding is acceptable, while the use recombinant techniques is not. There is, at best, a proposal for an argument here, and its completion needs a much more thorough and informed look at what has gone on in plant laboratories for the last 100 years.

Second, some think that agriculture itself is an affront to wild nature, and recombinant DNA technologies appear to be the next weapon in industrial agriculture's two-century-long assault on nature. This second tenet may be more widely held in North America, where environmentalists have long opposed water projects intended for agricultural use, as well as the grazing of livestock on fragile Western lands. Biotechnology is seen as the latest link in a chain of environmental disruption and habitat loss precipitated by the growth of industrial agriculture. As such, any use of biotechnology in agriculture is to be opposed as an extension of the basic commitment to preserve wild nature.[7]

It is clear that agriculture has an overwhelming impact on ecosystems in and near the areas where it is practiced. Habitat is irreversibly altered when new lands are given over to agriculture for the first time. When technical change permits dramatic change in land use, the land ethic must be deployed in any environmental ethics evaluation of such change. Crop biotechnology can facilitate such change, as modifications such as drought tolerance, salt tolerance, and virus resistance are often *intended* to expand the areas where crops can be grown. Indeed, a genetically engineered vaccine against sleeping sickness is currently raising questions about the environmental impact of pasturing cattle in the African veldt, where the presence of the tsetse fly has until now made this impossible.[8] What a careless preservationist fails to notice is that such questions in land ethics are not unique to biotechnology. They speak to a need to better integrate a land ethic into agricultural land-use decision making, without regard to whether biotechnology is involved.

The second preservationist argument may also conceal a pattern of reasoning antithetical to an agricultural land ethic. In viewing agriculture as *inimical* to nature preservation, the preservationist would seem to be committed to the twin goal of (1) minimizing the extent of land given over to agriculture, and (2) isolating agriculture from the rest of nature. This is the philosophical recipe for the most thoroughly industrialized agriculture we can possibly imagine. Water and solar energy could be captured in slime pits. Slurries could transport the biomass into factories where genetically engineered microbes turn it into edible protein. The futuristic food scenarios of Tomorrowland could become a reality, and it can all be self-contained in systems that recycle water and nutrients, reducing pollution to zero and leaving the most land "for nature" as is biologically possible.[9]

Extreme preservationism thus *supports* an ironic overstatement of the environmental ethics case for crop biotechnology. Furthermore, advocates of corporate agriculture have not overlooked this ploy for enlisting environmentalists in the industrialization of the food system.[10] While slime pits, slurries, and genetically modified bacteria are not consistent with my own conception of an environmentally friendly agriculture, further analysis of this theme would take the discussion in a direction other than the one I wish to pursue here. Preservationist environmental ethics needs to rethink its disposition toward agriculture. For now the point is simply that the preservationists' presumption *against* industrial agriculture can be transformed into an argument *for* the most aggressive industrialization of the food system, once biotechnical capacity to isolate and intensify the process of food production is taken into account. Wild nature preservationism is thus vulnerable to a reductio ad absurdem that environmental philosophers who might be tempted by this line of reasoning have thus far failed to address.

## The Teleological Argument

It is not difficult to imagine an explicitly religious argument against any form of biotechnology. If one presumes that a supernatural being designed the universe according to a detailed plan, and that human beings are endowed with a power to deviate from the plan, it is reasonable to understand the ability to move genes across species as a manifestation of that power. Indeed, many religious traditions involve beliefs about the purity of various plant and animal species that would suggest as much. It is also reasonable to think that God's wrath will be visited on those who spoil His handiwork in the form of great environmental disasters—flood, pestilence, and famine. If one drops the supernatural being (or takes a less confident view of His intentions), and presumes only that there is an implicit order within nature, the argument becomes a bit more difficult. If human beings are natural creatures, how is it that *their* actions are not simply a part of the larger plan? If disaster is not brought about by the hand of God, are we not brought back to the mechanisms that would be examined in a nonteleological view of nature?

There is, however, little point in pursing the metaphysical questions further. Gary Comstock has offered an extensive analysis of them, and readers seeking a more detailed discussion are encouraged to consult his work.[11] One application of metaphysical teleology that Comstock does not consider is whether such a teleological view of nature and environment ought to be taken seriously when assessing environmental risks.

Although authors such as Jeremy Narby and Jochen Bockemühl believe that it should, they admit that doing so also contradicts 200 years of science.[12] I would go farther: making a teleological argument abandons any hope that science can inform our understanding of biotechnology and its likely impact on the environment. If we abandon the molecular biologist's account of what gene transfer involves, who is to say that it challenges purity or design in the first place? Just think how the argument would go: Sure, biologists tell themselves an evolutionary/genetic story that omits reference to God or final purpose, but if we do not believe that the story is true, why should we believe that they have done what they say when the tell us that they have intermixed species? What if, instead, the biologists are mumbling meaningless gibberish while producing plant varieties wholly consistent with the larger plan? How could we tell the difference? When we adopt a throughgoing supernatural teleology, the argument against biotechnology evaporates.

This facetious conclusion illustrates that the teleological argument imagined above requires more subtlety than originally supposed. In fact, Narby and Bockemühl argue rather weakly for more open-mindedness in biology, rather than for a strong teleological view. Open-mindedness is always a virtue. Yet the larger problem for environmental ethics is that our conception of environmental risk from biotechnology presupposes the general metaphysical and scientific assumptions of evolutionary biology and ecology. Since these assumptions are decidedly nonteleological, direct appeals to teleology involve the use of claims that must be regarded as both arbitrary and as in potential conflict with the principles and reasoning on which our conception of environmental risk is based. The teleological argument against crop biotechnology requires us to accept what biology says about the role of DNA and genetics in the structure and function of organisms, but to reject what biology says about evolution and ecology.

Few critics make such a baldly teleological argument, and most religious groups have based concern about agricultural biotechnology on social and economic concerns. However, Laura Westra has interpreted ecological integrity in very teleological terms. For Westra, only and all unmanaged ecosystems have integrity, which includes goal directedness.[13] From her standpoint, natural phenomena are striving toward stable relationships of balance and harmony. Biotechnology is wrong because it violates integrity. It is on these grounds that Westra is among the most vociferous critics of biotechnology among environmental philosophers. Of course, *agriculture* violates integrity, so Westra seems to have the preservationist dilemma described above, as well. Agriculture might be excusable on the ground that humans must make some use of nature to survive, but if so, why not use biotechnology to leave the most room for nature?

Westra herself is aware that biologists—including both molecular geneticists and ecologists—are deeply reluctant to embrace a view of ecological integrity that posits goal directedness. Yet when this dimension is eliminated, ecological integrity must be interpreted in terms such as the genetic and species diversity in ecosystems, the interdependencies among these species, and the ecosystem's resilience: the capacity to maintain populations of resident species, especially under various forms of biological stress. Current thinking in ecology suggests that specific places have the potential to serve as habitat for very different configurations of resident organisms. Once established, an ecosystem configuration can be said to have or lack integrity in varying degrees, but a particular configuration might be succeeded by a different one. Evolutionary biology provides no basis for thinking that one succession has any moral or metaphysical superiority over any other.

The shift to a nonteleological form of argument does not rule out the potential for raising concerns about biotechnology's potential effect on ecological integrity. It merely places that concern squarely into our understanding of how ecosystems evolve and function. Furthermore, there is no reason why particular configurations of ecosystems *should not* be invested with moral significance, even if successional ecology does not provide the basis for such a judgment. The mere fact that an ecosystem exists in a given configuration could be a reason to invest that configuration with significance, and especially if the configuration has proven resilient and long-lived in the past. Indeed, it is in precisely these terms that I believe we *should* address the environmental risks associated with biotechnology. But note that we are now far from the teleological presumptions with which we began.

Before moving on, it is worth noting a speculation that a significant number of people who express concern about the environmental impact of biotechnology have a religious or quasi-religious teleology of nature at the root of their concerns. I have argued elsewhere that we have a political responsibility to respect their beliefs.[14] This means that we should not prevent them from acting on their beliefs by concealing the use of genetic engineering when people make daily food choices. It also means that they should be granted an opportunity to articulate the philosophical grounds for their concern in an open forum governed by standard rules of discourse. If a polity could reach closure on their opposition to crop biotechnology on cultural or religious grounds, that might well provide a reason to ban it. But *these* principles do not entail anything substantive about the actual impact that biotechnology might have on land use, wild species, or ecosystem processes.

## The Precautionary Argument

Both popular and scientific critiques of crop biotechnology note that we cannot, in truth, know the nature, probability, or extent of the detrimental impact from biotechnology. Yet our ignorance about the environmental impact of biotechnology is multitextured. On the one hand, it can be a function of poorly confirmed observations on specific and well-characterized phenomena such as gene flow or invasiveness. On the other hand, pleiotropy and epistasis are so poorly understood that we cannot even envision what their ecological consequences (or the likelihood of them) might be. In the former case, uncertainty derives from the fact that existing data provide little basis for predicting the likelihood of such events. In the latter case, a bland uncertainty—undifferentiated into possible event scenarios or their likelihood—pervades our thinking.[15] Dale Jamieson argues that it is inappropriate to classify such broad ignorance as "uncertainty," and like many critics who note our ignorance of the possible environmental implications of genetic engineering, he finds this sufficient reason to resist crop biotechnology.[16] This pattern of aggregating different forms of uncertainty in a generally cautionary argument has framed debate over a number of environmental issues.[17] Over the last decade, it has been associated with the precautionary principle.

Hans Jonas formulated the first statement of the precautionary principle in his book *The Imperative of Responsibility: In Search of Ethics for the Technological Age*. Jonas argued that when one can sketch a plausible scenario of events that would result in the extinction of autonomous reason, responsibility requires that we avoid any action that might contribute to such an outcome, without regard to the likelihood of the scenario. This means that when one can robustly imagine how technology might threaten the existence or quality of human life, one may not make cost/risk/benefit comparisons that incorporate an estimate of the probability that given events will occur. One must shun the technology in question entirely. Jonas clearly had several kinds of technology in mind. Foremost was nuclear energy, which through the development of nuclear weapons threatens the annihilation of all humanity. Jonas also feared technologies of behavior modification and social control. Such technologies threaten not the body but the soul, and it was the autonomous freedom of the moral agent that Jonas wished to enshrine. Jonas mentions that genetic engineering could become implicated in such a threat, but provides no details.[18]

Recent writings on the precautionary principle deviate substantially from Jonas's original argument. First, Jonas's precautionary principle had an implicitly anthropocentric orientation. It was activated by threats to a morally significant form of human consciousness. Certain kinds of devastating depletion of the natural order trigger the precautionary principle, but only because they threaten the annihilation of the human species. This restriction does not reflect a general anthropocentrism in Jonas's thought. He leaves open the possibility for duties to animals, ecosystems, and broader nature, but these duties do not follow from the precautionary principle, nor does his formulation of the precautionary principle apply to them. Second, Jonas limits the precautionary principle (he calls it the *principle of responsibility*) to risks of the most extreme and irreversible sort. Presumably, some sort of trade-off logic would continue to be used in evaluating the acceptability of technology that harms individuals (human or not) as well as ecosystems but that does not threaten the extinction of autonomous reason.

More recently, appeals to the precautionary principle apply its logic to a host of unwanted outcomes that fall short of annihilation, including detrimental impacts on nonhuman species and ecosystem integrity. Such applications suffer from one new problem and another problem inherited from Jonas. These recent applications weaken Jonas's approach in that they frame the precautionary principle as a general moral norm, one that would be applicable to virtually every action a moral agent might contemplate. In limiting its application to annihilating threats, Jonas limited the potential for a choice between two actions, both of which are prohibited by the precautionary principle. He also invoked an argument indebted to Kant: preservation of the potential for autonomous action is an overriding and universal moral responsibility. Failure to recognize this would be to deny (in one sense, annihilate) the very conditions that make freedom or autonomy possible. In contrast, extending the principle to virtually any circumstance that threatens detriment to animals (including individual humans), species, or ecosystems has implausible results. There cannot be many choices where this version of the precautionary principle does not either rule out every option, or revert to a rule of minimizing the chance of the worst case. The "minimax" rule has, of course, been the subject of intense discussion at least since the publication of John Rawls's *A Theory of Justice*. I believe we may assume that advocates of the precautionary principle have thought that they were offering something new.

It might be possible to redress this problem by limiting the precautionary principle to the evaluation of large-scale technological changes, but there is a further problem. Such an interpretation of the precautionary principle favors the status quo over any technological innovation. In doing so, one presumes that the status quo is not itself already part of a series of events leading to annihilation. It is not at all clear what the precautionary principle entails when agents are already embarked on a catastrophic course of action, and this is a problem not only for the expanded precautionary principle, but also for Jonas's original formulation. Surely one should try to get off this course, but what if one option with significant potential for averting foreseeable catastrophic consequences also involves a more remote chance of catastrophe itself? Is it clear that such options should be ruled out *tout court*, as the precautionary principle demands?

Many advocates of crop biotechnology see it precisely as an option of this general sort. They praise industrial agriculture for its capacity to produce prodigious amounts of food at prices that permit virtually every individual in the developed world to be fed. But industrial agriculture is engaged in a desperate game of continuous innovation to stave off resource depletion, population growth, and the tendency of plant pests to combat pesticides through adaptation. Advocates argue that giving up biotechnology would deprive agricultural scientists of a potentially important tool in their attempt to replace current practices with those that are both capable of meeting food needs and that are environmentally more benign. While by all accounts there *are* environmental risks associated with biotechnology, biotechnology's advocates argue that these risks must be evaluated in light of the risks associated with any course of action that continues present agricultural practices. This means that more conventional risk-benefit types of reasoning should be applied in place of the precautionary principle.[19]

My suspicion is that the precautionary sentiments reinforce preservationist and teleological proclivities. Each argument provides an initially plausible set of reasons for concern about crop biotechnology. Precaution and preservation support a conclusion consistent with a vague type of popular teleology, and they can be publicly asserted in secular democracies without ridicule. One suspects that honest religious concerns are being shunted off to precautionary and preservationist arguments that have been made familiar in environmental ethics. But since the true motivation for making these arguments is religious, there has been little to gain by working them out carefully. None of this is to say that there are not philosophically defensible interpretations of the precautionary principle that are applicable to crop biotechnology. The U.S. Department of Agriculture's Animal and Plant Health Inspection Service (APHIS) has arguably applied a minimax interpretation of the precautionary principle for nearly a century in its evaluation of imported agricultural products that might bring in plant and animal diseases. But though APHIS risk assessment does not weigh risk against benefit, it does pay close attention to the risk mechanisms that make plants invasive or weedy. In fact, my purpose in this chapter is to endorse such measured attempts to evaluate the risks of genetically engineered crops, but to do this is decidedly *not* to take a position of unilateral opposition based on a precautionary argument.

## Risk by Analogy

Advocates of biotechnology have suggested that transgenic crops are analogous to products of conventional plant breeding, while opponents have suggested an analogy to invasive species. These analogies lie at the heart of much confusion over the environmental risk of crop biotechnology. When environmental risks are evaluated in light of well-articulated mechanisms for producing a bad outcome, we broach a truly scientific understanding of the ecological impact that can be expected from the commercialization of transgenic crops. The ecology of invasive plants and animals provides a model that illustrates how transgenic crops might pose genuine environmental risk. It does so because the risk mechanisms revealed in the study of invasive species apply to *all* agricultural crops, irrespective of whether they have been developed conventionally or using recombinant DNA. Clearly, there *are* environmental risks associated with transgenic plants, as there are risks associated with all agricultural technology. Biologists have conducted research and debated the applicability of invasion ecology to the evaluation of environmental risk from transgenic plants (a bit more on this below). This debate is wholly proper and not without its own philosophical dimensions. Yet as it has percolated into the broader public sphere, it has reinforced a lay perception that the biological debate over risks from transgenic plants is analogous to the controversy over agrichemicals, global warming, or nuclear power.

Although it is easy to sympathize with this pattern of reasoning, on closer inspection it is a case of analogies run amok. First, transgenic crops are said by proponents to be like "ordinary plant breeding," but then we hear that some biologists are comparing them to species like kudzu, killer bees, and zebra mussels. So the analogy shifts not merely to that of invasive species, but to other problem children of the technological age, where scientific controversy is an indicator that profound environmental values are being threatened by Drs. Frankenstein, Strangelove, and their profit-seeking friends. In one sense, the promoters of biotechnology have gotten exactly what they deserve for having told the public that their technology is "nothing much new" at the same time they were telling their investors that it was revolutionary and unprecedented. But nonetheless, a philosophically respectable inference by analogy demands a closer look.

There are two kinds of inference by analogy at work. One analogizes transgenic crops to exotic species, the other to DDT, thalidomide, and Chernobyl. The analogies that a layperson might draw from the comparison to kudzu and zebra mussels need to be resituated in a more sophisticated understanding of what happens in ordinary plant breeding, and these issues are discussed below. But what about the analogy between transgenic crops and the bad-actor technologies of the twentieth century? This is an analogy that is explicitly suggested in some of the literature. Michelle Marvier puts her review of ecological risks from transgenic crops in context by reproducing a 1947 advertisement from *Time* singing the praises of DDT, and noting "Concerns that assurances of safety might end in a similar environmental calamity surely contribute to the current opposition to transgenic crops."[20] Here, too, it is important to know something about the history of agricultural technology. In fact, pesticides such as DDT are as much a result of ordinary plant breeding as of industrial chemistry. Commercial farmers find it easier to harvest crops that ripen and bloom simultaneously, and plant breeding has produced many varieties that do this. But these are also traits that make crops more vulnerable to insects, creating the need for some form of external control.

The larger philosophical point is that the issue with bad-actor technologies is rooted in the social context of science and technology. In the case of pesticides, academic departments of entomology became so closely wedded to the chemical paradigm that they lost their objectivity. There were always scientists within entomology who realized this problem, and there were research programs in biological control and integrated pest management (IPM) that provided alternatives. Such research programs needed (indeed, still need) the support of environmentalists, for both funding and implementation. But entomology departments in agricultural universities became isolated from environmental groups in the 1970s, and working scientists (as well as farmers) adopted a fortress mentality that stifled much of this work.[21] IPM has succeeded in displacing "spray-on-sight" thinking in crop protection research not because an environmentally concerned public called for it, but because a few savvy researchers realized that they would have to *dissociate* IPM (and themselves) from environmental groups in order to get farmers to accept it. Biological control continues to struggle for funding and for producer acceptance, though it is better established in the developing world.

These are not the sorts of circumstances that environmental philosophers should endorse. Other chapters in this book call for networking within democratic institutions to combat environmental problems, and note that polarizing philosophies make such action more difficult to accomplish. Here we are talking about a different kind of polarization, perhaps, but the message is still the same. What the analogy to bad-actor technologies should teach us is *not* to oppose biotechnology unilaterally, but to find the people within farming and agricultural research who are committed to objectivity and to environmental improvement. With them we should develop institutions that oppose the profit motive when it is inconsistent with environmental (and social) goals, but that do not oppose the profit motive when it can be deployed in pursuit of such goals. Here, I must simply say that my own experience indicates that such people can indeed be found, and that many of them see useful ways to employ rDNA techniques. Thus, like other arguments considered above, the argument from analogy to bad-actor technologies actually winds up being an argument *for* crop biotechnology, but here I say this without irony, without sarcasm. This is what I actually believe.

## The Environmental Benefits of Plant Biotechnology

Some of the legitimate risk issues associated with transgenic crops will be discussed in the following section, but we probably would not waste a lot of time thinking about risks if we did not have some presumptive reasons to think that plant biotechnology could do some good. What follows immediately is a brief conceptual and speculative discussion of some environmentally beneficial applications of gene transfer. The intent here is not to assert that such products are forthcoming or even that they are likely to be developed. In the case of benefits linked to products currently being used, the intent is not to suggest that they justify the use of these products, or to rebut criticisms based on their risks. Making either kind of argument would involve far more empirical detail than I care to provide. The point here is simply to give readers a picture of some of the reasons an environmentalist might find the prospect of transgenic plants appealing. Any adequate evaluation of these reasons would, of course, involve a great deal more discussion and debate.[22]

Indigenous farmers in the tropics have long tried to control insects by pulverizing the carcasses of dead individuals and distributing their remains on their crops. It works to the extent that it does because it redistributes a virus that killed one generation of insects to the next. It is considered a form of sustainable agriculture because such viruses are generally quite specific to a particular species and have no detectable effect on other organisms. However, even relatively poor farmers are switching to chemical agents because they provide a more reliable form of insect control. Biotechnology is currently being used to identify the DNA sequence of these insect viruses. It will be possible to insert that DNA into crop plants, though it is not clear that it will be an effective pest control agent. If it is, there will be no need for chemical pest control agents, and tropical farmers will have a more effective method of protecting their crops from insect damage than saving bug carcasses. There may also be applications in industrial agriculture, where chemical use continues to be rampant.

Although a handful of scientists are working on this approach to pest control, there is not yet any published work suggesting that this is a technology on any discernable horizon. And just imagine how a crop engineered to contain an insect virus would play on the evening news. Similar work has been proposed for scorpion toxins that may also be adapted for very specific applications, but a rumor circulates among crop protection researchers that a large corporation supporting scorpion toxin work pulled the funding because public relations officers did not relish the idea of defending this work in the media. This company will continue to invest its research dollars in chemical pesticides. It would be critical to do extensive risk work before bringing either of these approaches to the field, and it is entirely possible that their promise would not be realized. But the point here is that environmental philosophers should be supporting work to replace or avoid the adoption of broad-spectrum chemical pesticides, rather than opposing it unilaterally because it happens to involve biotechnology.

Another example is work to reduce lignin in woody plants that are to be used for paper products. Lignin must be removed from wood pulp, and its removal creates a significant environmental burden in the form of pollution and energy consumption. Again, it is neither certain that this can be done, nor that doing it will be compatible with environmental goals. What is relevant here is that lignin reduction could result in substantial improvement of environmental quality when compared to current practices.[23] Other scientists have suggested applications of tree biotechnology for reestablishing endangered populations,[24] and for timber production and forest management.[25] Yet genetic engineering of trees promises to be even more controversial than engineered crops. Beyond tree crops and pest-specific toxins, the list of potential products from biotechnology goes on and on. Obviously, not all of these are intended to have beneficial effects on the environment, but there is also work underway to use biotechnology to develop crops that will be effective agents for bioremediation of pollutants,[26] as well as the use of crops to produce nonfood materials that currently require energy consumptive and polluting industrial processes.[27]

If we turn to products currently being used, an environmental case can be made for virus-resistant squash and papaya. These crops have been transformed so that the plants produce antibodies that fight off the viral infection, much the way vaccination helps us fight off the flu. Plants attacked by the virus would eventually produce these antibodies naturally, but the fruit would have lost all value by that time. In the case of squash, the viral-resistance genes replace attempts to control fungal and insect vectors with chemicals. In the case of papaya, no cost-effective control was available prior to the creation of transgenic varieties. The environmental benefits in papaya are most dramatic in developing-country settings, where papaya is a backyard crop that is an important source of food. Poor people who cannot go out and harvest a papaya once or twice a week from their garden tree must find an alternative source of food. This is not only hard on people; it places added stress on land being used to grow the replacement crop.[28]

Clearly the most visible transgenic crops are those that confer resistance to the use of chemical herbicides and those that utilize bacillus thuringensis (Bt) to control caterpillars. Herbicide resistance has been used primarily for soybeans. Bt is used most notably in maize. Both are used in cotton. The net environmental impact of these two technologies is unclear. It is plausible to argue that they are at least no worse than the conventionally bred crop varieties they replaced, and there is mixed evidence suggesting that there may be some improvement. Recent research indicates that widespread planting of herbicide-resistant cotton and soybeans has led to an overall reduction of total herbicide use, and to the substitution of herbicides that degrade into nontoxic materials more quickly. Plantings of Bt cotton (but not maize) are also associated with a reduction in total insecticide use.[29]

We know that this is not the end of the story with either herbicide-tolerant or Bt crops. The most common type of herbicide-resistant crops are "Round-Up Ready," meaning that they are intended for use with the well-known Monsanto weedkiller. Round-Up is generally considered to be among the most environmentally benign herbicides, but as Round-Up Ready crops have become popular, other manufacturers have lowered the price of more toxic competitors. This may ironically be leading to an increase in the use of environmentally undesirable chemicals. Thus the net environmental impact of Round-Up Ready technology is an extremely vexed question, as is the matter of whether environmental effects brought about through such indirect economic causality really count against biotechnology on environmental grounds. Certainly similar kinds of economic causality would be observed if organic methods became effective competitors to chemical herbicides. It is thus important that this discussion of environmental benefits not be interpreted as an attempt to prove that current transgenic crops are unequivocal environmental winners. The point, rather, is simply to make a presumptive case for environmental benefits, one that would need to be followed up with empirical arguments that go far beyond the scope of this chapter. What the case of Round-Up Ready crops does illustrate is the way ecological and economic causes interact so as to make the assessment of environmental impacts and sustainability in agriculture exceedingly difficult.

The Bt case is also complex. As is well known, spreading insecticides throughout a field hastens the pace of acquired pesticide resistance, and can result in the creation of new and uncontrollable plant pests. This happens whether the insecticide is distributed by spray or by engineering the plant to synthesize the chemical in its own tissues. One problem is simply that pesticides become less effective as they are used. In the case of Bt toxin, which occurs in natural ecosystems, the rise of resistant insects could precipitate a cascade of ecological reactions that truly cannot be foreseen. Since we do not understand what ecological function Bt may be performing in natural ecosystems, it is impossible to predict what would happen if species of butterfly and moth acquire immunity to it. Nevertheless, it is true that Bt is nontoxic to mammals, and there is still a *presumptive* environmental benefit when mammalian toxins or broad-spectrum pesticides can be replaced with products that hit their intended target more precisely. Both herbicide-tolerant and Bt crops show that without a network of environmentally oriented citizens to monitor and adjust policy and economic incentives, even environmentally beneficial technologies can have disappointing results.

## The Environmental Risks of Transgenic Crops

Estimating the probability and degree of ecological harm from the introduction of a new crop or crop variety involves a complex mix of art, philosophy, and science. The complexity has already been illustrated in connection with varieties of maize that have been modified to produce the Bt toxin. These crops were introduced in the late 1990s and have been grown extensively throughout the United States since 1998. In 1999, a short paper documented the fact that pollen from Bt maize harms the larvae of Monarch butterflies.[30] The paper sketched a clear mechanism for environmental risk from Bt crops. Although Monarch butterfly larvae do not feed on maize, pollen from these crops could be consumed inadvertently. This discovery was widely covered in the press and often cited by opponents of crop biotechnology, but what was most important was the way it provided a conceptual basis for first estimating and then reducing the chance that this unwanted outcome might actually occur.

Such a response required several bits of knowledge that were either wholly unknown at the time that Bt maize appeared, or that could not be documented through published studies. One was the question of how pollen from Bt crops would move through the environment, and another was whether this pollen was likely to contaminate common food sources for Monarch larvae. Yet another was the general relationship between the life cycle of Monarch butterflies and maize production throughout the United States. A September 10, 2001, article in the *New York Times* reported that empirical research conducted since the 1999 paper by Losey, Raynor, and Carter was published indicates that Bt maize does not pose a significant risk to Monarch populations in the wild. One expects ecological risk issues to be addressed in this way: an initial description of a possible mechanism is followed by research to determine degree of risk. Had the empirical results been different in the Bt pollen case, action would be called for. That action might involve regulatory withdrawal, but in the case of Bt crops, one of the possible actions uses biotechnology: it is possible to regulate the expression of the Bt gene so that Bt is produced only in plant tissues consumed by larvae of the European corn borer, the main pest for which Bt maize was developed to control, but not in pollen.

A more developed literature exists for the question of whether transgenic crops should be regarded as invasive organisms, and it is thus a better example of how biologists respond to potential risks. Jane Rissler and Margaret Mellon argued that risks from transgenic plants are comparable to those of invasive organisms. The key risk mechanism identified in their argument is that gene flow to other plants could transfer characteristics that would make them become weedy (i.e., invasive). There is also the possibility that the crop itself could become weedy, though agricultural crops are almost without exception much less able to survive and reproduce without human assistance than wild types, which have considerably greater genetic diversity. While lay readers may have imagined the kudzu and zebra mussel scenarios alluded to above, the real significance of Rissler and Mellon's claim is that APHIS subjects all plants classified as noxious weeds to fairly strenuous regulatory review. Thus the underlying issue in this debate is whether APHIS regulators have adequately assessed the risks of invasiveness associated with currently commercialized transgenic crops.[31]

A U.S. National Academy of Sciences report (NRC 2002) addresses this question in detail. It notes that a number of hypotheses and specific traits have been proposed for evaluating a given species' potential for invasiveness in a given ecosystem, but that there is still a significant degree of debate among biologists as to the adequacy of these hypotheses, as well as their applicability to transgenic crops. Nevertheless, this literature has provided a basis for experimental evaluation of ecological risks associated not only with transgenic crops, but also with crops produced through conventional breeding. One experimental study indicates that currently grown transgenic crops have displayed no tendencies toward invasiveness, though the interpretation of data from these experiments is itself open to dispute.[32] In the present context, invasiveness itself is less critical than the question of whether the regulatory process, informal and reliant as it is on the professional activity not only of company scientists but also organizations such as Rissler and Mellon's Union of Concerned Scientists, represents an appropriate way to address ecological risks from transgenic plants. In considering this question, it is useful to reexamine the analogy between biotechnology and plant breeding one more time.

Virtually no one disputes the claim that there are ecological or environmental hazards that could be associated with crop biotechnology. What *is* claimed is that these hazards are comparable or similar in kind to those associated with plant breeding, and that the experience we have had with plant breeding provides a basis for estimating the probability that these hazards will be realized. Such a claim might be interpreted in several different ways. It is worth listing some of these possible interpretations.

1. Since conventional crops do not involve significant environmental risk, neither do transgenic crops.
2. The probability that any given transgenic crop will cause environmental damage is roughly equivalent to that of a typical crop produced from plant breeding.
3. Treated as classes, transgenic and conventional crops have roughly equivalent risks.
4. Transgenic methods of crop modification are comparable to conventional methods with respect to their capacity for introducing novel traits that pose environmental hazards.
5. The method for introducing novel traits into crops does not have significant bearing on the environmental risks that they pose.

Setting aside the fact that words such as *risk*, *hazard*, and *equivalent* are themselves open to interpretive difficulties, these five readings of the "comparable to plant breeding" claim map much of the contested turf.

It seems plausible to think that many nonbiologists, including some of the environmental philosophers who have opposed biotechnology, presume that the first interpretation is intended. Yet no competent biologist would claim that conventional crops involve no significant environmental risk. They might claim that the risks from plant breeding are acceptable, given the benefits, and that if this is so, biotechnology should also be acceptable. But it seems that the underlying source of confusion here may be in the failure of laypersons to recognize that products of plant breeding can have and have had detrimental ecological effects, especially when plant breeding occurs in conjunction with the development of other production technologies. In fact, conventional methods are capable of introducing a very high degree of genetic novelty into plants. When techniques such as radioactive bombardment and embryo rescue are included in the "conventional" category (and these techniques have been used for some time), it is not even true to say that products of conventional breeding contain only genes previously extant within a given plant family.

The second formulation is troublesome in that "any given transgenic crop" implies a statement that that is almost certainly too broad. Some transgenic crops may be too dangerous to grow under all but the most exacting conditions. Crops that will be used to produce toxic industrial materials, for example, should only be grown under conditions where we can be sure that the potential for gene flow is miniscule, if they should be grown at all.

The third formulation suffers from a similar kind of conceptual problem. On the one hand, scientists who make this sort of claim may be presuming that biotechnology will soon be utilized simply to increase the efficiency of breeding, hence many if not most uses of biotechnology will accomplish plant modifications that might have been done through plant breeding. If such a practice became routine, this third formulation might turn out to be true. On the other hand, if biotechnology is only used to accomplish what *cannot* be done through plant breeding, the class of transgenic crops will have a very different profile of phenotypic traits from that of conventional crops. The ecological risks may turn out to be incomparable in that event. The irony here is that public opposition to biotechnology may be leading plant researchers to avoid using it to accomplish things that might have been done (though in a more costly manner) through conventional methods. Hence the *reason* that products of crop biotechnology may be unlike those of crop breeding may depend more on social causality than on biology.

It is actually the fourth and fifth interpretations of the "just like plant breeding" claim that tell us most about the risks of crop biotechnology. Crop development can be characterized as involving two stages. The first is the introduction of a new, desired trait into the crop germplasm. Whether recombinant DNA methods are used, the typical result is a plant that is far from being useful to farmers. The second stage consists of backcrossing with established varieties and test plantings that are needed to develop and prove that one has a genetically stable and agronomically desirable variety containing the new trait. Biotechnology differs from classical approaches to crop development only with respect to the first of these stages. A transgenic crop can be judged comparable to a conventionally bred variety because, as with any conventionally bred variety, the process of backcrossing and test planting through six or more generations allows the research team to determine whether the crop is genetically stable, and whether phenotypes display any obvious unexpected traits (variations in color, foliation, height, fertility, and the like). This does *not* mean that when grown in an ecologically vulnerable area (near wild relatives, already disturbed ecosystems) or in conjunction with other agricultural technologies (pesticides, irrigation), the crop will not have unwanted environmental effects. It just means that what is needed (and this is what environmental philosophers should be working toward) is a general approach to the assessment of environmental impact from agricultural production practices, rather than an approach that singles out biotechnology.

There are biologists who dispute the "just like plant breeding" view. Writing for informed laypersons in the *Chronicle of Higher Education*, Stephen R. Palumbi argues that conventional breeding transfers not only the single functional gene (along with its promoters) but also a complex of genes that regulate the expression of the gene in different biological settings. All of the currently commercialized transgenic crops are single-gene transfers of the type Palumbi calls "brute force genetic engineering."[33] If Palumbi is right, even the fourth and fifth interpretations listed above have to be qualified. But even so, Palumbi (who is not a plant scientist) calls not for a ban on crop biotechnology but for a "stage for culling and sorting." Arguably, such a stage is already in place: the second phase of crop development described above. Many plant scientists would rebut Palumbi's claims by making two arguments. First, he may overstate the genetic stability of conventional breeding. Wide crosses can produce exceedingly dysfunctional plants. Second, problems in gene regulation become evident during the second phase of crop development, irrespective of the way novel traits were introduced. But in any case, readers must take note: this dispute between biologists is a substantially different kind of argument than the ones with which the chapter began.

Sometimes the objections turn out to be less substantive. Jane Rissler and Margaret Mellon also take the "just like plant breeding" claim to task in their book on ecological risks of transgenic plants by suggesting that transgenic crops could become invasive. In fact, Rissler and Mellon repeatedly cite research on conventional crop breeding in documenting the biological mechanisms that can create an environmental pest. Yet the inference they draw from this is not that all nonindigenous agricultural plant varieties can have environmental outcomes that deserve scrutiny, but that transgenic seeds in particular pose that risk. In effect, Rissler and Mellon insert the adjective *transgenic* into sentences that truthfully describe ecological risks of applied crop science in broad terms. Since all transgenic crops are examples of applied crop science, this inference preserves its truth value, but it would be equally permissible (and equally vacuous) to insert adjectives such as *green* or *perennial*. The most egregious instance of such adjectival insertion occurs in an extended quotation from a 1990 article titled "Gene Flow in Squash Species." Rissler and Mellon render Hugh Wilson's prose as follows: "Gene flow from [transgenic] cultivars to other elements of its gene pool, both domesticated and free-living, could provide a marked selective advantage to individual recipients . . . ," with the brackets appearing in Rissler and Mellon's text. Wilson's original statement offered as a general claim about gene flow in squash certainly *applies* to transgenic products of biotechnology, and indicates biological mechanisms that are highly relevant to the ecological risks of any nonindigenous plants. Rissler and Mellon's alteration makes it seem as if biotechnology contributes uniquely to this event.[34]

Although Rissler and Mellon call for regulatory responses to the risks of genetically engineered crops, their book actually presents the argument for regulatory attention to the ecological risks of *all* new crops having traits that would confer reproductive advantages on their progeny (or wild relatives) in unmanaged ecosystems. As with Palumbi's objections, rather than opposing crop biotechnology on environmental ethics grounds, we are now engaged in a much more focused discussion of protocols for developing and regulating transgenic plants. This is a debate that *presupposes* the central claim that I wanted to establish in this chapter, namely, that there is a general environmental ethics case favoring the use of this technology to develop more environmentally compatible agricultural practices. This claim does not contradict the observation that there are some transgenic crops that *could* be developed but ought not to be grown.

## Environmental Risk, Agriculture, and Environmental Philosophy

Farmers and agricultural scientists alike have neglected the environmental impact of new technology in the past. Today there are the awakenings of a new environmental ethic within both groups. Farmers especially will be slow to try strategies that threaten their economic viability, but most have accepted the need to embed agricultural production practices within a greater awareness of ecology and environmental impact. For their part agricultural scientists have embraced the goal of sustainable agriculture, though for some it is a cynical embrace. Progress toward a greater sensitivity to the environment in agriculture has been slower than many would like to see, but it has been significant. Realizing the possible benefit associated with responsible biotechnology presupposes that new crops are integrated with other practices in an ecologically sustainable food system, but there is no reason biotechnology should not be part of that system.

Techniques for the use of gene transfer in plants were proven in the mid-1980s, just at the point that the environmentally oriented critique of agriculture was gaining momentum. Almost to a man (and they *were* men), agricultural research administrators chose to invest in biotechnology rather than in technologies favored by those promoting a more environmentally sensitive view of agriculture. Research administrators openly admitted that new programs in gene transfer would form the basis for strong links to industry. Both public and private plant science institutions thus had a double financial interest in the success of gene transfer. First, both had invested their capital heavily in the form of molecular laboratories and staff expertise. Second, both envisioned future benefits tied equally to the science and to a public-private partnership. By the early 1990s, environmentalists interested in agriculture (there are few in North America) were made deeply suspicious by this turn of events.

The topics and goals of agricultural research have been contested since the very beginnings of organized agricultural science, and philosophy plays a large role in framing the controversy. Rather than being appreciated as crucial to our collective future, the debate over research allocations has resonated within lay publics, including environmental philosophers, as a source of mistrust in scientific institutions, generally. It is almost as if people think that to even talk about allocating dollars for technology development makes one venal and impure. Of course if one mistrusts plant scientists, one would quite rationally tend to regard their assurances about the ecological risks of transgenic plants with suspicion. Ironically, those assurances might even become the primary basis for judging the technology to be risky. The justification or defeat of this pattern of inference is a key normative issue for the risk politics of late industrial society.

Following suggestions made in Andrew Feenburg's book *Questioning Technology*, I submit that the most useful role for environmental philosophers is to take a more serious interest in the political economy of technology development.[35] This means that they should be engaged in networks with agricultural researchers who want to pursue environmental goals but who may lack either the right conceptual underpinnings or the right sociopolitical philosophy to do so effectively. One could go on and on documenting the weaknesses in scientists' worldview that contribute to failure here, but one example must suffice. An article by three University of North Carolina biologists concludes by remarking that "scientists must become more proactive in the public debate if agricultural biotechnology is to make a long-lasting and sustainable impact on improving food and fiber production and human health."[36] Unfortunately, they seem to think that this responsibility is fulfilled by talking to news reporters and publishing discussions of "misconceptions and public perception" in journals such as *BioTechniques*. They say nothing about networking with people in environmental studies (including philosophers) at their own universities, or with members of environmental or sustainable agriculture groups. There is apparently no need to talk with the people who actually care about the environment, or who are teaching and writing for students and informed lay audiences. However, the issue in the present context is to prevent environmental philosophers from making a complementary mistake. Democratic, participatory, and pragmatic environmental philosophy means working within social institutions (such as our own universities) to align research priorities in accordance with ameliorative approaches to environmental problems.[37] In the case of crop biotechnology, that probably means learning something about what plant scientists actually do before striking out against them.

There may be circumstances where lack of faith in scientists and their institutions warrants opposition to the technologies that their work has spawned. Indeed, transgenic technology may have begun under just such circumstances. If so, a very different set of issues in the ethics of environmental politics arise. Is it ethical to raise public concern over the food safety and environmental risk of biotechnology (two areas where scientific consensus indicates that risks are relatively low) in order to provoke reform of the research agenda and of the goals and incentives that influence both scientists and farmers? This is an important and worthy question for environmental philosophers and activists, but it can be raised clearly only if we do not deceive ourselves about the field-level, ecological risks of introducing genetically engineered plants into agriculture. A cavalier attitude toward ecological processes can be a source of poor judgment, and poor judgment can result in technology that poses significant ecological risk. But the issues here are whether the turn to molecular biology has increased the potential for poor judgment, and what can be done to shore up any deficiencies in scientists' ethic of responsibility.

In closing, it is worth noting that radical biologist Richard Lewontin, who with Richard Levin produced a devastating critique of agricultural research in the 1980s, has also found many of the critics' views on agricultural biotechnology rhetorically slanted and biologically uninformed.[38] Philosophers who would contend that crop biotechnology is either intrinsically or universally dangerous on environmental grounds do not have the arguments on their side. To the extent that philosophers participate in maintaining such a contention among themselves or among laypersons, they participate in a game of deception that obscures underlying issues about the structure and organization of science, and about the place of ecological integrity as a value for applied sciences in general. This is not to say that there could not be a serious and valuable debate about the adequacy of various regulatory regimes, or about the seriousness with which scientists and regulators have taken the ecological risks of crop biotechnology. Couched within a philosophical commitment to redeploy recombinant technology more resolutely in pursuit of ecological goals, such debate would be very healthy. But that is not the current context. After nearly twenty years of debate on the risks of agricultural genetic engineering, it would appear that our collective capacity to understand and discuss these risks is at an all-time low. Environmental philosophers should take the lead in refocusing that debate. To do so, they must understand that there *is* an environmental ethics case for crop biotechnology.

## Notes

1. Paul B. Thompson, "Food Biotechnology's Challenge to Cultural Integrity and Individual Consent," *Hastings Center Report* 27, no. 4 (1997): 34–38; Thompson, *Food Biotechnology in Ethical Perspective* (London: Chapman and Hall, 1997), 156–162.

2. Paul B. Thompson, "Science Policy and Moral Purity: The Case of Animal Biotechnology," *Agriculture and Human Values* 14 (1997): 11–27.

3. J. R. S. Fincham and J. R. Ravetz, *Genetically Engineered Organisms: Benefits and Risks* (Toronto: University of Toronto Press, 1991).

4. See David E. Cooper, "Intervention, Humility and Animal Integrity," in *Animal Biotechnology and Ethics*, ed. Alan Holland and Andrew Johnson (London: Chapman and Hall, 1998), 145–155.

5. Philipp Balzer, Klaus Peter Rippe, and Peter Schaber, "Two Concepts of Dignity for Humans and Non-Human Organisms in the Context of Genetic Engineering," *Journal of Agricultural and Environmental Ethics* 13 (2000): 7–27; Klaus Peter Rippe, "Dignity of Living Beings and the Possibility of a Non-Egalitarian Biocentrism," in *Intrinsic Value and Integrity of Plants in the Context of Genetic Engineering*, ed. David Heaf and Johnannes Wirz (Llanystumdwy, UK: International Forum for Genetic Engineering, 2001), 12–14; Henk Verhoog, "The Intrinsic Value of Animals: Its Implementation in Governmental Regulations in the Netherlands and Its Implication for Plants," in *Intrinsic Value and Integrity of Plants in the Context of Genetic Engineering*, ed. David Heaf and Johnannes Wirz (Llanystumdwy, UK: International Forum for Genetic Engineering, 2001), 15–18.

6. Robin Attfield, "Intrinsic Value and Transgenic Animals," in *Animal Biotechnology and Ethics*, ed. Alan Holland and Andrew Johnson (London: Chapman and Hall, 1998), 172–189.

7. Laura Westra, "A Transgenic Dinner: Social and Ethical Issues in Biotechnology and Agriculture," *Journal of Social Philosophy* 24, no. 3 (1993): 213–232.

8. Anne Rossignol and Phillipe A. Rossignol, "A Rift in the Rift Valley," *Reflections: Newsletter of the Program for Ethics, Science, and the Environment, Oregon State University* 5, no. 2 (1998): 2, 7.

9. See M. Rogoff, and S. M. Rawlins, "Food Security: A Technological Alternative," *BioScience* 37 (1987): 800–807.

10. Dennis T. Avery, "Why Biotechnology May Not Represent the Future in World Agriculture," in *World Food Security and Sustainability: The Impacts of Biotechnology and Industrial Consolidation*, NABC Report 11, ed. Donald P. Weeks, Jane Baker Segelken, and Ralph W. F. Hardy (Ithaca, NY: National Agricultural Biotechnology Council, 1999), 97–109.

11. Gary L. Comstock, *Vexing Nature: On the Ethical Case against Agricultural Biotechnology* (Dordrecht: Kluwer Academic Publishers, 2000), 183–215.

12. Jeremy Narby, "Shamans and Scientists," and Jochen Bockemühl, "A Goethean View of Plants: Unconventional Approaches," both in *Intrinsic Value and Integrity of Plants in the Context of Genetic Engineering*, ed. David Heaf and Johnannes Wirz (Llanystumdwy, UK: International Forum for Genetic Engineering, 2001), 26–31.

13. Laura Westra, *Living in Integrity* (Totowa, NJ: Rowman and Littlefield, 1997).

14. See note 1.

15. See Paul B. Thompson, "Risk, Ethics and Agriculture," *Journal of Environmental Systems* 13 (1998): 137–155.

16. Dale Jamieson, "Discourse and Moral Responsibility in Biotechnical Communication," *Science and Engineering Ethics* 6 (2000): 265–273.

17. See Paul B. Thompson, "Uncertainty Arguments in Environmental Issues," *Environmental Ethics* 8 no. 1 (1986): 59–75.

18. Hans Jonas, *The Imperative of Responsibility: In Search of Ethics for the Technological Age* (Chicago: University of Chicago Press, 1984).

19. Norman E. Borlaug, "Ending World Hunger: The Promise of Biotechnology and the Threat of Anti-Science Zealotry," *Plant Physiology* 123 (2000): 487–490; Edward Soule, "Assessing the Precautionary Principle," *Public Affairs Quarterly* 14 (2000): 309–328.

20. Michelle Marvier, "Ecology of Transgenic Crops," *American Scientist* 89 (2001): 160–167.

21. John Perkins, *Insects, Experts and the Insecticide Crisis* (New York: Plenum Press, 1982).

22. For a concise overview of the scientific literature, see L. L. Wolfenbarger and P. R. Phifer, "The Ecological Risks and Benefits of Genetically Engineered Plants," *Science* 290 (2000): 2088–2093.

23. T. Tzfira, A. Zuker, and A. Altman, "Forest-Tree Biotechnology: Genetic Transformation and Its Application to Future Forests," *Trends in Biotechnology* 16 (1998): 439–446.

24. Pew Initiative on Biotechnology, *Harvest on the Horizon: Future Uses of Agricultural Biotechnology* (Washington, DC: Pew Initiative on Biotechnology, 2001).

25. Pierre J. Charest, "Biotechnology in Forestry: Examples from the Forest Service," *Forestry Chronicle* 72, no. 1 (1996): 37–42.

26. V. Raboy, "Accumulation and Storage of Phosphate and Minerals," in *Cellular and Molecular Biology of Plant Seed Development*, ed. B. A. Larkins and I. K. Vasil (Dordrecht: Kluwer, 1997), 441–477.

27. C. R. Somerville, and B. Dario, "Plants as Factories for Technical Materials," *Plant Physiology* 125 (2001): 168–171.

28. Dennis Gonsalves, "Control of Papaya Ringspot Virus in Papaya: A Case Study," *Annual Review of Phytopathology* 36 (1998): 415–437. Rissler and Mellon discuss the possibility of gene flow from virus-resistant squash (see below). However, the controversy here stems less from ecological damage that would be associated with the movement of viral-resistance genes to wild relatives than with critics' belief that APHIS handling of regulatory approval for these squash varieties was flawed.

29. Janet E. Carpenter and Leonard P. Gianessi, *Agricultural Biotechnology: Updated Benefit Estimates* (Washington, DC: National Center for Food and Agricultural Policy, 2001).

30. J. E. Losey, S. Raynor, and M. E. Carter, "Transgenic Pollen Harms Monarch Larvae," *Nature* 399 (1999): 214.

31. Jane Rissler and Margaret Mellon, *The Ecological Risks of Transgenic Crops* (Cambridge, MA: MIT Press, 1997).

32. In the interest of conserving space, I have eliminated a number of citations for this paragraph and refer readers to National Research Council, *Ecological Risks of Transgenic Crops* (Washington, DC: National Academy of Sciences, 2002).

33. Stephen R. Palumbi, "The High Stakes Battle over Brute-Force Genetic Engineering," *Chronicle of Higher Education*, April 13, 2001, pp. B7–B9.

34. Rissler and Mellon, *The Ecological Risks of Transgenic Crops*; Hugh Wilson, "Gene Flow in Squash Species," *Bioscience* 40 (1990): 49–55.

35. Andrew Feenburg, *Questioning Technology* (London: Routledge, 1999).

36. C. Neal Stewart, Jr., Harold A. Richards IV, and Matthew D. Halfhill, "Transgenic Plants and Biosafety: Science, Misconceptions, and Public Perceptions," *BioTechniques: The Journal of Laboratory Technology for Bioresearch* 29 (2000): 832–843.

37. For an essay by a biotechnologist who sets a better example, see Robert Goodman, "Ensuring the Scientific Foundations for Agriculture's Future," in *Visions of American Agriculture*, ed. William Lockeretz (Ames: Iowa State University Press, 1997), 187–204.

38. Richard Lewontin, "Genes in the Food!", *New York Review of Books* 48, no. 10 (2001): 81–84.

8

# Yew Trees, Butterflies, Rotting Boots, and Washing Lines: The Importance of Narrative

Alan Holland and John O'Neill

We begin with a brief account of three walks we have taken, two together and one a solo walk, where we have encountered conservation problems of what we will call the "old world"[1] kind. In the second section we offer an old world suggestion about how these problems should be approached. In the third section we criticize certain "new world" approaches to such problems, and in the fourth section suggest that our "old world" approach has application to "new world" conservation problems also. Finally, we suggest that our preferred approach has explanatory power and is metaphysically sound.

## Ambulando: Three Walks

### Walk 1

The first walk was around an area called Little Langdale in the United Kingdom's Lake District National Park, and our guide was the regional manager for the National Trust, the conservation body that owns the land. Our attention was drawn to a number of problems typical of the region that organizations such as the National Trust face: how much grazing to permit; how to manage the small wooded areas; whether to fence off some of the higher slopes to allow the juniper, which still had a foothold there, to recover. Then our path turned through a farmyard and out the other side to a small mound. The mound, we learned, was a largely natural feature, slightly shaped at the edges by human hand. Recent archaeological investigation had established that this mound was once a "Thingmount," or Norse meeting place, and thus a site of some significance.[2] Part of the mound had been excavated unwittingly by the local farmer and was now buttressed by a concrete silage clamp. So, one question was whether the Trust should aim to "restore" the mound to its original condition by removing the clamp. The other question was raised by a more ephemeral adornment. Atop the mound and, as it were, its crowning glory, stood a thoroughly unabashed and utilitarian washing line. It was, after all, an excellent spot for drying clothes.

*Question*: Should the washing line be removed?

**Walk 2**

The second walk was around Arnside Knott, a small limestone outcrop in an area just south of the Lake District National Park known as Silverdale; this too was under the care of the National Trust. Once more we were fortunate to have the regional manager as our guide. Here, in recent memory, a certain butterfly had flourished—the High Brown Fritillary—which is relatively rare in the UK context. The colony was now much reduced and there was a considerable risk that it would disappear altogether. It had flourished because of local use of the land for grazing purposes; this created the limestone grassland that the insect requires for breeding purposes. The colony began to dwindle when the practice of grazing ceased. What the National Trust has recently done is to fence off a section of the land, cut down the naturally arriving yew and silver birch that had successfully begun to recolonize the land, and reintroduce grazing.

*Question*: Is this a defensible decision?

**Walk 3**

The third walk was around a disused slate quarry in North Wales. From a landscape perspective it would normally be judged something of an eyesore—a "scar"—and from an ecological point of view it would be judged relatively barren, showing little sign of life except for a few colonizing species. For both these reasons the local council decided to embark on a reclamation project that would involve landscaping the area. But as work started, there was local opposition. To carry through this project would be to bury the past: it would involve burying the history of the local community and the story of their engagement with the mountain—as revealed in the slate stairways, the hewn caverns, and the exposed slate face. Higher up, and most poignant of all, the workmen's huts were still in place, and inside the huts could be seen rows of decaying coats hanging above pairs of rotting boots, where the last men to work the quarry had left them.

*Question*: Should one let the quarry be?

**Andante Moderato: Adding Values?**

Our objective here is not to give precise solutions to the questions we have raised, but simply to offer a proposal about how they should be addressed. The Roman Stoic Epictetus[3] said that everything has two handles, one by which it can be carried and one by which it cannot, and that one should get hold of a thing by the handle by which it can be carried. In our view, one way above all *not* to get hold of these problems is to attempt to itemize and aggregate the "values" of the various items that feature in the situation—the Thingmount, the washing line, the butterflies, the rotting boots—and pursue a policy of "maximizing value." For besides making the implausible assumption that these values are in some way commensurable, this approach neglects their contextual nature. Nor, second, will any attempt to get hold of these problems be satisfactory that approaches them from an *a*temporal perspective. Rather, it is crucial that we pay attention to the *temporal*—the "diachronic"—dimension. Thus, the problem is, or should be construed as, the problem of *how best to continue the narrative*, and the question we should ask is: What would make the *most appropriate trajectory* from what has gone before? The value in these situations that we should be seeking to uphold lies in the way the constituent items and the places they occupy are intertwined with and embody the life history of the community of which they form a part. This perspective is in accord with a more general attempt to characterize (not define) the objectives of conservation in the following way: "Conservation is . . . about preserving the future *as a realization of the potential of the past*. . . . [It] is about negotiating the transition from past to future in such a way as to secure the transfer of maximum significance."[4]

What we are saying, then, is that time and history must enter our environmental valuations as constraints on our future decisions. It should be observed that many ethical theories fail in this regard, just because they have no place for time, narrative, and history in their accounts of how we should decide what is to be done. Utilitarianism, with its emphasis on future consequences, and existentialism, with its emphasis on the *un*constrained nature of human decision making, are notable culprits. While Rawlsian theory is not consequentialist, its impartial perspective stretches across time and is to that extent atemporal. Evaluations of specific history and processes form part of that body of knowledge of particulars of which those in the original position are ignorant: agents enter deliberation devoid of knowledge of the particular time and place in which they exist. And even theories that do introduce retrospective considerations do so in the wrong way. Some deontological theories, for example, make it a matter of some contract that has been entered into. But in the context of conservation we are not constrained by the past because of any promises we have made. The obligations we have to the past, if that is a proper way to speak here, are entirely nonvoluntary. Nor are our evaluations, and the constraints on our actions, about the legitimacy of the procedures that got us where we are. Robert Goodin is one of the few political theorists to have noted the importance that history and process have as a source of environmental value. However, he spoils his case by associating it with Nozick's "historical entitlement" theory of justice.[5] This appeal to Nozick to illustrate the point about history misses the mark entirely. The value a place may have, say as an ancient meadowland, has nothing at all to do with the justice or otherwise of any procedure that handed it down to us, and everything to do with the continuing historical process it encapsulates. In addition to these critical reflections on ethical theory, it should also be observed that a number of the currently proposed goals of environmental policy, such as "sustainability," "land health," and the "maintenance of biodiversity," as these are typically defined, fail similarly to incorporate the dimensions of time and history, and must be judged inadequate on that account.

At this point we will no doubt disappoint some of our readers by failing to give clear criteria for what exactly constitutes an *appropriate* trajectory from what has gone before, or what the *best* way of continuing a narrative might be. The main reason is that we believe this to be a matter for reasoned debate and reflective judgment on the part of those who have studied the situation carefully and thought hard about it: it is a matter, in short, of deliberative judgment, not a matter of algorithmic calculation according to some formula that we, or others, have supplied.[6] However, rather in the spirit in which Aristotle offers his principle of the "mean" to guide us in deciding on the right course of action in ethical matters,[7] we can offer some initial thoughts about what some of the guiding considerations might be, and hope in the process that we might preempt some of the reader's initial reservations.

First, the problems of nature conservation are not problems about change as such, or at least they need not be, but rather about the kinds of changes that are appropriate. Change can be too much or too little, not by any simple quantitative measure, but by a qualitative measure of degree of disruption to narrative significance. Some attempts at conservation can be disruptive precisely by virtue of stifling change and transforming the lived world into a museum piece. On the other hand, we are inclined to say that some rotting boots should be left to rot, and that some ancient monuments should continue to be decked with washing lines, rather than be removed from the intelligible temporal processes in which they feature. Some "histories" can even intelligibly incorporate "revolutionary" processes of fire and flood, but other dramatic forms of change are disfiguring, or worse. This is often because of their scale, pace, or source. Indeed, many of our conservation problems arise out of the fact that human-induced change generally has a faster pace than ecological change, thus preventing the numerous and subtle ecological checks and balances from operating as they might. The oil slick, for example, will invariably blight the potential of the marine and shoreline ecosystems that lie in its path. Cases involving so-called exotic species, on the other hand, are harder to call. They may inhibit, but can also release, potential, depending on circumstances.

Second, we have to acknowledge that the same site might embody quite *different* narratives that sometimes point to different trajectories between which we must adjudicate. The same mound of earth belongs at one and the same time to the story of an ancient meeting place, a need to dry clothes, and a farmer attempting to make a living in a world of unpredictable markets and state subsidies. There are different histories to which we have to be true—and there are histories that, when they are unearthed, change our perceptions of the nature of a place and what it embodies. The empty hills of highland Scotland embody not just a wild beauty but also the absence of those who were driven from their homes in the clearances. Their memory must also be respected. The argument over the fate of the 100-foot-high statue to the Duke of Sutherland on Beinn Bhraggie Hill near Golspie is about which history we choose to acknowledge.[8] We should perhaps support its removal not just on aesthetic grounds but also for what it represents to the local people, some of whom are descended from those who were driven out. For the same reasons, the now dilapidated cottages that their ancestors left behind should perhaps remain. That there is a problem about conflicting trajectories, often associated with differences of scale and pace between natural history and human history, we do not deny. But we hold that this is not a problem *for* our approach, but a problem revealed *by* our approach. It is not the task of analysis to make difficult problems appear easy, but to reveal difficult problems for what they are.

## Scherzo: New Worlds for Old?

An alternative, "new world," approach to all of the conservation problems we have mentioned might be to say that, given the chance, we should do what we can to restore a given site to its "natural state." But, however plausible this approach might seem at first sight, we believe that the attempt to apply it in the "old world" context is beset with problems.

Consider, to begin with, the site of our first walk, the semicultivated Little Langdale valley. The first question that needs to be asked here is: To *which* natural state should we attempt to restore it? To its Mesolithic state perhaps? Or to its natural state during the glacial, or interglacial period? To *which* interglacial period, exactly? And why this one rather than that? But besides the arbitrariness implicit in such a suggestion, there is also the point that restoration in this sense would be inappropriate, in view of the ecological changes that have no doubt taken place in the meantime.

A natural response to such objections is to propose that we "restore" the site, so far as possible, to "what it would *now* be like if there had been no human intervention." We will not dwell on the practical difficulties lurking behind the phrase "so far as possible," for we believe that this idea is *in principle* misconceived. More precisely, we suggest that this definition of the "natural state" as "how things would be, if humans were abstracted" is either incoherent or radically indeterminable.

If, first, we *simply* imagine humans removed, we have an incoherent situation: because humans *were* in fact there, we should have to try to imagine a vacuum in nature.

Accordingly, if we imagine humans removed, we have to imagine something there to replace them. But this, we argue, must be a radically indeterminable state of affairs, due to the radical contingency of natural processes and the arbitrary choice of starting point. Given that the slightest event may have the most far-reaching consequences—consider, for instance, the difference that may flow from the arrival or nonarrival of a particular species at a particular site at a particular time—then one can only specify what the situation *would* be like, if all potentially relevant variables are assigned determinate values. But it would be impossible to complete such an assignment, and even if it were possible, the assignments would be bound to be arbitrary, except on the assumption of a completely deterministic universe. Further, we have to ask at what point one would begin this "alternative" natural history, for at whatever point one chose to begin, a different history would unfold. One is reminded of the conundrum sometimes posed in graduate logic classes, where counterfactual examples are used to suggest that two incompatible statements might both be true. (This is a conundrum because if two statements are incompatible, at least one of them ought to be false.) Two such statements are: "If Caesar had been alive today, he would have used arrows" and "If Caesar had been alive today, he would have used the atom bomb." The truth is that, because of the radical contingency referred to, *there is simply no saying* what Caesar would have done. To take an example closer to home, the situation is exactly analogous to attempting to speak of "what Langdale would be like if there had been no humans." The fact is: *there is simply no saying* what Langdale would be like under that supposition.

Perhaps the best response to the difficulty we have just outlined is to settle for the weaker notion of what a site *might* be like (rather than what it *would* be like), save for the human presence. But this has some disadvantages. One is that it is not clear what it excludes; perhaps a site *might* (naturally) have come to be a pile of ashes. Another is that it becomes less clear that we have any incentive to bring about a situation just because it *might* have become like that naturally. This supposes that we should be prepared to endorse *any* possible natural world, and it is not clear that we should be so prepared. We will return to the point later.

## Vivace: An "Old World" Symphony

So far, our stance has been merely defensive. We have defended an "old world" approach to certain "old world" problems, and found a "new world" approach wanting. But now we propose to go onto the offensive. We are aware that the inhabitants of "new world" countries, such as the United States or Australasia, are likely to find these "old world" conservation problems quaint or parochial. They hardly concern big global issues, and they hardly concern the problems of conservation in large wild places, the areas of so-called minimal human influence. Or so it may seem. But in fact, we want to suggest that, far from it being helpful to bring over "new world" concepts of nature to help solve the quaint and parochial problems of the old world—and we have just attempted to show that it is not helpful—on the contrary, what would be more helpful would be to take the perspective we have just suggested for dealing with the old world's conservation problems out into the new world's wildernesses.

The first of our reasons for favoring the old world perspective is empirical: the "new" world is much more like the "old" world than it likes to pretend. In short, we do not quite buy the "wilderness story." Our preferred version of that story would be this: that emigrants from the "old" world arrived in the "new" world and mistakenly thought it to be wilderness; this folk memory has lived on . . . and on. In fact, what these immigrants were encountering as new was another people's old world, another people's home. The Indian and the Aborigine had already radically transformed their land. Although (as previously argued) we may not know, because it is impossible to say, what that land would have been like "naturally," we may at least be sure that the presence of these indigenous peoples will have made some considerable difference.

The failure to recognize this has itself been the source of problems in the treatment of the ecology of the "new" world. In particular, in both

Australia and the United States the treatment of nature as a primitive wilderness led to a failure to appreciate the ecological impact of native land management practices, especially those involving burning. Thus, consider the history of the management of one of the great symbols of American wilderness, Yosemite National Park. In the influential report of the Leopold Committee, *Wildlife Management in the National Parks*, we find the following statement of objectives for parks: "As a primary goal we would recommend that the biotic associations within each park be maintained, or where necessary be recreated, as nearly as possible in the condition that prevailed when the area was first visited by the white man. A national park should represent a vignette of primitive America."[9] What was that state? The first white visitors represent the area thus: "When the forty niners poured over the Sierra Nevada into California, those who kept diaries spoke almost to a man of the wide-space columns of mature trees that grew on the lower western slope in gigantic magnificence. The ground was a grass parkland, in springtime carpeted with wildflowers. Deer and bears were abundant."[10]

However, this "original" and "primitive" state was not a wilderness but a cultural landscape with its own history. The "grass parkland" was the result of the pastoral practices of the Native Americans who had used fire to promote pasture for game, black oak for acorns, and so on. After the Ahwahneechee Indians were driven from their lands by Major Savage's military expedition of 1851, "Indian-style" burning techniques were discontinued and fire-suppression controls were introduced. The consequence was the decline in meadowlands under increasing areas of bush. When Totuya, the granddaughter of chief Tenaya and sole survivor of the Ahwahneechee Indians who had been evicted from the valley, returned in 1929, she remarked on the landscape she found that it was "too dirty; too much bushy."[11] It was not just the landscape that had changed. In the Giant Sequoia groves, the growth of litter on the forest floor, dead branches, and competitive vegetation inhibited the growth of new Sequoia and threatened more destructive fires. Following the Leopold report, both cutting and burning were used to "restore" Yosemite back to its "primitive" state.

Such talk, however, simply disguises the nature of the problems that, we suggest, are much better approached from our "old world" perspective. Reference to wilderness suppresses one part of the story that can be told of the landscape. The non-European native occupants of the land are themselves treated as part of the "natural scheme," of the "wilderness." Their history as dwellers in a landscape that embodies their own cultural history is made invisible. Moreover, such language also disguises the way the history of the landscape is being frozen at a particular point in time: "The goal of managing the national parks and monuments should be to preserve, or where necessary recreate, the ecologic scene as viewed by the first European visitors."[12] To refer to "natural" or "wilderness" states avoids the obvious question, "Why choose that moment to freeze the landscape?" There are equally obvious answers to that question, but they have more to do with the attempt to create an American national culture than with ecological considerations.

This position invites two criticisms. The first criticism appeals to the fact that there were large tracts of the Americas and Australasia that the Indian and the Aboriginal peoples did not, in fact, settle on. So here, at any rate, the wilderness concept remains applicable. Without disputing the factual claim, we will respond to this challenge presently by arguing that the old world perspective not only provides a better account of the value of human ecological systems but of natural ones too. The second criticism appeals to a distinction between low-level human impact and "technohuman" impact, and argues that it is only the second kind of impact that results in a seriously degraded and "unnatural" environment. Our response to this criticism would be to insist that, even though the Indian and Aboriginal peoples may not have wrought any great devastation, they will nevertheless have made a great *difference* to their environments, and that therefore, in broad terms, the conservation problem has to be about how to continue the story in which they have been involved, rather than about how to construct a different story in which we imagine they were never there, because their presence is somehow assimilated within the natural world.

The second of our reasons for favoring the "old world" perspective is, in the broad sense, a moral one, to do with the question of what goals and objectives are desirable. The question is most simply approached by asking, "What is so good about a nature characterized by minimal human influence?" The most plausible answer is to connect the defense of nature, understood in this sense, with resistance to biological and ecological *impoverishment*. To defend nature is to hold the line against biological and ecological impoverishment. Two questions arise.

The first question is: Won't the goal of ecological *health* serve this purpose just as well as the pursuit of minimal human influence? The point behind the question is that those who pursue the latter goal believe that it will serve to justify the protection of wild places in particular. At the same time, they seem willing to grant that ecological *health* is in principle as applicable to human ecological systems as to natural ones. Indeed, this is why the goal of minimal human influence is preferred. But if biotic "impoverishment" alone is the villain of the piece, and we can insure against such impoverishment by maintaining ecological health, we seem so far to lack a justification for the protection of wild places in particular.

The second question is: What gives us the right to be speaking of biotic *impoverishment* in the first place? To be sure, *if* biotic impoverishment threatens, we should no doubt do our best to avoid it, but have we earned the right to such language? As instances of biotic impoverishment, attention is often drawn to the following kinds of process: habitat destruction, "pest" outbreaks, increases in exotic species, and the depletion of natural resources. But so far as the first three of these are concerned, one might ask why, exactly, are they not just "changes in the pattern of flourishing"? So-called habitat destruction might alternatively be described as a "change in the pattern of niches." "Pest" outbreaks and increases in the presence of exotics, it might be said, are no more than "changing distributions of plant and animal species." The question that needs to be pressed is: Why are these *bad* changes?

The reason we press the question is that, although we are inclined to agree that many such changes are indeed bad, we are disinclined to accept that they are always so. Consider another short walk at Reposaari on the southwest coast of Finland taken by one of us. This was a walk on what is in a quite straightforward sense a new land. The land is a recent postglacial uplift from the sea—Finland's area is increasing by about 1000 sq km every hundred years. The land on which Reposaari stands emerged over the last 1000 years.[13] The nearest large town, Pori, was founded as a port on the coast in 1558. The port moved with the land uplift. Pori is now an inland town, 30 km from the harbor of

Reposaari. Reposaari's status as a harbor is witnessed by the quite remarkable and sometimes beautiful graffiti carved into the rock from the last century. Its history as a port is also embodied in the very particular biology of the area. The export trade from the harbor was predominantly timber; the imports generally were lighter cargoes, among them spices from further south in Europe. The pattern of heavy export cargoes and lighter imports meant that the ships of previous centuries arrived at Reposaari carrying ballast—soil from southern Europe—and left without it. The ballast soil deposited on the new land contained the seeds of a variety of "exotic" species of plant that were able to flourish in the coastal climate of Finland. The result is a flora unique to the area.[14] The history of human activity is part of the narrative of the natural history of the area. The idea that the proper way to continue that narrative is to cleanse the area of any seed of human origin in the name of biological or ecological integrity strikes us as quite wrong, and would be properly resisted by local biologists and inhabitants of the area. The ballast flora forms as much a part of the ecology of the area as do the seeds distributed by migrating birds. The position we are urging instead is that in considering whether an "introduced" species constitutes harm or good we need to consider the specific narrative we can tell about it, not whether its origins happen to be human or not.

A final instance of biotic impoverishment involves the depletion of natural resources. We have two reservations about this kind of example. The first is that it fails to differentiate between the concept of "natural resources" and the concept of the "natural world." The distinction is important because, even if it can be demonstrated that the natural *world* is disappearing fast, it does not follow that natural *resources* are dwindling. The reason is that natural resources are understood as comprising the natural world only insofar as it is capable of supplying human needs, and possibly those of other selected species. It is well recognized that human-made capital, and technology in particular, can enhance the value of the natural world in this sense, thus enabling resources to be maintained even while the natural world is dwindling. Indeed, in some cases, the increase of natural resources *requires* the depletion of the natural world. Our second reservation stems from the suspicion that what most fuels concern about the depletion of natural resources is indeed concern for resources rather than for the natural world as such. However natural hurricanes and locusts might be, they are more often conceived as a curse than as a blessing. Should this suspicion prove to have some foundation, and should natural *capital*, as distinct from the natural world, prove relatively robust, then this aspect of the case for claiming that biotic impoverishment results from human impact would remain to be made.

But even where there is agreement in judgment as to what will count as biotic impoverishment or loss, the question remains whether the new world approach provides an adequate account of the justification for such judgments. This is the subject of our final section.

**Presto**

We want to suggest, finally, that the "old world" historical perspective we are proposing has considerable explanatory power not only with respect to old world problems but *with respect to new world conservation problems also*, in the sense that it does a great deal to explain the claim that we feel the natural world has on us and the source of its value. For the natural world, just as much as human culture, has a particular history that is part of our history and part of our context, both explaining and giving significance to our lives. Thus, what it is that we value about an ancient human habitation has much more in common with what it is that we value about the natural world than the new world accounts of such value would allow. The value of the natural world should be measured not in terms of the degree of freedom from human impact, but in terms of a continuity true to the historical processes of natural selection that it embodies. Moreover, and paradoxically, the new world accounts do not always carry value as far out into the natural world as does the old world perspective that we are recommending. A test case is the hurricane. Hurricanes are a very evident feature of the natural world. But they wreak much havoc, and it is hard to think of many perspectives from which they would not be judged to promote rather than hinder processes of "impoverishment." The historical approach we are advocating would view things somewhat differently. Of course there are various human and natural ecological systems whose histories are ruptured by hurricanes, which are therefore unwelcome. But, as the practice of naming hurricanes perhaps intimates, this is not the whole story. They too have their own history and play their own awesome role in the histories of others, so that they come to have their own significance in the narratives of human and natural events.

We suggest two areas in particular where the virtues of our account show through. The first is in diagnosing cases of conflict and the second is in diagnosing the tragedy of environmental loss. Taking conflict first, if we return to the conservation problems with which we started, we see that part of the tension between human history and natural history arises from the different paces of change in the two. And the history of the natural world and the objects it contains matter just as much as do those from human history, for the "natural," too, has its own narrative dimension, its own "natural history." It is the fact of their embodying a particular history that blocks the substitutivity of natural objects by human equivalents, rather than, for example, the inability to replicate their function. While "natural resources" may be substituted for one another and by human equivalents—they have value by virtue of what they do for us—natural objects have value for what they are, and specifically for the particular history they embody. The block on "faking" nature[15] lies not just in the origin of natural objects but in the history that takes us from their origin. Were we in 500 years time to release into some natural bamboo-plantation pandas developed from embryos that were naturally conceived but then frozen, their natural origins would be unlikely to confer a "natural" status on the result. Thus, "naturalness" is a question of both origin and history, and conservation problems are frequently associated with conflicting historical narratives.

Our second illustration concerns the "tragic" dimension of environmental loss. We do not here challenge the appropriateness of applying terms such as *impoverishment* and *loss* to certain environmental situations, but we do question whether reference to the ecological characteristics of these situations is alone adequate to convey the gravity of such applications. We find a hiatus here. If you look in the larder and find that you are running out of sugar, you might speak of this as a tragedy, but in doing so you would be conscious of using hyperbole. There is a shortfall between saying of a situation that it is unsustainable, and saying that it is tragic. The ecological characterization fails, in our view, to capture what is at stake—fails to capture the element of tragedy that environmentalists feel. A sense of tragedy requires that there be a story—

and the story can be fiction or, as Colin Macleod argues in his essay on "Thucydides and Tragedy," fact. "Thucydides," Macleod writes, "can certainly be said to have constructed his history and interpreted events, in a strict sense of the term, tragically. This is not at all contrary to his aims as a historian. History is something lived through."[16] We agree. It is only in the context of a *history* of nature, we submit, that the sense of something akin to an environmental tragedy can find adequate expression.

## Coda

It is worth noting that the perspective offered here has some sound metaphysical backing. For what we are in effect proposing is that the term *nature* should be taken in a sense that the philosopher Saul Kripke has identified as that of a "rigid designator."[17] That is, we should be taken to be using *nature* in the manner of a (proper) *name*, as referring to a particular historically identifiable individual. We should not be taken to be using the term descriptively as referring to "whatever is, or might be, natural." This coincides with the way Kripke himself has proposed construing terms referring to constituents of the natural world, such as organic and inorganic kinds, and is consonant with the theoretical position adopted by some leading biologists who construe species (ontologically) as individuals rather than classes.[18] We are claiming, accordingly, that our evaluative attachment is to *this* natural world, not to any possible natural world. For there could be no guarantee that any possible natural world would be good. Some possible natural worlds could turn out to be horrific, like certain medieval depictions of hell. A corollary of this position is that there is no such thing as a state or condition of something that constitutes its "being natural," or an identifiable set of characteristics that makes any item or event "natural." Being natural is, and is only, determined by origin and by history: it is a spatiotemporal concept, not a descriptive one.

## Notes

1. We recognize that the distinction between "old" and "new" world perspectives is something of a (gentle) caricature. It is intended only to register a

"tendency." In fact, many of the sharpest critiques of the wilderness concept are by *new world* authors. See, for example, the discussions in J. Baird Callicott and M. Nelson, eds., *The Great Wilderness Debate* (Athens: University of Georgia Press, 1998), and William Cronon, ed., *Uncommon Ground: Toward Reinventing Nature* (New York: Norton, 1995).

2. In fact, this is thought to be the only site of its kind on the UK mainland, although another one has been found on the Isle of Man.

3. Epictetus, *Enchiridion* XLIII, from *The Moral Discourses of Epictetus*, trans. Elizabeth Carter (London: Dent, 1910), 270.

4. Alan Holland and Kate Rawles, *The Ethics of Conservation*, Report presented to The Countryside Council for Wales, Thingmount Series no. 1 (Lancaster: Department of Philosophy, Lancaster University, 1994), 37. We would suggest that the "significance" or "making sense" referred to here should form a key element of the "coherence" that characterizes "public reflective equilibrium" (see the editors' introduction to this volume).

5. Robert Goodin, *Green Political Theory* (Cambridge: Polity Press, 1992), 26–30.

6. An approach broadly endorsed by the UK's old Nature Conservancy Council: "The standards of nature conservation value thus became established through practice and precedents based on collective wisdom" (Nature Conservancy Council, *Guidelines for Selection of Biological Sites of Special Scientific Interest* (Peterborough: Nature Conservancy Council, 1989), 13).

7. Aristotle, *Nicomachean Ethics*, Book II.

8. Between 1814 and 1819, the "Black Duke," as he was known, played a leading role in evicting the indigenous people from their homes. See David Craig, *On the Crofters' Trail* (London: Jonathan Cape, 1990).

9. Aldo S. Leopold et al., *Wildlife Management in the National Parks*, U.S. Department of the Interior, Advisory Board on Wildlife Management, Report to the Secretary, March 4, 1963, p. 4; cited in A. Runte, *National Parks: The American Experience*, 2nd ed. (Lincoln: University of Nebraska Press, 1987), 198–199.

10. Leopold et al., *Wildlife Management in the National Parks*, 6; cited in Runte, *National Parks*, 205.

11. Kenneth Olwig, "Reinventing Common Nature: Yosemite and Mt. Rushmore—A Meandering Tale of a Double Nature," in *Uncommon Ground: Toward Reinventing Nature*, ed. William Cronon (New York: Norton, 1995), 396.

12. Leopold et al., *Wildlife Management in the National Parks*, 21; cited in Runte, *National Parks*, 200.

13. Michael Jones, *Finland: Daughter of the Sea* (Folkestone: Dawson, 1977), 15, 59.

14. J. Suomin, "The Grain Immigrant Flora of Finland," *Acta Botannica Fennica* 111 (1979): 1–108; H. Jutila, "The Seed Bank of Ballast Area in Reposaari, SW Finland," unpublished ms.

15. Robert Elliot, "Faking Nature," *Inquiry* 25 (1982): 81–93.

16. Colin Macleod, *Collected Essays* (Oxford: Clarendon Press, 1983), 145–146.

17. Saul Kripke, *Naming and Necessity* (Oxford: Blackwell, 1980), 3–4.

18. See, for example, Michael Ghiselin, "Species Concepts, Individuality, and Objectivity," *Biology and Philosophy* 2 (1987): 127–143; Ernst Mayr, "The Ontological Status of Species," *Biology and Philosophy* 2 (1987): 145–166.

# III

# Rethinking Philosophy through Environmental Practice

# 9

# The Role of Cases in Moral Reasoning: What Environmental Ethics Can Learn from Biomedical Ethics

Robert Hood

How do cases function in reasoning about moral practices concerning the environment? At the heart of the question about the role of cases in environmental ethics is the issue of what kind of understanding environmental ethics provides. Is ethics something like a scientific discussion that shows how moral perceptions exemplify rules and laws that are more general and more certain than the particular practices in which we are engaged? Or is ethics more of a practical discussion informed by particular experiences and prudence? While there has been discussion of these questions in all areas of applied ethics, I think that biomedical ethics is the most successful in sorting through the role of cases and creating a nonacademic role for ethicists. I suggest some ways environmental ethicists might draw on the insights of bioethicists and develop a "clinical approach" to environmental ethics by supplementing theoretical discussion with a practical role focusing on cases.

Environmental ethics as an academic discipline emerged after Earth Day in 1970, at about the same time as other applied ethics disciplines, including biomedical ethics, legal ethics, business ethics, and the study of ethics in the professions. Each of these disciplines has produced a substantial theoretical literature legitimating its object of study. Whereas other applied fields have also produced substantial literatures focusing on cases and devoted to helping practitioners solve particular problems, a comparably substantial body of literature has yet to emerge in environmental ethics. To the extent that there has been some writing about cases in environmental ethics, it has not been accompanied by the creation of institutions and roles for environmental ethicists to work as an integral part of the environmental management team. Environmental ethics remains a largely academic discipline, unlike other areas of applied ethics, where there are also roles for ethicists in business and professional environments.[1]

Perhaps the most successful applied ethics discipline, both in terms of developing an applied, case-based focus and in creating a nonacademic role for ethicists, is biomedical ethics. The clinical side of biomedical ethics is notable for its departure from "normal" philosophical ethics and its success in engaging interest and respect from the health care professions. Clinical biomedical ethicists are employed by many hospitals and participate as integral members of health care teams. As yet there is no analog in environmental ethics to the practical or clinical side of clinical biomedical ethics, where philosophers work as part of the care team side by side with physicians and nurses and patients in a clinical setting. In sum, whereas environmental ethics by and large still is focused on the theoretical issues, biomedical ethics has both a theoretical and a clinical side. To understand what environmental ethics can learn from bioethics, I next review a debate in bioethics.[2]

## The Debate about Cases in Bioethics

Medicine has had longstanding norms of ethical conduct. However, urgent ethical problems such as those arising after World War II—including assisted suicide, cloning, euthanasia, and the question of how to prioritize scarce medical resources such as organ transplants—have forced scientists, medical professionals, policymakers, and the public to grapple with bioethical issues. The ensuing debates have focused in part on how best to approach the emerging field of clinical biomedical ethics—in terms of principles, or in terms of cases or casuistry.[3]

According to Albert Jonsen and Stephen Toulmin, two proponents of a case-based approach, their own commitments on the matter emerged out of their experience participating in a U.S. government commission charged with studying bioethics issues. As they worked with people on the commission, they noticed that there was disagreement about justifications and matters of theory. Nevertheless, although the various commission members had different academic, religious, and philosophical perspectives, the commissioners could still reach consensus by approaching moral problems using cases. The commission participants bracketed their differences on matters of principle; they began with clear cases or paradigm cases where there was agreement, then explored more complex cases posed by biomedical research. Using arguments from analogy, precedents, and counterexamples, in a process similar to that found in common law, Jonsen and Toulmin report they triangulated their way across the complex terrain of moral life, gradually extending their analysis of relatively straightforward problems to issues requiring a much more delicate balancing of competing values.[4]

As Jonsen and Toulmin tell the story, the call for ethicists to focus on cases is at odds with the dominant traditions in ethics in the last several centuries, which they characterize as being focused on principles. Regardless of whether we agree with all of Jonsen's and Toulmin's history of ethics, nevertheless their heretofore-overlooked discussion of case-based ethics is a fruitful source for a discussion of the role of cases in environmental ethics. In addition to their defense of the use of cases in moral reasoning, they provide a detailed model of what a case-driven approach to moral and political philosophy looks like. Through their work in medical ethics they provide an analogous model or direction of how case-driven environmental ethics might develop.[5]

Jonsen's and Toulmin's argument for a case-based approach appeals to a broadly construed Aristotelian picture of moral reasoning. They characterize their method as being

> the analysis of moral issues, using procedures of reasoning based on paradigms and analogies, leading to the formulation of expert opinions about the existence and stringency of particular moral obligations, framed in terms of rules or maxims that are general but not universal or variable, since they hold good with certainty only in the typical conditions of the agent and circumstances of action.[6]

Their approach includes the following characteristics, among others:

1. Similar type cases ("paradigms") serve as final objects of reference in moral arguments, creating initial "presumptions" that carry conclusive weight, absent "exceptional" circumstances.
2. In particular cases the first task is to decide which paradigms are directly relevant to the issues each raises.
3. Substantive difficulties arise, for example, if the paradigms fit current cases only ambiguously, so that presumptions they create are open to serious challenge, and when two or more paradigms apply in conflicting ways, which must be mediated.
4. The social and cultural history of moral practice reveals a progressive clarification of the "exceptions" admitted as rebutting the initial

moral presumptions and a progressive elucidation of the recognized type cases themselves.[7]

I discuss each of these issues in turn.

### The Priority of Cases

A case-driven approach involves a reliance on similar type cases or "paradigms." A paradigm, as Jonsen and Toulmin use the term, is a clear case all participants in a debate or moral decision-making group would agree on. This case-based approach to moral philosophy starts with the actual practices in which we are enmeshed. Cases are both descriptive and normative accounts of these practices. Jonsen and Toulmin are skeptical of the ability to abstract away from the details of practices, and provide three arguments for why ethicists should focus on detailed examples.

The first aspect of the priority of cases concerns the details of moral life. Moral phenomena are sufficiently particular, context dependent, and unique that only by attending to the details of cases can we hope to find appropriately nuanced answers to moral problems. From the case-based perspective, moral phenomena are of a kind that requires detailed understanding; it is only by knowing the details of particular situations that we can hope to make competent choices. Through rich descriptions of cases we can appreciate and learn from the deliberations of others. Details flesh out the situation, providing evidence needed to determine the context, circumstances, and motivations of the protagonists. While knowing the details of a case is not sufficient for proper choice, details are necessary.

An implication of this emphasis on the details of moral life is that case-driven reasoning in environmental ethics must draw on thick descriptions. Consider Aldo Leopold's account of his change of mind and recognition of the value of biological diversity for its own sake. Over the course of his career, Leopold came to emphasize the importance of biological diversity, both because of his belief that biological diversity resulted in greater stability and land health, and because he came to believe diversity is important for its own sake, independent of the functional, stabilizing role it plays in systems. Yet Leopold did not always provide an argument per se, but communicated using detailed narratives. Concerning his recognition of the value of biodiversity, Leopold recounted that his "own conviction on this score dates from the day I saw a wolf die." Leopold and others were eating lunch when they spied some wolves. "In those days we had never heard of passing up a chance to kill a wolf. In a second we were pumping lead into the pack." Among the kill was an old wolf: "We reached the wolf in time to watch the fierce green fire dying in her eyes. I realized then, and have known ever since, that there was something new to me in those eyes—something known only to her and the mountain." This event was a pivotal one in Leopold's intellectual and personal development. When a friend urged Leopold that *Sand County Almanac* should contain material that illustrated Leopold's own transformation, Leopold noted the story, including this passage:

> I was young then, and full of trigger itch; I thought that because fewer wolves meant more deer, that no wolves meant hunters' paradise. But after seeing the green fire die, I sensed that neither the wolf nor the mountain agreed with such a view.
>
> Since then I have lived to see state after state extirpate its wolves. I have watched the face of many a newly wolfless mountain, and seen the south-facing slows wrinkle with a maze of new deer trails. I have seen every edible bush and seedling browsed, first to anemic desuetude, and then to death. I have seen every edible tree defoliated to the height of a saddlehorn. Such a mountain looks as if someone had given God a new pruning shears, and forbidden Him all other exercise. In the end, the starved bones of the hoped-for-deer herd, dead of its own too-much, bleach with the bones of the dead sage, or molder under the high-lined junipers.[8]

Leopold's description of this event reveals the degree to which his position had changed from his early work in the Forest Service, where he viewed the wolf as "vermin." He had come to the understanding that land health just *is* biological diversity, and for health to be maintained native biological diversity must be maintained. However, he communicated this in terms of this narrative, which allows the audience to accompany him on his moral transformation.[9]

Such detailed cases are an extension of life not only horizontally, bringing the reader into contact with events or locations or situations or problems not previously met, but also vertically, giving the reader experience that is deeper, sharper, and more precise than much of what takes place in life. Given the fact that many people lack firsthand understandings of natural areas and systems, detailed descriptions such as Leopold's, which include aesthetic and ecological factors, help develop a moral imagination, thereby training judgment and encouraging more competent choices. Additionally, this story includes emotive appeal and an absorbing plottedness, and confronts us with the variety and indeterminacy of life in its context of a chance encounter with a pack of wolves. Leopold's account has an additional advantage of drawing the reader in as a participant and coevaluator in a community. As this example illustrates, morally competent choice requires appropriately detailed knowledge of the circumstances.

A second argument for the priority of cases is that we are confronted with practical matters that are mutable, lacking in fixity. Jonsen and Toulmin draw on William James's view of the dynamic nature of ethical problems: "There is no such thing possible as an ethical philosophy dogmatically made up in advance. . . . There can be no final truth in ethics any more than in physics, until the last man has had his experience and his say."[10] Jonsen and Toulmin claim that case-based reasoning accounts for changing circumstances in a way that principles cannot. A system of rules can encompass only the sorts of things that have been seen before. However, the world confronts us with ever-new configurations of circumstances, sufficiently novel such that rules will apply. Moreover, rules are applied with reference to a set of cases and circumstances. Insofar as the circumstances that define the interpretation of rules change, the application and meaning of rules may change as well. Concrete ethical cases may contain particular, novel, and nonrepeatable elements. An ethical system with "principles" this context-specific would be a vast and infinitely extensible series of principles. Thus, because of context dependence, and the possibility of ultimately novel and particular elements, principles are insufficient for understanding moral phenomena.

A third advantage of cases is that outside the context of the generally accepted understandings of practitioners, we would not know which principles were relevant to a given case. Practices "frame" the relevance of principles. Practices license the application of principles—that is, we would not know why this or that principle would be relevant in abstraction from the accepted practice, from the embedded understandings of a practice. The way common law emerged out of the embedded practices of society is an example of this point.

In sum, because moral choices require detailed understandings of the circumstances, motivations, and other factors of concrete cases; because moral phenomena change in ways that prevent any general understanding; and because of the way particular details frame rule-following behavior itself, a focus on cases holds particular promise for applied ethics.

**Paradigms, Analogies, and Cases**

According to Jonsen and Toulmin, moral disputes are resolved by appeal to paradigm cases—clear cases all participants in a debate agree on. When a situation arises about which there is no agreement, the first step is to decide which paradigms are directly relevant to the issue that the case raises—that is, to identify the case as a case. A case under moral consideration is then situated in a family or taxonomy of related cases, where similarities and differences between them can be compared. The context of an individual case and how its conflicting maxims appear within that particular context are the raw materials of the case-comparison method. Then in succession, cases are proposed that move away from the paradigm case by introducing variations of circumstances and motives that made the offense in question less apparent. As Jonsen and Toulmin note, the "gradual movement from clear and simple cases to the more complex and obscure ones was standard procedure for the casuist; indeed, it might be said to be the essence of the casuistic mode of thinking." The relative weight of conflicting maxims in an individual case is ascertained by comparison to analogous cases. With case-based moral reasoning, moral guidance is provided by an ever-growing taxonomy of paradigm cases that represent unambiguous instances in which moral consensus is obtained.[11]

**Moral Disagreement**

The third element of case-based moral reasoning is an account of how to evaluate problem cases against clear paradigm cases. Disputes are settled by identifying relevant similarities among cases, and figuring out by a series of arguments from analogy where the problematic case fits within the established paradigmatic case taxonomy. The circumstances of particular cases—the details acquired by specifying *who, what, where, by what means, why, how, when, and about which or whom*—are used to locate a problematic case in the taxonomy. Weighing the circumstances, motivations, and degrees of probability in a case are examples of the sorts of things that are brought to bear on a cumulative assessment of the situation; it is the cumulative weight of all evidence bearing on a case that is important. This process is not, strictly speaking, a formal one but a practical sort of wisdom involving judgment and discernment.

### Elucidation over Time

The final element of case-based moral reasoning is the belief that the process of sorting out moral problems becomes more sophisticated over time. As additional cases arise the taxonomy becomes more comprehensive and informative, although the process of sorting out moral disputes is intrinsically incomplete. New circumstances are always presenting themselves, and it is impossible to specify in advance what principles would govern them. Over time the substantive areas of moral concern are elucidated, but so also is the process of case-driven reasoning itself. Just as the development of the common law has produced a more nuanced and precise view over time, so too has this "common law morality" become more sophisticated over time.

In sum, the kind of understanding ethics provides is a practical one. Moral problems are resolved by locating them within a taxonomy of paradigm cases. The process of handling moral disagreement is not formal, in the sense that solutions are inferred from principles, but a process of working through what to do based on analogies with similar cases and past precedent. A refinement of moral understanding occurs over time as a richer, more detailed taxonomy of cases is developed against which to compare problematic ones.

## Implications for Environmental Ethics

Jonsen and Toulmin argue that medical ethics, and any applied field, should be case based. Focusing on cases can facilitate agreement about treatment and policy even when there is disagreement over principles. This suggestion that different justifications can nevertheless "converge" on the same policy or plan of action has also been made, for different reasons, by Bryan Norton. Indeed, it would seem that there are a number of similarities between biomedical ethics and environmental ethics. Bioethics emerged to cope with novel problems due to technological, economic, and social changes in medicine such as euthanasia and organ transplantation. So, too, did environmental ethics emerge out of a context of technological, economic, and social changes. Early thinkers such as Gifford Pinchot, John Muir, and Aldo Leopold all tried to address the increased scale and rate of resource depletion. Just as preexisting moral norms in medicine were not adequate to cope with the new problems in bioethics, so existing ethics and law in areas such as property law, common law, the law of torts, and other areas of environmental law also have not been adequate to cope with genuinely novel environmental problems that emerged after World War II—such as those posed by nuclear energy, chemical toxins in the environment, genetically modified organisms, and the global crisis of biodiversity. Just as bioethics grew out of a need to address urgent moral problems, so too did the rise of modern environmental ethics in the 1970s grow out of a desire by philosophers to say something relevant concerning environmental degradation.[12]

Notwithstanding these similarities, I think that differences between medicine and environmental management suggest that developing a case-based approach in environmental ethics may be more difficult than in bioethics. As observed above, a case-based approach in environmental ethics has not yet emerged as in medical ethics. As medical ethics has matured over the last thirty years to become institutionalized as part of medical training, philosophers have come to teach in medical schools and participate with physicians and nurses on rounds in hospitals. There is not yet a similar institutional representation of environmental ethicists outside academic institutions. If cases are to play a similar role in environmental ethics, I think the following issues need to be addressed.

One difference between biomedical ethics and environmental ethics is that, while there is widespread agreement on the moral status of persons and that of patients in medical treatment, there is less agreement on the moral status of the environment or of the environment as an object of management. The value of individual human lives is taken as a given in the health care profession, and there is recognition that trade-offs concerning human lives need justification. Exceptions to the practice are regarded as failures to live up to these norms, and the seriousness with which such failures are regarded indicates the extent of this shared understanding. For example, public knowledge of the Tuskegee Syphilis Study led to widespread changes in law and policy.[13]

In contrast to medicine, there is not widespread cultural agreement on the nature and kind of attitudes that should be held with regard to the environment. Whether the environment has value in and of itself is a controversial question within environmental ethics as well as for the general public, as is the question of whether or under what conditions environmental trade-offs might be permissible. In addition, the moral status of animals and their exploitation remains contested. In sum, environmental ethics has, since its inception, been a field where there is much less agreement than in bioethics, and one where arguably the range of disagreement is wider. To the extent that clear cases or paradigm cases of agreement might be more difficult to locate in environmental ethics, a case-based approach would seem to face greater difficulty than in bioethics.[14]

A second difference between medicine and environmental management is that historically medicine has been recognized and esteemed as a moral profession. The call to update the medical tradition to accommodate new technologies and practices was well within the moral norms of that profession. In contrast, resource management has involved an ongoing attempt to define itself as a moral profession and to develop a moral focus. For example, several influential architects of environmental management felt they had to make the case that environmental management should be recognized as a profession. Moreover, it is a revealing admission of the gap between the esteem of medicine and environmental management that reformers tried to legitimate environmental management by comparing it to medicine. Thus Gifford Pinchot, one of the first professional foresters in the United States, compared foresters to physicians both in terms of the requirements of their moral character and in terms of their skills. In his attempt to make forestry into a profession, Pinchot emphasized the analogy with medicine, writing that "it is just as essential a part of the Forester's equipment to be able to see what is wrong with a piece of forest, and what is required for its improvement, as it is necessary for a physician to be able to diagnose a disease and to prescribe the remedy."[15]

Aldo Leopold, the founder of wildlife management, also drew comparisons to the medical profession to make his case. Leopold spoke elegantly and at length, using a medical analogy for understanding the land. He developed the concept of land health and suggested a model of land doctoring, though he cautioned that land doctoring was still so new that actions "must not be confused with cures. The art of land-doctoring is being practiced with vigor, but the science of land-health is a job for the future."[16] However, even when there is recognition within the management professions of normative issues, there is no widespread cultural recognition and esteem for environmental management as engaged in a moral endeavor. Unlike medical ethics, where the medical profession was already seen as engaged in a moral endeavor, in environmental ethics we have had to make the case that the problems truly are moral ones. In addition, environmental ethics calls for huge changes at the core of the resource management fields, not just incremental ones. Whereas bioethics never questioned the ends of medicine, environmental ethics has raised searching questions about the ends of management and technology.[17]

A third difference concerns the institutional contexts of medical and environmental decision making. In medicine, the decision team is relatively small, and usually there are clear channels of authority. In contrast, much environmental decision making in the United States is public in the sense that government agencies have requirements to solicit public comment on policy matters and specific land-use decisions. The public context of environmental decision making means that environmental decisions are made according to democratic or representational paradigms, where notions of satisfying the aggregate or average preferences may be decisive.

A fourth difference between environmental ethics and biomedical ethics concerns the discourse and practice of medicine and environmental management. Even though there has been debate about justification in bioethics, nevertheless there is recognition that the day-to-day discourse and practice of the medical profession is organized in terms of cases. Medical professionals approach patient care in terms of a narrative of a *case*—questions are asked about the patient's history and symptoms, recommendations are made concerning treatment, and the case is concluded when the patient is cured. Interesting cases are the subject of professional debate and discussion *as cases*—they are published as case studies in medical journals and the case is the subject of general rounds in teaching hospitals. Moreover, the education and socialization of doctors and nurses involves exposing them to clinical cases and teaching them to reason in terms of cases. This focus on cases as an aspect of the sociology of the health care profession in turn influences the debate in bioethics. The success of bioethics is in part due to its capacity to craft a discourse that effectively communicates within the existing rhetorical expectations of clinical practitioners, namely, to communicate in terms of cases.

In contrast, environmental management and environmental ethics are not practiced or taught in terms of cases. Textbooks in environmental ethics emphasize theoretical understandings involving, for example, animal rights, anthropocentrism, biocentrism, ecocentrism, ecofeminism, deep ecology, and the like, and the majority of courses in environmental ethics similarly reflect this theoretical orientation. Perhaps more significantly, environmental management historically has not been taught or practiced in terms of cases. A challenge for environmental ethicists is to create an effective discourse that addresses the concerns of practitioners.[18]

To the extent these differences exist, strategies used to cope with disagreement in biomedical ethics, in particular the case-based approach, may prove fruitful in developing a case-based clinical approach to environmental ethics. The case-based approach suggests that environmental ethicists should focus on developing a taxonomy of clear cases about which there is agreement. A case-based approach recognizes that the identification of clear cases and their progressive elucidation over time exists within a certain community of discourse.

Given the public context of environmental decision making in the United States, it is possible that clear cases might be defined in terms of minimal standards like *avoiding the creation of environmental disaster.* That is, it is possible that there will be greater agreement that one should refrain from harming the environment than that one should be required to help restore the environment. Rather than a positive vision for the protection of biodiversity or wilderness, it is possible that what will be identified as a clear case is just the minimum condition that we should avoid causing anything like the extinction of the Passenger Pigeon, or the creation of another Love Canal, or another Bhopal, or another Chernobyl, or the slaughter of another Mad Cow. Be this as it may, as I illustrate with an example below, a clinical approach to environmental ethics would involve using analogical and case-based reasoning to move from clear cases to more contested terrains.

## An Example Case: Sea Lamprey Control in the Great Lakes

I will sketch how the case-based process might work with the following example. Clinical environmental ethics cases, when viewed as ethical problems, might be analyzed in terms of the following four questions and topics:

1. Ecosystem condition and status: What is the condition of the ecosystem and what management interventions should be done to protect, maintain, or improve the condition?
2. Preferences of stakeholders: What are the goals to which management should be directed according to the preferences of stakeholders?
3. Ecosystem health: What are the goals to which management should be directed based on the health of the ecosystem?
4. Contextual features: What social, economic, legal, and administrative factors influence management?

Exploring these four topics, in order, provides a structured way of investigating and resolving ethical issues given the condition of the ecosystem, the management goals and facts about the ecosystem, and the constraints imposed by the specific context. These topics could help environmental managers understand where the moral principles meet the circumstances of the clinical case. Asking these four questions helps organize information so that clear cases and analogies can be developed. Consider the following case:

The successful control of sea lampreys (*Petromyzon marinus*) in the Great Lakes is one of the major success stories in fisheries management. Sea lampreys are parasitic eel-like fish native to the Atlantic Ocean. They entered the Great Lakes in the early 19$^{th}$ century after the construction of a shipping canal, and then migrated throughout the Great Lakes and tributaries. They are an exotic species and lack significant predation in the Great Lakes. By the 1950s they were largely responsible for the severe decline of lake trout and whitefish populations in the Great Lakes, as well as for effects on the ecosystem as a whole. In the late 1950s a lampricide chemical, TFM (3-trifluoromethyl-4-nitrophenol), was discovered which kills larval lamprey and, at the levels used for control, is non-toxic or has minimal and temporary effects on aquatic plants, invertebrates, fish and waterfowl. It is also non-toxic to humans and other mammals. TCM is reapplied every 3–10 years in streams infested with larval lamprey, and has resulting in a recovery of fish populations and the sport and commercial fishing industries. Even though the lampreys have not been eradicated, they have been managed and controlled.[19]

I think this case exemplifies clinical environmental ethics, not because it shows an environmental problem, but because the ethics have gone as smoothly as the management. The ecosystem conditions and status were well enough known; the preferences of stakeholders were in agreement, the treatment or management options were actually effective and caused no side effects or new problems in the ecosystem; and there were no specific contextual features that complicated the situation. However, in some cases, ethical aspects become ethical problems. This case would have been complicated if the lampricide also resulted in the elimination of a species, or if a well-organized contingent of people blocked use of the treatment in the courts. As I discuss elements of the case below I indicate ways it might deviate from being a paradigm case and suggest the general direction of a case-based approach.

The first topic in a case concerns the details of the ecosystem conditions. A clear view of the situation and of the possible benefits of intervention is the first step in assessing the ethical aspects of a case. This might seem backward, particularly when so much uncertainty exists concerning ecosystems. Yet knowing the factual basis gives needed context to stakeholder preferences and management goals. While the circumstances of the case do not, by themselves, determine what ought be done, they constrain what is possible. For example, it is important to this case that it was known at the outset the lamprey is an exotic species. This case would have been complicated, in different ways, if the species in question was native to the area, or if the lamprey had resulted in only minor ecosystem changes—say, only affecting noncharismatic species that were of relatively less interest to people than commercial and sport species. Each of these examples moves the situation away from being a clear case, and makes it less obvious what should be done. Who should mediate such a dispute? What sort of standard should be used? In the case-based view, answers to such questions would emerge in the details of working it out in a specific context, and would be constrained by precedents.

A second issue concerns the preferences of stakeholders. In all management the ethical preferences of the stakeholders are ethically relevant.

The preferences of stakeholders in this case were to remove the species. The case would have been complicated if commercial interests and recreational interests, for example, had clashed. On commercial grounds, on recreational grounds, and in terms of protecting the value of biodiversity, in this case all stakeholders could agree that this exotic species should be removed even though they appealed to different reasons. However, it is worth nothing that, in contrast to the situation in biomedical ethics, the role of stakeholders in environmental ethics is significantly complicated by the fact that some stakeholders have yet to be born, or as nonhumans are not capable of indicating their interests. This case would have become more complicated if the lampreys were not exotic, more so if they were endangered. In such a case, the precedents of the Endangered Species Act would help provide a context for deciding what to do here.

A third issue concerns the goals of management from the perspective of the system, which in this case means the removal of the lamprey species or a substantial reduction in its numbers, thus creating conditions for the return of the system to its previous level of diversity. The management goals turned out to be realized when the lampricide was found to be effective. Also, the case would have been complicated if the treatment had had side effects—for example, if it caused significant reduction or expansion in another species.

Finally, every management case is embedded within a wider social, legal, political, economic, and ecosystemic context. Several contextual features were relevant to this case. For instance, the Great Lakes are governed by two countries; if the United States and Canada did not enjoy cooperative relations, management of this exotic species would have been more difficult. However, although the lampricide chemical TCM has been shown to be safe, a certain percentage of the public prefers that chemical pest control measures not be used, even if they are safe. As a result, experiments have been done using mild electric shocks to deter lampreys in a way that does not affect other fish. In addition, if this public concern were stronger or better organized, it would complicate the successful management plan currently in place. Finally, TCM is relatively expensive, and that expense would be of serious concern in some parts of the world.

This cursory exploration of a simple case—simple in the sense that the ethics and the management worked out together—illustrates two points. First, a case-based method can be used to evaluate environmental practices that work, and need not be reserved for unusually challenging cases. More detailed work using the case-based approach may increase our understanding of well-functioning environmental practices and enrich theoretical environmental ethics, just as clinical bioethics has informed theoretical ethics. Second, a case-based approach would be significantly more complicated if there was disagreement about management goals, or if the treatment was not as successful. However, if it is clear that a case-based approach would have difficulty with more complicated situations, I hope it is apparent that theoretical approaches would likely do no better, and might do worse to the extent that they did not take into account the nuances and details of the particular case.

## Conclusion

In contrast to biomedical ethics, environmental ethics currently does not have much of a discourse focusing on cases, or an established "clinical" practice of environmental ethics in management. The case-based approach outlined here suggests how this might proceed. We should add the study of certain cases to the study of theoretical work in environmental ethics. However, even if the casuist claim that all applied ethics should be case based is too strong, environmental ethics could benefit from exploring the role of cases as in medical ethics. Differences between biomedical ethics and environmental ethics suggest a number of difficulties in developing a case-based clinical approach. However, we should seek the creation of institutions for applied case-based consultations regarding environmental matters. As in bioethics, there should be a complementary wing of environmental ethics engaged in clinical environmental ethics.

## Notes

1. Examples of the use of cases in environmental ethics include Alan R. Beckenstein et al., *Stakeholder Negotiations: Exercises in Sustainable Development* (Chicago: Irwin, 1996); Brian Furze, "Ecologically Sustainable Rural Development and the Difficulty of Social Change," *Environmental Values* 2, no. 2 (1992): 141–156; Eric Katz, "A Pragmatic Reconsideration of Anthropocentrism," *Environmental Ethics* 21 (1999): 377–390; Lisa Newton and Catherine Dillingham, *Watersheds II: Ten Cases in Environmental Ethics* (New York:

Wadsworth, 1996); Lilly-Marlene Russow, "Why Do Species Matter?", *Environmental Ethics* 3 (1981): 101–112; Kristin Shrader-Frechette and Earl D. McCoy, *Method in Ecology: Strategies for Conservation* (New York: Cambridge University Press, 1993). For environmental pragmatism, see Andrew Light and Eric Katz, *Environmental Pragmatism* (New York: Routledge, 1996).

2. For an overview of clinical biomedical ethics, see Albert R. Jonsen, Mark Siegler, and William J. Winslade, *Clinical Ethics : A Practical Approach to Ethical Decisions in Clinical Medicine*, 4th ed. (New York: McGraw-Hill Health Professions Division, 1998); for a clinical approach to environmental ethics, see David Rapport, "What Is Clinical Ecology?", in *Ecosystem Health: New Goals for Environmental Management*, ed. Robert Costanza, Bryan G. Norton, and Benjamin D. Haskell (Washington, DC: Island Press, 1992), and Eugene C. Hargrove, "Environmental Therapeutic Nihilism," also in *Ecosystem Health.*

3. For a history of bioethics, see Albert R. Jonsen, *The Birth of Bioethics* (New York: Oxford University Press, 1998). For principles-based approaches, see Tom Beachamp and James Childress, *Principles of Medical Ethics* (Oxford: Oxford University Press, 1994); Tom Beachamp, "Principalism and Its Alleged Competitors," *Kennedy Institute Ethics Journal* 4, no. 3 (1995): 181–198; James Childress, *Practical Reasoning in Bioethics* (Bloomington: Indiana University Press, 1997). For case-based approaches, see John D. Arras, "Getting Down to Cases: The Revival of Casuistry in Bioethics," *Journal of Philosophy and Medicine* 16 (1991): 29–51; Jonsen, Siegler, and Winslade, *Clinical Ethics*; Albert R. Jonsen and Stephen Toulmin, *The Abuse of Casuistry: A History of Moral Reasoning* (Berkeley: University of California Press, 1988).

4. Jonsen and Toulmin, *The Abuse of Casuistry*, 16–19.

5. A complete taxonomy is not possible here, but it is worth noting that there are several ways of rejecting theory and emphasizing the particular in environmental ethics that are distinct from that of case-based reasoning, including an emphasis on narrative; see Max Oelschlaeger, "Earth-Talk: Conservation and the Ecology of Language," in *Wild Ideas*, ed. David Rothenberg (Minneapolis: University of Minnesota Press, 1995). On postmodernism, see Jim Cheney, "Postmodern Environmental Ethics: Ethics as Bioregional Narrative," *Environmental Ethics* 11 (1989): 117–134; Robert Frodeman, "Radical Environmentalism and the Political Roots of Postmodernism," *Environmental Ethics* 14, no. 4 (1992): 307–320; Max Oelschlaeger, ed., *Post-Modern Environmental Ethics* (Albany, NY: SUNY Press, 1995); Peter Quigley, "Rethinking Resistance: Environmentalism, Literature, and Poststructural Theory," *Environmental Ethics* 14, no. 4 (1992): 291–306; On virtue, see Geoffrey B. Frasz, "Environmental Virtue Ethics: A New Direction for Environmental Ethics," *Environmental Ethics* 15 (1993): 259–274; Thomas E. Hill, Jr., "Ideals of Human Excellence and Preserving Natural Environments," *Environmental Ethics* 5 (1983): 211–224; Kenneth M. Sayer, "An Alternative View of Environmental Ethics," *Environmental Ethics* 13 (1991): 195–213; David Schmidtz, "The Problem with Preservation," *Environmental Values* 6 (1997): 327–340. On pragmatism, see Light and Katz, *Environmental Pragmatism*; Bryan Norton, *Toward Unity among Environmentalists* (New York: Oxford University Press, 1991).

6. Jonsen and Toulmin, *The Abuse of Casuistry*, 257.

7. Jonsen and Toulmin, *The Abuse of Casuistry*, 306–307.

8. Aldo Leopold, *A Sand County Almanac* (New York: Oxford University Press, [1949] 1993), 138.

9. Susan L. Flader, *Thinking Like a Mountain* (Madison: University of Wisconsin Press, 1974); Robert L. Hood, "Ecosystem Health: A Critical Analysis," unpublished doctoral dissertation, Bowling Green State University, 1998; Curt Meine, *Aldo Leopold: His Life and Work* (Madison: University of Wisconsin Press, 1988).

10. Jonsen and Toulmin, *The Abuse of Casuistry*, 282.

11. Jonsen and Toulmin, *The Abuse of Casuistry*, 252.

12. Norton, *Toward Unity Among Environmentalists.*

13. Allan Brandt, "Racism and Research: The Case of the Tuskegee Syphilis Study," *Hastings Center Report* 8, no. 6 (1978): 21–29; W. J. Curran, "The Tuskegee Syphilis Study," *New England Journal of Medicine* 289, no. 14 (1973): 730–731; James Jones, "The Tuskegee Legacy: AIDS and the Black Community," *Hastings Center Report* 22, no. 6 (1992): 38–40.

14. Jennifer R. Wolch and Jody Emel, *Animal Geographies: Place, Politics, and Identity in the Nature-Culture Borderlands* (New York: Verso, 1998).

15. Gifford Pinchot, *The Training of a Forester* (Philadelphia: Lippincott, 1914), 66. See also Pinchot, *Breaking New Ground* (New York: Harcourt, Brace and Company, 1947); Pinchot, *The Fight for Conservation* (New York: Doubleday, Page and Company, 1910).

16. Aldo Leopold, "Report to the American Game Conference on an American Game Policy," in *The River of the Mother of God and Other Essays by Aldo Leopold*, ed. Susan L. Flader and J. Baird Callicott (Madison: University of Wisconsin Press, 1991), 153. Leopold's writings on land health have been collected in J. Baird Callicott and Eric T. Freyfogle, eds., *For the Health of the Land: Previously Unpublished Essays and Other Writings of Aldo Leopold* (Washington, DC: Island Press, 1999).

17. For conservation biology, see J. Baird Callicott, "Conservation Values and Ethics," in *Principles of Conservation Biology*, ed. Gary K. Meffe (Sunderland, MA: Sinauer Associates, 1994); Michael E. Soulé, ed., *Conservation Biology: The Science of Scarcity and Diversity* (Sunderland, MA: Sinauer Associates, 1986). For restoration, see Susan Power Bratton, "Alternative Models of Ecosystem Restoration," in *Ecosystem Health: New Goals for Environmental Management*, ed. Robert Costanza, Bryan G. Norton, and Benjamin D. Haskell (Washington, DC: Island Press, 1992); Andrew Light and Eric Higgs, "The Politics of Ecological Restoration," *Environmental Ethics*, 18 (1996): 227–248; Donald Scherer, "Evolution, Human Living, and the Practice of Ecological Restoration," *Environmental Ethics* 17, no. 3 (1995): 359–380. For ecosystem management, see Norman L. Christensen et al., "The Report of the Ecological Society of America Committee on the Scientific Basis for Ecosystem Management," *Ecological Applications* 6, no. 3 (1996): 665–691; Michael P. Dombeck,

"Thinking Like a Mountain: BLM's Approach to Ecosystem Management," *Ecological Applications* 6, no. 3 (1996): 699–702; Jerry F. Franklin, "Ecosystem Management: An Overview," in *Ecosystem Management: Applications for Sustainable Forest and Wildlife Resources*, ed. Alan Haney and Mark S. Boyce (New Haven, CT: Yale University Press, 1997); Sherri W. Goodman, "Ecosystem Management at the Department of Defense," *Ecological Applications* 6, no. 3 (1996): 706–707; Steven L. Yafee et al., *Ecosystem Management in the United States: An Assessment of Current Experience* (Washington, DC: Island Press, 1996).

18. Newton and Dillingham, *Watersheds II*. For information about environmental ethics courses, see the Environmental Ethics Syllabus Project on the World Wide Web at http://appliedphilosophy.mtsu.edu/syllabusproject/index.html.

19. Robert A. Daniels, "Untested Assumptions: The Role of Canals in the Dispersal of Sea Lamprey, Alewife, and Other Fishes in the Eastern United States," *Environmental Biology of Fishes* 60, no. 4 (2000): 309–329; J. R. M. Kelso and K. I. Cullis, "The Linkage among Ecosystem Perturbations, Remediation, and the Success of the Nipigon Bay Fishery," *Canadian Journal of Fisheries and Aquatic Sciences* 53 (1996): 67–78; James F. Kitchell et al., "Sustainability of the Lake Superior Fish Community: Interactions in a Food Web Context," *Ecosystems* 3, no. 6 (2000): 545–560; C. P. Schneider et al., "Predation by Sea Lamprey (Petromyzon Marinus) on Lake Trout (Salvelinus Namaycush) in Southern Lake Ontario, 1982–1992," *Canadian Journal of Fisheries and Aquatic Sciences* 53 (1996): 1921–1932; Shawn P. Sitar et al., "Lake Trout Mortality and Abundance in Southern Lake Huron," *North American Journal of Fisheries Management* 19, no. 4 (1999): 881–900; B. R. Smith and J. J. Tibbles, "Sea Lamprey in Lakes Huron, Michigan, and Superior: History of Invasion and Control, 1936–1978," *Canadian Journal of Fisheries and Aquatic Sciences* 37 (1980): 1780–1801; G. R. Spangler and J. J. Collins, "Response of Lake Whitefish (Coregonus Clupeaformis) to the Control of Sea Lamprey (Petromyzon Marinus) in Lake Huron," *Canadian Journal of Fisheries and Aquatic Sciences* 37, no. 11 (1980): 2039–2046; William D. Swink, "Effectiveness of an Electrical Barrier in Blocking a Sea Lamprey Spawning Migration on the Jordan River, Michigan," *North American Journal of Fisheries Management* 19, no. 2 (1999): 397–405.

# 10

# Grab Bag Ethics and Policymaking for Leaded Gasoline: A Pragmatist's View

Vivian E. Thomson

In *Little Dorrit*, Charles Dickens mocked nineteenth-century British politicians and public employees as devotees of one "sublime principle": "how not to do it." Dickens's scorn knew few bounds: "From the moment when a general election was over, every returned man who had been raving on hustings because it hadn't been done . . . and who had been asserting that it must be done, and who had pledging himself that it should be done, began to devise, *How it was not to be done*."[1] Failed political promises, bad intentions, and administrative unresponsiveness (e.g., the "Circumlocution Office") were the stuff of "how not to do it." Dickens's cynical remarks would find a receptive audience today in the United States. Political observers of all ideological stripes lambaste government policymaking as an exercise in inefficiency and ineptitude designed to thwart any change that does not benefit self-interested politicians and slothful bureaucrats.

Yet, sometimes government policies work well. That is, sometimes good intentions—the conviction that something "must be done"—translate into effective political action. A case in point is federal policy to rid gasoline of lead additives. Lead air emissions in the United States plummeted by 98 percent between 1970 and 1990, after government policymakers regulated the lead content of gasoline under authority provided by the Clean Air Act Amendments. Decreasing lead in gasoline had a palpable effect on blood lead levels, which dropped by 78 percent between the late 1970s and the late 1980s.[2]

Governments are supposed to help us make decisions that we might not make as individuals—to help us discover the kind of moral community we want to be. From this principle flow the central questions of this chapter: What values did policymakers use in lowering lead levels in gasoline, and are these values that we can applaud? In short, was this a good example of "how to do it," or are the results tainted by the means used to obtain them?

I examine, in turn, the effectiveness of lead-in-gasoline policies and the ethical underpinnings of the government's actions. I show first that this policymaking effort exemplified "how to do it" when measured by two of its most important results: federal policies markedly decreased United States lead air emissions and public blood lead levels, and the policy process engaged a broad segment of the general public. However, government officials used different, even contradictory values in response to changing political pressures—that is, they employed "grab bag ethics." I evaluate this tangle of ethical and political considerations from a pragmatist's perspective and I conclude that lead-in-gasoline policymaking was a success, albeit a qualified one.

This analysis reinforces themes articulated by Andrew Light and Avner de-Shalit in their introduction to this book. In particular, the case of leaded gasoline links theory with practice by illuminating the specific kinds of ethical discourse that are explicit or, more often, implicit in environmental policymaking. Further, one can see a clear need for philosophers in decisions like that of leaded gasoline, which are dominated by technical experts who often fail to articulate the contestable normative assumptions hidden behind elaborate analytical facades. However, environmental ethicists will not become integral to public debates until they acknowledge the worth of, and become fluent in, the language that policymakers use. As the case of leaded gasoline demonstrates all too clearly, that language is messy, inconsistent, and largely anthropocentric. Many environmental ethicists may understandably prefer to express themselves through the elegance of abstract theoretical arguments, but in so doing they will continue to consign themselves to the margins of environmental policymaking and many valuable ideas will never enter the public sphere.

## Getting the Lead Out

It is all too tempting for Americans to look back from the vantage point of 2002 and to proclaim the "inevitability" of leaded gasoline's demise. But in so doing we would revise history, for the reduction and subsequent elimination of lead in gasoline in the United States were anything but inevitable. As long ago as the 1920s, petroleum refiners and lead additive manufacturers successfully overcame the warnings of public health advocates. And throughout the 1970s and early 1980s health advocates, opposing business interests, members of the general public, and representatives of government institutions fought pitched battles over lead in gasoline. The opposition to removing lead from gasoline was steadfast and vehement. As a result, federal policymaking for leaded gasoline spanned twenty-five years from start (with the 1970 Clean Air Act Amendments) to finish (when leaded gasoline was banned for highway vehicles starting on January 1, 1996).

The inventors of tetraethyl lead hailed it as a "gift of God" because its octane-enhancing properties ensured that powerful, high-compression engines would run smoothly.[3] Leaded gasoline was first introduced in 1923, when federal regulation of environmental hazards was minimal. Nonetheless, public health officials immediately raised concerns because of lead's notorious reputation as a poison, and their worst fears were realized when lead additive workers suffered violent deaths and bizarre, incapacitating injuries. At one New Jersey factory five workers died over a five-day period and forty-nine others were seriously injured. Three hundred workers succumbed to lead poisoning at another lead additive factory that was nicknamed the "House of Butterflies" because of hallucinations induced by lead exposure ("The Victim pauses . . . gazes intently at space and snatches at something not there"[4]). Industry representatives attempted to downplay such episodes, claiming that "these men probably went insane because they worked too hard" or that the workers themselves were to blame.[5]

Several city and state governments temporarily banned leaded gasoline. The Ethyl Corporation suspended production of lead additives, and a task force appointed by the Surgeon General investigated the potential public health dangers. After several months of study the task force recommended that the Surgeon General regulate leaded gasoline and conduct long-term health studies. Leaded gasoline production resumed. However, no federal standards or long-term studies were forthcoming. Instead, lead manufacturers simply adopted more precautionary workplace standards.[6] Environmental levels remained unregulated and over the next several decades leaded gasoline became a staple for United

States motorists. By 1970, 98 percent of gasoline consumed contained lead.[7]

The policymaking process by which lead would finally be eliminated as a gasoline additive began with the 1970 Clean Air Act Amendments. Between 1970 and 1985 the United States Environmental Protection Agency (EPA) regulated the lead content of gasoline in two distinct ways under Section 211 of the Clean Air Act. First, the Agency required in 1973 that many service stations sell unleaded gasoline, since new cars would be equipped with catalytic converters to reduce emissions of smog-forming pollutants and carbon monoxide, and converters are ruined if exposed to only a few tankfuls of leaded gasoline. Second, on three separate occasions (in 1974, 1982, and 1985) the Agency decreased the amount of lead permitted in leaded gasoline because of public health concerns. This stepwise reduction in lead concentrations was the "lead phasedown."

Early congressional hearings on these two initiatives reflected a marked dissensus that pitted oil refiners and lead additive manufacturers against the auto industry and public health advocates. Witnesses testifying on the 1970 Clean Air Act Amendments' fuel provisions and on President Richard Nixon's proposal to tax lead additives disagreed on such basic questions as: Would new, low-emission cars require unleaded gasoline? Should the federal government force the production of unleaded gasoline? Did the health data compel federal action on leaded gasoline? Predictably, the petroleum industry opposed the administration's proposals out of economic self-interest. Standard Oil President Robert Gunness complained that manufacturing and distributing unleaded gasoline would require a $5 billion investment. The Ashland Petroleum Company not only objected to added production costs but also attacked the scientific evidence for health risks. For its part, General Motors expressed doubt that the new auto emission standards could be met with leaded fuel. Public health advocates supported some form of regulation but conceded the lack of definitive evidence linking leaded gasoline with adverse health effects.[8]

The dissensus and uncertainty evident in these 1970 hearings persisted through much of the 1970s. In the initial phase of the lead phasedown EPA could not establish definitively that gasoline additives, as opposed to lead in food or drinking water, contributed significantly to lead exposures.[9] In a 21 February 1997 letter to the editor of the *Washington*

*Post*, former EPA official Robert Sansom recalled that the Agency's recommendation to reduce lead in leaded gasoline for public health protection "was among the most difficult decisions Administrator William Ruckelshaus ever made, opposed by the White House, John Ehrlichman, Treasury Secretary George Schultz, OMB Director Caspar Weinberger, even by the Department of Health, Education, and Welfare." At the height of the energy crisis oil refiners placed full-page ads in the 27 November 1973 and 19 March 1974 *Washington Post* in which they predicted direly that removing lead from gasoline "could have the net effect of dumping 1 million barrels of crude oil every day." The Circuit Court for the District of Columbia at first overturned, and then barely upheld, EPA's rulemaking to reduce lead in leaded gasoline (*Ethyl Corp. v. EPA*, 541 F. 2d 1 (D. C. Cir. 1976), cert. denied 426 U. S. 941 (1976)).

But by the early 1980s elite attitudes toward lead in gasoline had shifted, as became evident in the backlash against a short-lived attempt to undermine the lead phasedown. The Reagan Administration sought to eliminate government "interference" in the marketplace, and at the urging of the White House Task Force on Regulatory Relief, EPA proposed relaxing or rescinding the lead phasedown.[10] Even more disturbingly, EPA Administrator Anne Gorsuch met with representatives of a small refinery and effectively promised them she would not enforce the existing leaded gasoline limits. One EPA analyst who attended that astonishing meeting recalled that the Administrator "stated that she was not going to enforce the regulations on something that was going to be drastically changed or abolished."[11]

Elite reactions to these actions were overwhelmingly negative. Members of Congress and representatives of the medical community condemned any effort to increase gasoline lead concentrations.[12] The *Washington Post* (22 March and 27 May 1982) and the *New York Times* (18 April 1982) editorialized against "backroom deals" and relaxation of the lead rules. Syndicated columnist Jack Anderson and even conservative commentator George Will strongly endorsed the lead phasedown.[13] Large refiners also protested, claiming that EPA's proposal would perpetuate an unfair competitive advantage for smaller refiners.[14] Under fire for their proposals, EPA's political appointees quickly reversed course and ordered further reductions in gasoline lead concentrations in 1982 and again in 1985.

Senate hearings in 1984 on a proposed leaded gasoline ban illustrate the emerging consensus that removing lead from gasoline was right and proper. Ashland Oil testified that

> [We] were skeptical when lead particulate emissions from automobile exhaust pipes were first suggested as a severe health hazard. We then argued that the evidence from the health studies available at that time was not conclusive enough to justify curtailing or banning the use of lead. . . . Further medical studies confirm . . . that airborne lead particles from automobile exhausts are the principal contribution to lead in our environment. New evidence also suggests an even stronger association between leaded gasoline use and severe public health problems than was first believed. . . . *We feel that EPA, in making its decisions in 1973 and 1982, properly responded to the harmful effects of lead in gasoline* (italics added). . . . "Get the lead out" is an idea whose time has come! In fact, it is past due.[15]

Ashland's motives were not entirely altruistic: the company was also concerned about recouping its investment on facilities that would manufacture unleaded gasoline and ethanol, a lead substitute. Even so, the company's definitive public acknowledgment of lead's health impact marked a dramatic change from its position in 1970. Many other refiners supported EPA's 1984 proposal to reduce lead gasoline concentrations to negligible levels, even though some argued about the appropriate timing thereof. State and local government representatives, health officials, and conservation groups overwhelmingly advocated further reductions in lead levels or an outright ban on leaded gasoline.[16]

Strikingly, many citizens—individually and in groups—also commented directly on these actions. EPA received over 1000 and 1500 written comments on its 1982 and 1985 rules, respectively. Among the groups offering these comments were a predictable range of Washington insiders—refiners, lead manufacturers, national and local environmental organizations, state and local governments, trade groups, farming organizations. But hundreds of comments came from citizens outside the Washington policy community. Included in this latter group was an enormous hodgepodge of citizen organizations, local government representatives, and businesses, such as the Studebaker Drivers Club of Central Oklahoma, the Northwestern Boating Council, the Six Rivers Racing Association, the Ear, Nose and Throat Associates of Spokane, the New York City Office of the Comptroller, the City of Avon Lake, Ohio, the Tahoe Fracture and Orthopedic Medical Clinic, the Small Car Shop of

Rosehill, Kansas, Sid's Auto Parts and Hardware Store (no city given), Yankee Wood Saw of Hoosick Falls, New York, the Heart of Dixie Mustang Club, Hack'd Magazine of Portland, Oregon, Bacon's Sugar House of Jaffrey Center, New Hampshire, and the Fairville, New York Volunteer Fire Department. Others listed no group affiliation, implying that they were offering input as individuals. Comments came from virtually every state, showing that concern over leaded gasoline's regulation extended widely and was not limited to particular geographic areas.[17] And the leaded gasoline rule making attracted far more comments than comparable Clean Air Act rules completed in this same period. Between 1983 and 1988 other major, high-profile Clean Air Act actions garnered from 60 to 500 comments.[18]

Why were so many citizens from across the country activated over policymaking for leaded gasoline? EPA records do not reveal the nature of each individual's remarks. But the summaries reveal that many worried about increased gasoline prices and about whether their boats, cars, and farm equipment would run effectively on low-lead gasoline. If public concern over environmental issues is inversely related to public confidence in the federal government's actions, as one observer has speculated,[19] then some commenters may have been provoked into participating by the political firestorm that eventually consumed many of President Reagan's controversial environmental appointees.

However, many commented on EPA's proposals because of their concern over lead's public health effects.[20] Survey data indicate that this may have been a longstanding concern for many Americans. As far back as the early 1970s two national polls showed that 60 to 70 percent of respondents said they "often" or "sometimes" worried about leaded gasoline exhaust.[21] Reasonably frequent media articles undoubtedly kept these concerns alive. Throughout the 1970s and 1980s popular, wide-circulation magazines like *Parents Magazine, Psychology Today*, and the *Saturday Evening Post* reported often on lead, frequently highlighting the special risks for children.[22] More elite publications were no less attentive: the *New York Times* published 274 stories between 1970 and 1990 on the subjects of lead air pollution or leaded gasoline, with pulses of articles in 1970 (63 stories), 1982 (20 stories), and 1984 (24 stories). As a result of this sustained media coverage, many members of the general public must have been at least vaguely aware that products with lead were hazardous and that leaded gasoline was the subject of policy discussions.

Those debates ended with the 1990 Clean Air Act Amendments, which banned leaded gasoline for highway sale after 1995 and which also forbade the production or sale of engines requiring leaded gasoline after model year 1992.[23] The policymaking system that, at best, provided mixed signals in 1970 regarding the need for, and advisability of, unleaded or low-lead gasoline had legitimized drastic reductions in lead additives by the mid-1980s and a ban by 1990.

A variety of evidence testifies to the dramatic environmental impact of removing lead from gasoline. By 1990 leaded gasoline comprised only 5 percent of total gasoline sold in the United States.[24] Between 1970 and 1996 estimated national lead air emissions plummeted by over 98 percent, from 220,869 to 3869 tons annually.[25] Outdoor air monitoring data from a wide variety of urban areas show that lead concentrations dropped by over 90 percent between 1976 and 1992.[26] Lead can be transported for great distances via atmospheric currents and lead levels in Greenland snow decreased in tandem with U.S. gasoline lead consumption.[27]

A key question was whether lowering lead in gasoline would have an impact on blood lead levels, given that humans are exposed to lead from sundry sources (e.g., lead paint, lead solder in cans, lead in pottery and drinking water). Several studies have confirmed a strong link between gasoline lead additive production and blood lead levels. National blood lead levels and lead additive production moved in parallel: even seasonal shifts in lead additive production were mirrored by similar changes in national mean blood lead levels.[28] Between 1976 and 1991 mean national blood lead levels and automobile lead emissions dropped by 78 percent and by over 95 percent, respectively.[29] Despite early questions about whether regulating gasoline lead additives would have a measurable effect, these analyses put such doubts to rest.

Although the growing use of catalytic converters would have spurred refiners to supply some unleaded gasoline, the evidence clearly shows that gasoline lead levels would not have dropped as precipitously without federal intervention. Industry and public resistance to the advent of unleaded and lower-lead gasoline was initially quite strong for several reasons: adding lead was the cheapest method for raising octane levels;

unleaded gasoline required more crude oil to produce than leaded, which was an important consideration during the 1970s energy crisis; and it was unclear whether automakers would use catalytic converters or some other method to meet federal auto emission standards. Even as they produced more unleaded gasoline, refiners introduced more lead into leaded gasoline because of its octane-enhancing qualities. And about 10 to 15 percent of motorists driving cars with catalytic converters "misfueled" their cars with leaded gasoline, for a variety of reasons.[30] As a result, consumption of leaded gasoline and lead additive in the early 1980s remained much higher than anticipated.[31]

In summary, policymaking for lead in gasoline worked. The evidence shows that lead air emissions dropped markedly after the government mandated the widespread availability of unleaded gasoline and as the amount of lead in leaded gasoline was decreased through government regulation; monitored air concentrations of lead decreased substantially in tandem with lead emissions; and, as gasoline lead levels dropped, blood lead levels fell substantially. Other important measures of policymaking success were the widespread consensus that gradually coalesced in the policymaking community and the fact that, at least in the 1980s, leaded gasoline policy actions engaged a hundreds of individual citizens and citizens' groups who were scattered widely across the country.

### What Ethical Benchmarks Did Decision Makers Use?

Environmental laws generally delineate an ethical playing field for subsequent policy decisions.[32] For example, in one portion of the Clean Air Act EPA is instructed to set ambient air quality standards to protect the public health with "an adequate margin of safety" while in another section national industrial emission standards are to be based on best available technology. These sections exemplify two commonly employed moral standards for environmental policymaking: the right to safety, on the one hand, and "the best we can do technologically," on the other. When establishing safety-based standards decisionmakers make further ethical judgments: they might try to protect the population at large or concentrate on particularly vulnerable subpopulations—for example, people living near a polluting facility or those who are overly susceptible to a given pollutant's effects. Some American statutes (e.g., the

Toxic Substances Control Act) employ cost-benefit analysis, which rests on the welfare economics principle that we should provide only as much environmental protection as society is willing to buy, thereby achieving an efficient allocation of social resources. These principles—safety, the best we can do technically, and economic efficiency—represent contradictory sets of values that often point to very different policy solutions.

Which principles were invoked as lead was removed from gasoline? EPA's rationale for requiring service stations to sell unleaded gasoline was that its widespread availability could only be assured through government fiat.[33] Although refiners took EPA to court over this rulemaking, the Agency prevailed (*Amoco Oil v. EPA*, 501 f.2d 722, 737–739 (D. C. Circuit 1974)). In requiring the marketing of unleaded gasoline the government relied on one empirical and two moral claims. EPA observed that markets sometimes do not respond quickly to consumer demand (the Agency's empirical claim), thus justifying government intervention to guarantee that consumers could purchase unleaded gasoline easily and conveniently. In requiring the sale of unleaded gasoline EPA signaled the depth of its commitment to reducing automobile exhaust, which threatened the health of the public at large (its first moral claim). The government also indicated its willingness to help protect automakers' investment in converters (its second moral claim). Having forced reluctant automakers to install emissions controls, the executive branch effectively sheltered that investment by requiring the widespread sale of unleaded gasoline.

When we examine the ethical underpinnings for the lead phasedown, we find that they varied over time. EPA initially proposed to establish a safety-based standard.[34] Only eleven months after this proposal was published, however, EPA rejected a safety-based rationale because of the many uncertainties over lead's health effects and exposure pathways. The Agency confessed that there was no conclusive evidence indicating that airborne lead by itself presented a public health problem but, nonetheless, asserted that "it would be prudent to reduce *preventable* [italics added] lead exposures from automobile-emitted airborne lead to the fullest extent possible." The key term "preventable" went undefined, although EPA indicated vaguely that the standard "reasonably" protected public health and was "reasonable . . . from the standpoint of economic and technological feasibility." While observing that much of the population was overexposed to lead, EPA emphasized the special need to lower children's exposures.[35]

Despite the twists and turns in EPA's rationale, the standard finally established in 1973 achieved the same emission reductions (60 to 65 percent) associated with the original safety-based standard. This implies that safety—that is, the uniform right to protection against unsafe levels of lead—was a predominant concern in the early phasedown, with technical feasibility employed as a secondary consideration. Economic factors seemed to play a tertiary role. Between proposed and final rule makings EPA undertook an economic analysis, which confirmed that the phasedown would affect crude oil use and consumer costs only slightly. The phasedown was also stretched out over five years rather than four, apparently in response to the combined pressures of the Office of Management and Budget, the Department of Interior, and the White House.[36] Despite these acknowledgments of the lead phasedown's economic impacts no formal cost-benefit analysis was conducted.

The evidence thus indicates that a precautionary impulse to secure public safety lay at the heart of EPA's initial decision to lower lead levels in leaded gasoline. Even though leaded gasoline's health effects had been called into question and even though EPA could not predict the phase-down's precise impacts, it made sense to act, in light of the potential for airborne lead to cause adverse health effects and considering the particular susceptibility of children to lead exposures. Emission reductions achieved through the phasedown would help lower ambient lead levels below those deemed harmful and would not have serious economic repercussions.

Refiners and lead additive manufacturers asserted that EPA had not found that lead additives "will endanger" the public health, a required statutory finding, and challenged EPA's action in a court battle that would delay the lead phasedown by over two years. Although a three-judge panel of the D. C. Circuit Court of Appeals initially agreed with the industry petitioners, the entire court vacated that judgment by a slim, five to four margin. The Court ruled that the Administrator's preventive approach was entitled to "great deference," that the statute is "precautionary in nature and does not require proof of actual harm before regulation is appropriate" (*Ethyl Corp. v. EPA*, 541 F.2d 1 (1976), cert. denied 426 U. S. 941 (1976)).

Despite EPA's emphasis on precautionary health protection the Agency was willing to be flexible about exactly how refiners met the phasedown standards. For example, according to a 24 September 1976 press release EPA extended regulatory deadlines by nine months so that refiners could install additional processing facilities. Further, each ruling allowed averaging. That is, at no point did refiners have to demonstrate that every gallon of fuel met federal lead restrictions; rather, they could average across gasoline batches. Later rules allowed refiners to trade or bank lead rights.[37]

Although EPA's 1982 rule making for leaded gasoline ultimately rested on the same rationale used in 1973[38] the Agency's rationale in 1985 pivoted around a very different ethical principle, that of economic efficiency. In a detailed cost-benefit analysis EPA estimated that the social benefits of lowering lead limits once again would far exceed the costs, even when unquantifiable benefits were excluded. Net monetized benefits ranged from $264 million in 1985 to $1.08 billion in 1992 and included improvements in children's health, lower car maintenance costs (lead corrodes engines and exhaust systems), increased fuel economy (lead fouls spark plugs and oxygen sensors), and lower emissions of pollutants controlled by catalytic converters (the rule would increase leaded gasoline prices, thus decreasing fuel switching).[39] EPA stopped short of banning leaded gasoline because some farm equipment allegedly required some lead for proper valve operation. In light of this concern Congress prevented EPA from promulgating any standards lower than 0.1 grams/gallon before 1988.[40]

The 1990 Clean Air Act Amendments drove the final nail into leaded gasoline's coffin. Congressional reports offer only a brief justification for the ban, referring briefly to the increasing evidence on lead's effects at low levels, to EPA's cost-benefit analysis showing net benefits in the hundreds of millions or billions of dollars, and to evidence that off-road vehicles could operate well without leaded gasoline.

Thus the ethical foundations of the federal government's decisions on lead in gasoline changed from initial action to eventual ban. Early actions stressed not only the government's willingness to protect automakers' investment in catalytic converters, but precautionary action in the face of uncertainty, a related unwillingness to wait until health and economic impacts could be precisely quantified, and feasibility. But the United

States energy situation and the politics of environmental policymaking were substantially altered by the early 1980s. Consequently, later decisions hinged more on detailed estimates of industry impacts and on cost-benefit tests.

## Assessing the Removal of Lead from Gasoline

Was the removal of lead from gasoline a good example of "how to do it?" A reasonable observer might applaud this policymaking effort, given that policy discourse changed from a marked dissensus to a widespread consensus, that hundreds of individual citizens became directly engaged in the policymaking process, and that policymaking for lead-in-gasoline yielded measurable, dramatic reductions in lead emissions, ambient levels, and blood lead levels. It is unlikely that anyone would wish away these impressive accomplishments.

But close scrutiny of the ethical principles involved might give us pause. EPA modified the ethical principles invoked as a function of the politics-of-the-day—that is, the Agency indulged in grab bag ethics. Some observers might be dismayed at this moral muddle and lack of clearly drawn, consistently applied principles. Those who are skeptical of policy analysis techniques might applaud the qualitative basis for early lead-in-gasoline policymaking while cringing at the analytical "quantomania" characterizing the 1985 phasedown decision. Environmental rights advocates might praise the precautionary ethic underlying EPA's early decisions and might condemn EPA for its later reliance on economic efficiency, while cost-benefit adherents would react in opposite fashion. Some environmental ethicists might stop short of endorsing the lead phasedown as an exemplary case of "how to do it" because the associated policy decisions rested on a pluralist foundation in which policymakers not only failed to embrace any consistent set of principles, they also ignored potential effects on non-humans and they explicitly valued human lives.[41]

However, examining policymakers' grab bag ethics through a pragmatist's lens yields a more favorable assessment of the values employed in policymaking for leaded gasoline. "Pragmatism" refers to a distinctly American philosophical movement that Richard Rorty has called the "chief glory of our country's intellectual tradition."[42] Pragmatism's original exponents included John Dewey, William James, and Charles Pierce and, while the movement fell out of fashion in the mid-twentieth century, it has enjoyed a recent resurgence of popularity among political theorists. Many of the early pragmatists were public philosophers who wrote on the pressing political issues of their day, thereby establishing pragmatism as philosophy in the practical service of politics.

Pragmatism is not easily reduced to one summary description, for it encompasses many views on the place of ethics, science, and the public in politics.[43] However, pragmatists commonly advocate incorporating a wide variety of public values into policy decisions and they reject the idea that any one particular ethical principle can apply to all situations.[44] For the pragmatist meaning and values are drawn out of experience rather than determined a priori.[45] Pragmatists endorse resistance to hierarchy, entrenched power, and expertise.[46] The pragmatist would argue that we must always illuminate a policy's ethical implications and that there is no single, neutral method for deciding whether that decision is efficient or fair or just. Rather, the allegedly objective algorithms used to justify safe or efficient or technologically best policy solutions inevitably mask political judgments about which goals—and people—should count more than others. Pragmatists agree with policy scholars who worry about relying on complicated technical jargon because it unjustifiably privileges the views of experts and excludes the uninitiated from policymaking discourse.[47]

While pragmatists challenge the idea that technical experts should dominate policymaking discourse, they invite the use of scientific information in policy decisions. To the pragmatist scientific inquiry can help us understand the world, although such knowledge is necessarily corrigible and should not be regarded as immutable truth.[48] The key lies with the distribution of scientific knowledge: experts should have no monopoly over its use. That is, policymaking inquiry should be democratized. We should aspire for an egalitarian "politics of inclusion" that makes a place for the lay public even in (perhaps especially in) decision making involving technical matters.[49]

A final crucial element in the pragmatist's worldview is action.[50] In the first place, action facilitates our sense of connection with, and responsibility to, the communities we inhabit.[51] Pragmatists believe that we come to understand ourselves only through our relationships with nature and with other people—that is, knowledge is contextual. In the second place, action and meaning are inseparable for the pragmatist.[52] That is, we discover the world through action, not observation, "what a thing means is simply what habits it involves."[53] Another way of putting this is that pragmatists want to understand the world in order to engage it directly; they are not interested in abstract reflection about truth.[54]

What does this mean for how a pragmatist would regard policymaking for leaded gasoline? Many environmental ethicists would criticize the 1985 lead phasedown decision because it rested on cost-benefit analysis and its attendant ideological dedication to quantification, monetization, discounting, individualism, and the notions of utilitarianism and consumer sovereignty, and they would probably agree with economist Charles Lindblom's unvarnished criticisms of microeconomists' analytical techniques:

> They give undue weight to variables that can be quantified and undervalue others. They treat noncomparables as comparable. . . . They employ dubious tactics such as treating guesses as facts. . . . They pretend to a neutrality that they do not in fact achieve. . . . They are often sufficiently esoteric to confuse and intimidate clients. . . . They are antidemocratic. They oversimplify. They mistakenly reduce problems of policy choice to problems in resource allocation.[55]

While these formidable objections would resonate with a pragmatist, she would point out that policy experts pursuing goals other than efficiency inevitably stumble into many of the analytical quagmires that Lindblom condemns. Pragmatists are skeptical of an infatuation with quantification and esoteric technical languages no matter what principles are served, given the many value judgments disguised behind allegedly objective techniques like quantitative risk assessment.[56] And a pragmatist could not commit herself to one authoritative definition of "fairness" or "equality" in environmental policymaking, since to do so yields decisionmaking to one authoritative scheme.[57]

The pragmatist's perspective leads us away from fixed ethical benchmarks and towards questions of political legitimacy as we try to ascertain the "rightness" of particular decisions. A pragmatist would argue that a legitimate policy consensus is shared, not imposed; it is one in which experts (government and nongovernment), politicians, and the public are integrally involved; and it considers a wide variety of public values and the interests of the less powerful. In this view experience and professional dedication should count in policymaking discourse and we should encourage political deliberation and reflection. But expert opinions, judgments, and languages should not be unduly privileged. The general public must also legitimize environmental policies.

Within this version of the pragmatist's framework, policymaking for lead-in-gasoline represents a qualified case of "how to do it." A pragmatist would applaud the fact that a wide variety of public values—precautionary and equitable public health protection, technical feasibility, industry and consumer costs—played an important role in policy deliberations, especially since the interests of the less powerful (minority children) were a central concern. Policymakers' motivation for employing these varying standards—namely, to act effectively despite changing sets of political and social constraints—would also be commended. A pragmatist would note approvingly that by the mid-1980s expert and public attitudes toward lead additives seem to have been transformed. In effect we seem to have achieved a new, hard-won collective understanding that was partially based on the general public's concern over lead's adverse health effects.

Several considerations temper this approbation. The pragmatist would be troubled that, while average citizens could react to policy decisions, they were not instrumental in formulating those decisions. In effect, policymakers did not promote public learning and informed inquiry. Political scientist Jane Mansbridge notes that differences in wealth, education, and access inevitably "lace deliberation with power," and policymaking for leaded gasoline is no exception to that generalization in that industry interests cast a long shadow over policy decisions.[59] Pragmatists would object to the technocratic, efficiency-based ideology that dominated EPA decision making in the 1980s, as would most environmental ethicists.[60]

Through a pragmatist's lens we can appreciate the flexibility policymakers displayed in making decisions regarding leaded gasoline, and we can applaud their willingness to act in the public interest. But we will still feel impelled to press for policymaking practices that are less obsessed with unachievable analytical precision and exclusive technical discourse, that consistently engage a wider set of public values, and that include members of the general public as an integral, not accidental, part of the policy process.

## Notes

1. Charles Dickens, *Little Dorrit*, ed. Harvey Peter Sucksmith (Oxford: Clarendon Press, 1979), 101, emphasis added.

2. Michael P. Walsh, "Phasing Lead Out of Gasoline: The Experience with Different Policy Approaches in Different Countries," Draft Issue Paper written for the United Nations Environment Programme and the Organisation for Economic Cooperation and Development, 1998.

3. Jack Lewis, "Lead Poisoning: A Historical Perspective," *EPA Journal* 4 (1985): 15; David Rosner and Gerald Markowitz, "A 'Gift of God'?: The Public Health Controversy over Leaded Gasoline during the 1920s," *American Journal of Public Health* 75, no. 4 (1985): 344.

4. Rosner and Markowitz, "A 'Gift of God'?", 347.

5. Rosner and Markowitz, "A 'Gift of God'?", 347.

6. P. B. Hammond, "Lead Poisoning: An Old Problem with a New Dimension," in United States Senate Committee on Public Works and Committee on Commerce, *Joint Hearings before the Subcommittee on Air and Water Pollution of the Committee on Public Works and the Committee on Commerce: Air Pollution—1970*, 91st Cong., 2d sess., 24 and 25 March 1970, 1151–1177.

7. "Behind the Hubbub over Leaded Gas," *Changing Times* 25, no. 10 (1971): 6–9.

8. United States House of Representatives Committee on Ways and Means, *Hearings before the Committee on Ways and Means on the Subject of the Tax Recommendations of the President*, 91st Cong., 2d sess., 9, 10, 14, 15, 16, and 17 September 1970; United States Senate Committee on Public Works and Committee on Commerce, *Joint Hearings before the Subcommittee on Air and Water Pollution of the Committee on Public Works and Committee on Commerce: Air Pollution—1970, Part 3*, 91st Cong., 2d sess., 24 and 25 March 1970.

9. United States Environmental Protection Agency, *EPA's Position on the Health Implications of Environmental Lead* (Washington, DC: U.S. Environmental Protection Agency, 1973).

10. Felicity Barringer, "Debate over Lead in Gasoline Revs Up Again," *Washington Post*, 5 October 1981, sec. A.

11. United States House Committee on Government Operations, *Hearing before a Subcommittee of the Committee on Government Operations: Lead in Gasoline: Public Health Dangers*, 97th Cong., 2d sess., 14 April 1982, 83.

12. United States House Committee on Government Operations, *Hearing before a Subcommittee of the Committee on Government Operations.*

13. Jack Anderson, "Danger to Children and Other Living Things," *Washington Post*, 1 August 1982, sec. B, George Will, "The Poison Poor Children Breathe," *Washington Post*, 16 September 1982, sec. A.

14. Sandra Sugawara, "EPA Trying to Ease Out of a Leaden Box," *Washington Post*, 19 May 1982, sec. A.

15. United States Senate Committee on Environment and Public Works, *Hearing before the Committee on Environment and Public Works: Airborne Lead Reduction Act of 1984*, 98th Cong., 2d sess., 22 June 1984, 203, 204, 206.

16. United States Senate Committee on Environment and Public Works, *Hearing before the Committee on Environment and Public Works*; ESI International, "Summary of Comments on the Environmental Protection Agency's August 2, 1984, Proposal on Regulation of Fuel and Fuel Additives (Lead Phasedown)," prepared for the Office of Mobile Sources, Environmental Protection Agency, 1984.

17. United States Environmental Protection Agency Docket no. A-81-36; United States Environmental Protection Agency, "Regulation of Fuels and Fuel Additives," *Federal Register* 50, no. 45 (7 March 1985): 9386.

18. This survey was performed on LEXIS/NEXIS (Envirn library, Fedreg file) on 14 April 1999. The search was designed to capture all Clean Air Act rules classified as major by the Office of Management and Budget (i.e., those estimated to cost more than $100 million annually) that were finalized between 1983 and 1988. Even the 1987 action limiting the use of ozone-depleting chemicals attracted far fewer comments (500) than the leaded gasoline rules.

19. Riley Dunlap, "Public Opinion and Environmental Policy," in *Environmental Politics and Policy: Theories and Evidence*, ed. James P. Lester (Durham, NC: Duke University Press, 1989).

20. Environmental Strategies, Inc., "Summary of Comments on the Environmental Protection Agency's February 22, 1982, Proposal on Regulation of Fuel and Fuel Additives (Lead Phasedown)," prepared for the Office of Air, Noise and Radiation Enforcement, Environmental Protection Agency, 5 October 1982.

21. Louis Harris and Associates, March 1971 Harris Survey; the 1972 Virginia Slims American Women's Opinion Poll; both sets of results obtained from the Public Opinion Location Library, a Roper Center database available through LEXIS/NEXIS.

22. See, for example, Max L. Fogel, "Warning: Auto Fumes May Lower Your Kid's IQ," *Psychology Today* 13, no. 8 (January 1980): 108, and Claude A. Frazier, "Suffer Little Children," *Saturday Evening Post* 252, no. 7 (October 1980): 72.

23. United States Senate Committee on Environment and Public Works, *A Legislative History of the Clean Air Act Amendments of 1990*, Committee Print (Washington, DC: Government Printing Office, 1993).

24. United States Environmental Protection Agency, *Motor Vehicle Tampering Survey 1990* (Washington, DC: Office of Air and Radiation, 1993).

25. United States Environmental Protection Agency, *National Air Pollutant Emission Trends, 1900 to 1996* (Research Triangle Park, NC: Office of Air Quality Planning and Standards, 1997).

26. United States Environmental Protection Agency, *National Air Quality and Emissions Trends Report, 1992* (Research Triangle Park, NC: Office of Air

Quality Planning and Standards, 1993); Richard G. Kozlowski, "Revisiting the Lead Phasedown," *EPA Journal* 13, no. 8 (October 1987): 28.

27. K. J. R. Rosman et al., "Isotopic Evidence for the Source of Lead in Greenland Snows Since the Late 1960s," *Nature* 362 (1993): 333–335.

28. United States Environmental Protection Agency, *Costs and Benefits of Reducing Lead in Gasoline: Final Regulatory Impact Analysis* (Washington, DC: U.S. Environmental Protection Agency, 1985).

29. Stephen J. Rothenberg et al., "Blood Lead Levels in Children in South Central Los Angeles," *Archives of Environmental Health* 5, no. 51 (1996): 383, obtained from LEXIS/NEXIS ENVIRN library, CURNWS file; United States Environmental Protection Agency, *National Air Pollutant Emission Trends*, 1997.

30. United States Senate Committee on Energy and Natural Resources, *Hearing before the Committee on Energy and Natural Resources: Price Differential between Leaded and Unleaded Gasoline*, 95th Cong., 2d sess., 11 December 1978; Sobotka and Company, "An Analysis of the Factors Leading to the Use of Leaded Gasoline in Automobiles Requiring Unleaded Gasoline," in United States House Committee on Interstate and Foreign Commerce, *Hearings before the Subcommittee on Oversight and Investigations: Environmental Effect of the Gasoline Tilt Rule*, 96th Cong., 1st sess., 12 and 13 March 1979; United States Environmental Protection Agency, *Costs and Benefits of Reducing Lead in Gasoline*, 1985; United States Environmental Protection Agency, *Motor Vehicle Tampering Survey* (Denver: National Enforcement Investigations Center, 1983).

31. United States Senate Committee on Environment and Public Works, *Airborne Lead Reduction Act*, 1984.

32. In my use of the term *ethical* I follow VanDeVeer and Pierce, who define an "ethic" as a "moral claim . . . about what someone . . . ought or ought not to do" (Donald VanDeVeer and Christine Pierce, *The Environmental Ethics and Policy Book* (Belmont, CA: Wadsworth, 1994)).

33. United States Environmental Protection Agency, "Regulation of Fuels and Fuel Additives," *Federal Register* 38, no. 6 (10 January 1973): 1251.

34. United States Environmental Protection Agency, "Regulation of Fuels and Fuel Additives," *Federal Register* 37, no. 36 (23 February 1972): 3882.

35. United States Environmental Protection Agency, "Regulation of Fuels and Fuel Additives," *Federal Register* 38, no. 6 (10 January 1973): 1258.

36. John Quarles, *Cleaning Up America: An Insider's View of the Environmental Protection Agency* (Boston: Houghton Mifflin, 1976).

37. Barry D. Nussbaum, "Phasing Down Lead in Gasoline through Economic Instruments," *Journal of Energy Engineering* 117, no. 3 (1991): 115–124.

38. United States Environmental Protection Agency, "Regulation of Fuels and Fuel Additives," *Federal Register* 47, no. 27 (August 1982): 38078.

39. United States Environmental Protection Agency, *Costs and Benefits of Reducing Lead in Gasoline*, 1985; United States Environmental Protection

Agency, "Regulation of Fuels and Fuel Additives," *Federal Register* 50, no. 45 (7 March 1985): 9386.

40. David T. Deal, "Mobile Source Fuels and Fuel Additives," in *Clean Air Law and Regulation*, ed. Timothy A. Vandever (Washington, DC: Bureau of National Affairs, 1992).

41. Andrew Light, "Contemporary Environmental Ethics: From Metaethics to Public Philosophy," forthcoming in *Metaphilosophy* (2002).

42. Richard Rorty, "Pragmatism, Relativism, and Irrationalism," in Richard Rorty, *Consequences of Pragmatism (Essays: 1972–1980)* (Minneapolis: University of Minnesota Press, 1982).

43. Andrew Light and Eric Katz, "Environmental Pragmatism and Environmental Ethics as Contested Terrain," in *Environmental Pragmatism*, ed. Andrew Light and Eric Katz (London: Routledge, 1996); Matthew Festenstein, *Pragmatism and Political Theory from Dewey to Rorty* (Chicago: University of Chicago Press, 1997).

44. Kelly A. Parker, "Pragmatism and Environmental Thought," in Light and Katz, *Environmental Pragmatism.*

45. Ian Shapiro, *Political Criticism* (Berkeley: University of California Press, 1990).

46. Shapiro, *Political Criticism*; Light and Katz, "Environmental Pragmatism and Environmental Ethics as Contested Terrain."

47. Deborah A. Stone, *Policy Paradox: The Art of Political Decision Making* (New York: Norton, 1997); Bruce A. Williams and Albert R. Matheny, *Democracy, Dialogue and Environmental Disputes: The Contested Languages of Social Regulation* (New Haven, CT: Yale University Press, 1995).

48. Shapiro, *Political Criticism.*

49. James Bohman, "Democracy as Inquiry, Inquiry as Democratic: Pragmatism, Social Science, and the Cognitive Division of Labor," *American Journal of Political Science* 43, no. 2 (1999): 590–607.

50. Morris Dickstein, "Introduction: Pragmatism Now and Then," in *The Revival of Pragmatism: New Essays on Social Thought, Law, and Culture*, ed. Morris Dickstein (Durham, NC: Duke University Press, 1998).

51. Alan Ryan, *John Dewey and the High Tide of American Liberalism* (New York: Norton, 1995); Paul B. Thompson, "Pragmatism and Policy: The Case of Water," in *Environmental Pragmatism*, ed. Andrew Light and Eric Katz (London: Routledge, 1996).

52. Sandra B. Rosenthal and Rogene A. Buchholz, "How Pragmatism *Is* an Environmental Ethic," in *Environmental Pragmatism*, ed. Andrew Light and Eric Katz (London: Routledge, 1996).

53. Dickstein, "Introduction: Pragmatism Now and Then," 3.

54. Shapiro, *Political Criticism*; Ryan, *John Dewey and the High Tide of American Liberalism.*

55. Charles Lindblom, *Inquiry and Change: The Troubled Attempt to Understand and Shape Society* (New Haven, CT: Yale University Press, 1990), 277n35.

56. Sheila Jasanoff, "Acceptable Evidence in a Pluralistic Society," in *Acceptable Evidence: Science and Values in Risk Management*, ed. Deborah G. Mayo and Rachelle D. Hollander (New York: Oxford University Press, 1991); K. S. Shrader-Frechette, *Burying Uncertainty: Risk and the Case against Geological Disposal of Nuclear Waste* (Berkeley: University of California Press, 1993).

57. John Dewey, "Philosophy and Democracy," in *John Dewey: The Political Writings*, ed. Debra Morris and Ian Shapiro (Indianapolis: Hackett, 1993).

58. Cass Peterson, "How the EPA Reversed around the Gas Pumps," *Washington Post*, 1 August 1984, sec. A.

59. Jane Mansbridge, "Public Spirit in Political Systems," in *Values and Public Policy*, ed. Henry J. Aaron, Thomas E. Mann, and Timothy Taylor (Washington, DC: Brookings Institution, 1994), 157.

60. See, for example, Mark Sagoff, *The Economy of the Earth: Philosophy, Law, and the Environment* (Cambridge: Cambridge University Press, 1988).

# 11

# Animals, Power, and Ethics: The Case of Foxhunting

Clare Palmer and Francis O'Gorman

Foxhunting—"the pursuit of the wild fox with a pack of hounds"[1]—has been the subject of widespread popular debate in the United Kingdom for many years and has been in the public forum again recently. Indeed, in the United Kingdom the question whether foxhunting should be made illegal has emerged as one of the key issues in current political discussion. It has recently been banned in Scotland, and discussions about a ban in England have led to popular demonstrations in favor of its continuance. Hunting, more generally, has also been the subject of considerable scholarly debate among those working in environmental and animal ethics.[2] Much of the popular debate about foxhunting has turned on questions relating to class, a putative rural-urban divide, and whether foxhunting causes unnecessary suffering to foxes. The scholarly debate has tended to focus on the "naturalness" or otherwise of human hunting, the idea that foxhunting is an exhibition of human domination over nature, the place of human hunting in ecocentric ethical systems, and questions about the intrinsic value of sentient life.

In this chapter, we will be exploring foxhunting from the perspective of human-animal power relations that might be involved, and looking more closely at the idea of human "domination" in this context.[3] We hope that this might suggest different angles on some of the debates, at least, about foxhunting and ethics.

## Humans, Nonhumans, and Power

The small amount of existing work relating to human-animal power relations generally conceptualizes human power as a monolithic dominating force *over* animals. This is the approach adopted, for instance, by Yi-Fu

Tuan in his influential book *Dominance and Affection: The Making of Pets*.[4] In particular, sport hunting—alongside meat eating—in Western societies is often viewed as primarily an exhibition of human abilities to control and dominate the natural world.[5] Drawing on a broadly Foucauldian approach to power, however, we suggest that power relations assume widely varied forms and strategies—not excluding, but not limited to, domination.[6] Foxhunting—involving as it does domesticated horses and dogs and wild foxes—provides a particularly apt case study of very different kinds of power relations. We will move on to look at these more particular relations shortly. First we will consider—albeit very briefly—some more general aspects of the human-animal power relations on which we will be concentrating.

Central in this context is the body, as a material, biological phenomenon.[7] Since animals do not share a discursive community with humans, the different ways in which (primarily) human power bears on animal bodies is important when exploring the power relations involved in foxhunting. This may involve thinking about the way power creates particular kinds of bodies and identities, rather than only considering the way power represses them. Alongside this stress on the creativity of power, we will also be focusing on reactivity or resistance to it. We use *resistance* in a specifically Foucauldian sense, emphasizing resistant practices and actions rather than the intentions that may lie behind such behaviors—that is, we are interested in what Certeau describes as responsive *tactics* to strategies of power.[8] In particular, how far are the humans and nonhumans involved in foxhunting able to react to or to resist the power of others? Foucault himself argued that power relations passed over into regimes of domination only when all kinds of resistance to power become impossible—and he believed there are remarkably few occasions where this is the case.[9] Viewed in this way, is foxhunting an exercise of human "domination over nature"?—and if so, in what senses?

## The Context of the Hunt

Before considering the course of the hunt itself, we want to look at the broader locus of human-animal power relations in which the hunt is located. Hunting a fox is, after all, a complex, interspecies activity where members of three species are in pursuit of a member of one.

For this human-driven pursuit to occur, the cooperation of dogs (both foxhounds and terriers) and horses must be obtained, and this cooperation forms the backdrop against which hunting takes place. Let's think, for instance, about the processes involved in producing a foxhound ready to take part in a hunt (although the production of a hunting horse or terrier is somewhat different, many of the fundamental features remain the same).

First, the foxhound is a pedigreed, domesticated dog. Thus its very body, its genetic makeup, biological constitution, appearance, and temperament are inscribed with human choice. Over generations, desirable characteristics have been optimalized while undesirable ones have been "bred out." Sexual partners have been carefully selected by humans to produce an individual hound displaying the qualities of "intelligence, speed, stamina, voice and nose."[10] Of course, this process may produce "imperfections," as Foucault argues: "Differences, peculiarities, deviance and eccentricities are even more highlighted in a regime concerned to seek them out."[11] It is also possible, given the relatively limited human knowledge of genetics, that the pedigree process may create more serious genetic vulnerabilities among the hounds apparent only at some future date. But however flawed the results, the very constitution of even the "imperfect" foxhound has been shaped by humans.

Once born, the hound puppy enters a world where its subjectivity is constructed and developed by those both human and nonhuman around it. It will learn something about rank and pecking order among its fellow dogs. However, it will also be involved in a series of relations with humans that will affect its ability to act. These may be restrictive on the hound. Patton characterizes such restrictions, in the human context, as "external limits to the kinds of act which may be carried out or . . . internal limits to the kinds of action the agent is capable of undertaking."[12]

The "external limits" in the case of a dog may take a variety of forms. Control over available space, kind of space, and mobility is one form of limit; much of the time the hound will be spatially confined in kennels; when taken out it may be restrained. It may also be isolated from other members of its species for periods of time. Such spatial and social constraints form the context in which the hound develops. At the same time, the hound is learning internal limits to its actions. It will be taught particular kinds of behavior relating to its body—when and where to defecate, what sort of behaviors are appropriate in response to which sounds and verbal commands. This knowledge is psychologically internalized from various external human practices relating to the body, such as the use of violence or rewards of food. In learning to behave as an adult foxhound ready for a hunt, the hound must deeply internalize human disciplinary practices. Viewed in this way, the adult foxhound is profoundly inscribed with human power practices: the very constitution and form of the hound is shaped by human desires; the space in which the hound may move and the social world with which it interacts is constructed by human decision; and the rules it has internalized and the behavioral responses it demonstrates are shaped by human hands. Indeed, the production of domesticated animals is one example of the *creativity* of human power practices.

However, the hound is not entirely a "passive surface" shaped by humans. It may resist elements of this process, to some extent at least: otherwise we would have no concept of a dog as being "disobedient." It may howl for hours when left in a place it does not want to be, ignore human commands, escape from confined spaces, fight other dogs, eat what is not intended for it. But for the domesticated dog, these forms of resistance are entirely located within a framework of human domination; ultimately dog resistance is what Foucault describes in a different context as "no more than a certain number of tricks which never brought about a reversal of the situation."[13] Selective breeding cannot be "resisted" by animals; it is a dominating practice that produces, even where an animal is purposefully bred for ferocity, a *docile body*, in that it is a body created to do what humans want of it. It represents "a form of capture of the other's own power or capacities."[14]

The fox, in contrast, can be characterized as having an *unruly body*. This might apply in several respects. For instance, foxes are rarely bred by humans for hunting. Where this does happen, particular favored characteristics are not selected (there is no attempt to produce bushier brushes, longer legs, pointier snouts). Thus, even on the occasions where they are bred, foxes retain what we might think of as a constitutive wildness, understanding wildness here to mean that their genetic constitution is not deliberately selected for by human practices. Foxes' bodily practices are also regarded as unruly (especially in urban and suburban areas). Urination and defecation are uncontrolled by humans; their lairs smell; they rummage in trash cans and eat garbage—transgressing widely accepted Western human rules about purity and contamination (even if some popular domesticated animals such as cats, on occasion, transgress these too).

More generally, foxlike behavior—characterized as cunning or wily—is often constructed as exhibiting resistance to human control. Foxes enter places from which humans intend to exclude them; they eat what they are not intended to eat. In particular, they kill members of domesticated species—lambs and chickens—thus disrupting established human-animal power relations and interfering with human decisions about when and how such animals should die and for what their bodies should be used. Worse still, they kill gratuitously—not for food but for pleasure (it is almost impossible not to comment on the irony of this objection to foxes in the context of foxhunting). Indeed, the fox is energetically recruited by pro-hunt discourse as comparable to rats, urban pigeons, or cockroaches as a creature, in the United Kingdom at least, which resists human power. This kind of comparison, for instance, prompted the MP James Grey in a House of Commons debate on foxhunting to comment: "Anyone who has . . . seen a fox gnawing at a recently born calf . . . will know what vermin they are. In that context, foxes are precisely the same as rats and should be dealt with in the same way. We think nothing of hitting a rat on the head with a shovel in the farmyard, and we should have the same feeling about dealing similarly with a fox."[15]

The hunt, then, is located in a context of diverse yet interrelated human-animal power relationships. First, there are the human relationships with the domesticated animals—horses and dogs—with which they work. Second, there are human relationships with the rural fox, that liminal figure slipping in and out of human farms and habitations. Third, there is the negative impact on humans of the fox's own power relationships with other kinds of domesticated animals—chickens, lambs, and calves—where foxes are seen as usurping roles reserved for humans. This complex of power relations comes together at the time of the hunt itself.

## Power Relations in the Hunt

In the United Kingdom, there are two different forms of foxhunting: the autumn hunt, or cub hunting from August to October, and the main winter hunt season from November to April. Autumn hunting takes place when the season's cubs are maturing, and entire families of foxes are together in woodland coverts (which have sometimes been planted by humans for this purpose). These coverts are encircled by the humans and horses of the hunt, while the dogs are sent into the woodland and encouraged to remain there. Dogs within the wood hunt foxes until foxes are killed or break cover and disperse in different directions. The reduction of fox numbers and the dispersion of fox families is one of the purposes of cub hunting. In addition, it provides humans with the opportunity to train and discipline young foxhounds "to distinguish between, for instance, foxes and deer, and only to hunt foxes; to know the sound of the horn and what the various calls mean."[16]

In contrast to the autumn hunt, the full foxhunting season may entail dogs, horses, and humans chasing a fox for considerable periods of time across open countryside. Although the average chase is only seventeen minutes before a fox is killed or its scent lost, the Worcestershire Hunt recently claimed to have covered ninety miles following one fox (and several horses died of exhaustion). The hounds are encouraged to chase the fox by being deprived of food prior to the hunt; however, if the fox is caught, it is supposed to be killed by the lead hound biting it at the back of the neck rather than by being torn apart by the hounds en masse. Its body, however, is frequently given back to the hounds to eat. If a chased fox runs into an unstopped earth, terriers are often used to flush foxes out into the open again, although sometimes the terriers kill the fox underground.

What then can we make of human-animal power relations in the course of foxhunts? A number of different descriptions exist in hunting literature; we can only focus on a few here.[17] One characteristic frequently discussed is the imagined "bonding" of horses/dogs with humans in the course of the hunt. Scruton, for instance, comments that "hunting with hounds is the noblest form of hunting. And this is because it is the form in which our kindred nature with the animals is most vividly present to our feelings. The pleasure that we feel in this kind of hunting is borrowed from the animals who are really doing it—the hounds who pursue and the horses who excitedly follow them."[18] Certainly, humans, horses, and dogs must cooperate in the course of a hunt. There is a common purpose (at least between the humans and the dogs) and some degree of mutual dependency—the humans are dependent on both horses and dogs to chase the fox; the dogs and horses at various times rely on human guidance. But what is this sense of bonding and kindred nature that is felt with dogs and horses while hunting—a sense that does not extend to the fox? It is only possible because humans have shaped domesticated animals so that they are doing what humans want when humans want it. (The imagined bonding soon breaks, presumably, if the horse refuses to jump a fence or the dogs lose the fox and chase a rabbit.) Bonding with the fox does not occur not just because the hunters are trying to *kill* the fox (in some cultures an imagined bonding with hunted animals is ritually necessary) but because, in one sense at least, the fox is *defying* a human project (even though, in another sense, the pursuit is necessary for the hunt to happen).[19]

In exploring these power relationships further, it is helpful to look more closely at the (usually unwritten) rules of hunting—in particular the idea of the "fair chase." This is widely discussed in hunting literature and has a long historical heritage.[20] Luke suggests that the idea of the fair chase means that animals are not pursued when their flight is restricted by water, deep snow, or fencing; when motorized vehicles are used to track them; when they are electronically tagged; or when they are placed outside their natural habitat.[21] That is to say, a fair chase is possible when the odds are not so overwhelmingly against animals that they have no possibility of escape. These rules seem generally to be observed in the case of British foxhunting, although the chances of foxes escaping are lessened by the practice of blocking fox earths before dawn so that they cannot take refuge during the course of the hunt.[22] However, that the fox has an opportunity to escape is an important part of the hunt. Scruton maintains that a hunt should ensure that "the fox or stag has the best chance of saving himself," while the great hunting advocate, Ortega, comments: "If these countermeasures did not exist, if the inferiority of the animal were absolute, the opportunity to put the activities of hunting it into effect would not have occurred."[23]

This raises a number of questions about the power relationships between humans and foxes. In the course of the hunt, foxes have a genuine opportunity to escape from the hounds. Their flight could be successful and frequently is. In Foucault's terms, the fox's attempt to escape from the hunt can be seen as animal resistance to human power strategies. The fox produces a series of "responses, reactions, results and possible inventions" that represent "insubordination and a certain essential obstinacy on the part of the principles of freedom."[24] We might also conceptualize this in terms of Certeau's understanding of *tactics*. He sees tactics as the reactions of those placed in weak positions to the strategies of those in strong positions, performed where "there is no option of planning general strategy" and "based on the chance offerings of the moment."[25] Clearly, the hunted fox is in just this position: unable to "strategize" (for many reasons) about its position, it resorts to a series of tactics such as doubling back on itself, crossing water, and going underground in order that the hounds should lose the trail.

However, further questions are raised here. The rules of fair chase mean that humans deliberately construct a space for the fox to resist in order for the hunt to happen at all. If humans choose voluntarily to refrain from exercising some of their power (e.g., by shooting/gassing/chasing in motorized vehicles), how far is the fox's behavior genuine resistance? It is, after all, *permitted* resistance: that which the fox is allowed to display in order to give maximum enjoyment to the hunter. As Luke comments, "The *point* of fair chase is to preserve the hunting experience for the hunter, in particular to maintain hunting as the development and application of certain skills in distinction from effortless killing via high technology."[26]

A Foucauldian response to such a question would be to shift the focus from human intentions to actual practices. Theoretically, humans certainly could (and do) kill many more foxes by other means (the Countryside Alliance (1999) maintains that gamekeepers kill 150,000 foxes a year, while hunts kill only 16,000). And certainly, to create a proper chase, humans intend to allow the fox to elude the hunt, at least for a while. But if we focus narrowly on the actual event of the hunt—the sweat, the scent, the twisting and turning of the chase—from the fox's perspective, there is a real chance that the fox may employ tactics, however unconscious, that enable it to escape altogether. In practice, for that seventeen minutes, the outcome is not predetermined, as it would be with shooting or gassing. It is only at the point where the leading hound catches the fox and leaps on it to bite it, that the fox has no more tactical responses. At that point, the dogs (rather than the humans, directly) have domination over the fox, because its dying struggles, however violent, have no chance of success. Only here, then, does the hunt demonstrate complete human dominance over the fox, and then only through the agency of the hounds.

Having explored just some of the many human-animal power relations entangled in foxhunting, we want to move on to think about how some of this might be useful in addressing ethical questions relating to foxhunting.

## Ethics, Power, and Foxhunting

There are, of course, many ethical arguments both for and against hunting. What ethical decisions are made in this context depend on the ethical framework and methodology adopted. This chapter does not attempt to address questions of this kind, and we are not intentionally proposing either a prohunt or antihunt position. However, we do suggest that thinking about the whole context of human-animal power relations surrounding the hunt, as well as during the hunt itself, may indicate new perspectives on a few elements of the ethical debate.

One of the arguments that constantly resurfaces in ethical debates about hunting is what hunting says about human relationships with nature. Much of this turns on different uses of the highly contested term *nature*. One strand of antihunting discourse argues that hunting is an exhibition of human domination over nature and for this reason should be ethically condemned. In contrast, one strand of prohunting discourse maintains that hunting is natural, allowing humans to feel "oneness with nature" and for this reason should be ethically applauded.

Here, of course, *nature* is being used in different senses. Within antihunting discourse the idea that hunting is an exhibition of human domination over nature is usually underpinned by the idea that nature is what is "not human," or as Cartmill puts it in his recent book on hunting, "what is not part of the human domain."[27] The prohunting argument that hunting is "natural" in part relates to an idea of nature as the expression of, in Ellen's terms, "inner essence"—hunting as "doing what humans do" (or perhaps "doing what male humans do"), before and in a more fundamental way than engaging in the "artifice" of human culture.[28] As Scruton describes it, hunting expresses the hunter-gatherer in all of us (or all males at least: on some occasions Scruton's "us" excludes women). But it also depends on an idea of "wild" nature. A nature-culture divide still exists (that is, the term *nature* is not extended to mean "everything that is") but hunting is a practice in which, by expressing their inner essence, (male) humans can identify or feel at one with (wild) nature rather than with (human) culture. Based on (quite widely held) ideas that domination is bad or that feelings of oneness with nature are good, these views underpin some, at least, of the ethical discourses on hunting.

But the exploration of human-animal power relationships in this chapter would suggest that there are difficulties with both these arguments, in relation specifically to foxhunting at least. For instance, we have suggested that there are, indeed, elements of human domination in the hunt. But if domination is understood, in a Foucauldian sense, as power that the "other" is unable to resist, it is not so much the fox that is dominated in the hunt as the horses and dogs bred for it. The fox, at least, has an "unruly body," not shaped by human desires; it lives a life largely independent of human provision and interaction; and the whole idea of the fair chase means that the death of the fox is not inevitable, however much its chances of escape are limited. The fox can resist human power to the end. The very constitutions of the dogs and horses, however, have been constructed by human beings in ways they could never resist. They survive due to human provision; their behavior has been constructed in relation to humans; the nature and timing of their deaths are usually decided by humans—and once decided, will inevitably follow. While some degree of day-to-day resistance is possible, this resistance is located within a framework of overall human domination. If it is domination that is viewed as ethically problematic, objections to the foxhunt are merely the tip of an extremely large iceberg; it is domestication that should be the primary focus of moral disagreement, not foxhunting.

But the presence of large numbers of domesticated animals on a foxhunt also provides one particular argument (among others) for undermining pro-foxhunting perceptions of "naturalness" and "oneness with nature."[29] Even if one were to concede that in some sense hunting is an expression of inner human essence (and this is, of course, profoundly difficult to maintain[30]), it could hardly be argued that foxhunting is outside human culture. It is a practice dependent on large numbers of domestic animals bred, disciplined, trained, and (one might say) enculturated by humans. The stamp of such human practices of power and domination are found all over the hunt. And this leads us back to the strangeness of Scruton's assertion both that hunting allows humans to express "primordial relations" with nature and that humans imaginatively bond during a hunt with horses and dogs with whom we have "kindred nature." A bond with the horses and dogs, who are doing what humans have bred and trained them to do, is a bond with human culture, rather than with nature, if nature is understood as the wild and "not human." This is not to deny that hunters may feel bonded in this way, and that they may interpret this bonding as bonding with nature. Our argument here is that the idea that such bonding is with "wild nature" is based on a mistake—on the invisibility of human power practices in relation to domesticated animals. And this undermines ethical arguments that foxhunting allows the expression of the natural in humans and promotes their identification with wild nature.

In this chapter we have tried to explore some of the human-animal power relations that underpin the complex human practice of foxhunting. We have suggested that an exploration of the whole context of the hunt, including the role of domesticated animals as well as the fox, indicates that many different power relations are involved, and that thinking about these relations more closely is relevant when studying ethical discourses about hunting.

## Notes

1. Countryside Alliance definition: http://www.countryside-alliance.org/country/edu/edu2-3-5fox.htm.

2. See, for instance, P. V. Moriarty and Mark Woods, "Hunting Does Not Equal Predation," *Environmental Ethics* 19, no. 4 (1997): 391–404; C. J. List, "Is Hunting a Right Thing?", *Environmental Ethics* 19, no. 4 (1997): 405–416; Brian Luke, "A Critical Analysis of Hunters' Ethics," *Environmental Ethics* 19,

no. 1 (1997): 25–44; Roger J. H. King, "Environmental Ethics and the Case for Hunting," *Environmental Ethics* 13, no. 1 (1991): 59–85; Gary Bekoff and Dale Jamieson, "Sport Hunting as Instinct—Another Evolutionary Just-So Story," *Environmental Ethics* 13, no. 4 (1991): 375–378; Thomas Vitali, "Sport Hunting—Moral or Immoral?", *Environmental Ethics* 12, no. 1 (1990): 69–82.

3. *Animal* is used here in the traditional sense, understood as nonhuman. We acknowledge that the use of the opposition *human-animal* suggests (inaccurately) that humans are not animals; it is adopted merely as convenience.

4. Yi-Fu Tuan, *Dominance and Affection: The Making of Pets* (New Haven, CT: Yale University Press, 1984). Tuan understands power as dominance and as a consciously held possession.

5. See, for instance, Andrée Collard and Joyce Contrucci, *Rape of the Wild* (London: The Women's Press, 1988), 48; Nick Fiddes, *Meat: A Natural Symbol* (London: Routledge, 1993), 173; Matt Cartmill, *A View to a Death in the Morning* (Cambridge, MA: Harvard University Press, 1993), 135. A. Franklin, in "On Fox-Hunting and Angling: Norbert Elias and the 'Sportisation' Process," *Journal of Historical Sociology* 9, no. 4 (1996): 454, also argues that, in the past at least, foxhunting was a broader demonstration of "power over rural England."

6. There are a number of difficulties involved in applying Foucauldian approaches to animals—in particular, Foucault's interest in discursive forms of power. However, ways some of these difficulties may be approached have been addressed in Clare Palmer, " 'Taming the Wild Profusion of Existing Things?' A study of Foucault, Power and Animals," *Environmental Ethics* 23, no. 4 (2001): 339–358.

7. Bryan S. Turner, *The Body and Society* (London: Sage, 1993), 79.

8. Michel de Certeau, *The Practice of Everyday Life* (Berkeley: University of California Press, 1984), xx.

9. Michel Foucault discusses this in some detail in "The Subject and Power," included in *Beyond Structuralism and Hermeneutics*, ed. Hubert Dreyfus and Paul Rabinow (Chicago: University of Chicago Press, 1982), 208–226. The idea is also discussed by Paul Patton in "Taylor and Foucault on Power and Freedom," *Political Studies* 37 (1989): 260–276.

10. Countryside Alliance, *Hunting the Facts* (1999), available at http://www.countryside-alliance.org.

11. Alec McHoul and Wendy Grace, *A Foucault Primer: Discourse, Power and the Subject* (Melbourne: Melbourne University Press, 1993), 72.

12. Patton, "Taylor and Foucault on Power and Freedom," 262.

13. Michel Foucault, "Interview with Herodote," in *Power/Knowledge: Selected Interviews and Other Writings 1972–1977*, ed. C. Gordon (Brighton: Harvester, 1980), 89.

14. Paul Patton, "Taylor and Foucault on Power and Freedom," Political Studies 37 (1989): 260–276.

15. A comment that has uncomfortable resonance with the slave overseer who said, "Why, I wouldn't mind killing a nigger more than I would a dog." Reported in K. Jacoby, "Slaves by Nature? Domestic Animals and Human Slaves," *Slavery and Abolition* 15, no. 1 (1984): 89–99. Attributed to James Gray, United Kingdom House of Commons Hansard Debates, 29 October 1997, Part 9, Column 847.

16. Countryside Alliance, 1999.

17. Norbert Elias, in "An Essay on Sport and Violence," in *Quest for Excitement*, ed. Norbert Elias and Eric Dunning (Oxford: Blackwell, 1986), argues that the relationships involved in foxhunting have also changed over time—that in early days humans were more closely involved in the kill, but that later, with "sportization," the hounds became the principal actors and responsible for the "violence." If Elias is right, this suggests (unsurprisingly) that the human-nonhuman relations involved in hunting are not universally constant, but change over time.

18. Roger Scruton, "From a View to a Death: Culture, Nature and the Huntsman's Art," *Environmental Values* 6, no. 4 (1997): 471–482.

19. In contrast, though, some antihunt discourses also rely on a sense of "bonding" and "kindred nature"—but with the fox rather than the hounds. The fox is seen as humanlike in the sense of being sentient and intelligent, and the discourse depends on some sense of personal identification (and hence bonding) with "what it would be like to be that fox." Here the fox is constructed in terms of familiarity and harmony with humans, rather than otherness and defiance.

20. See Anne Rooney, *Hunting in Middle English Literature* (Cambridge: Brewer, 1993); Mira Friedman, *Hunting Scenes in the Art of the Middle Ages and the Renaissance*, 2 vols. (Tel Aviv: Tel Aviv University Press, 1978). Edward of Norwich, 2nd Duke of York's *The Master of the Game: The Oldest English Book on Hunting* (London: Chatto and Windus, 1909) is a useful starting point in any consideration of the sizable amount of primary literature on medieval hunting and its rules.

21. Luke, "A Critical Analysis of Hunters' Ethics," 27.

22. See D. Itzkowitz, *Peculiar Privilege: A Social History of English Fox-Hunting 1753–1885* (Brighton: Harvester, 1977), 3; United Kingdom House of Commons Hansard Debates November 29 1997 Part 3 Column 1206.

23. José Ortega y Gasset. *Meditations on Hunting* (New York: Charles Scribner's Sons, 1972), 49.

24. Michel Foucault, "Governmentality," in *The Foucault Effect: Studies in Governmentality*, ed. G. Burchill, C. Gordon, and P. Miller (Sussex: Harvester, 1993), 220; Foucault, "The Subject and Power," 225.

25. Certeau, *The Practice of Everyday Life*, 37.

26. Luke, "A Critical Analysis of Hunters' Ethics," 28.

27. Cartmill, *A View to a Death in the Morning*, 29.

28. Roy F. Ellen, "The Cognitive Geometry of Nature: A Contextual Approach," in *Nature and Society: An thropological Perspectives*, ed. Philippe Descola and Gisli Palsson (London: Routledge, 1996), 111.

29. For other arguments against this, see Bekoff and Jamieson, "Sport Hunting as Instinct—Another Evolutionary Just-So Story," and Moriarty and Woods, "Hunting Does Not Equal Predation."

30. See Cartmill, *A View to a Death in the Morning*, and Moriarty and Woods, "Hunting Does Not Equal Predation."

# 12

# Ethics, Politics, Biodiversity: A View from the South

Niraja Gopal Jayal

## Moral Arguments for Biodiversity Conservation

This chapter explores the ethical and political issues that arise from environmental practice in the area of biodiversity conservation. The central concern of environmental ethics has been the relationship between human beings and nature: providing moral justification for the value of the nonhuman natural environment; exploring the limits of the rights of humans vis-à-vis nature; positing the rights that nature could be said to have against humans, and the considerations of justice that place duties on human beings in their interactions with nature. On the other hand, no such unity of purpose characterizes environmental practices that emanate from a variety of sources—ranging from popular practices to environmental activism and from government regulation to international agreements—and therefore speak quite different languages.

Biodiversity conservation is an area where the gap between environmental ethics and environmental practice is arguably the widest. Drawing attention to it is important for two reasons. First, the chief policy and indeed political questions that have been at the center of debates on biodiversity in recent times are underpinned by rival ethical assumptions that call for explication. At the bottom of every policy and political statement on the subject, there lies implicit a normative position about the relationships between nature and human beings, as well as the diverse social, political, and moral communities to which individuals belong.

Second, there appears to be a certain inadequacy in the traditional way mainstream environmental ethics has looked at the human-nature relationship. Even as they recognize the differentiatedness of nature (in terms of animals, plants, and other living organisms) and seek to account for the treatment of whole species as well as of individual members of these species, environmental ethicists—with a few notable exceptions like Murray Bookchin—frequently tend not to differentiate between human beings as they relate to nature. For the purposes of environmental ethics, therefore, a merchant banker in New York is, in her actual practices and her moral responsibilities toward nature, no different from a tribal person displaced from her forest habitat by a dam project in western India. Implicit in such a view is a certain universalism, and nowhere does this become more problematic than in the consideration of biodiversity, where we may observe a vast range of human-nature interactions. Such interactions may go beyond a purely instrumentalist view of the environment in terms of natural resource use (which would include the consumers as well as the conservers of nature), to encompass relationships with the environment that enter into the self-perception of individuals and even into the very constitution of their individual personhood and their social self.

An undifferentiated view of the category of the human person in environmental ethics may therefore be unhelpful if we are to explore the different ethical values implicitly invoked in claims of rights or justice in relation to the natural environment. The validity of such an approach may be justified in two distinct ways, one theoretical and the other empirical. For the first, theoretical justification, let us take the argument for welfare rights as an example. Such an argument is premised on a recognition of special needs in situations of social and economic inequality, which justifies special provisioning by the state, on the grounds that the universalist ascription of civil and political rights to all citizens is an insufficient guarantee of their equality.[1] Similarly, we need to go beyond the rather flat standard account of the human person that, assuming the task of equipping individuals with rights to be complete, seeks to extend these to animals, plants, and even nonhuman, nonsentient, living beings. Such a project does not take into account how some people are more closely and intimately linked with nature, such that their individual and social self is constituted by that relationship. The following excerpt from a letter written by a tribal person in the Narmada Valley in western India—facing displacement on account of the Sardar Sarovar dam—to the Chief Minister of the state government, illustrates this:

> You tell us to take compensation. What is the state compensating us for? For our land, for our fields, for the trees along our fields. But we don't live only by this. Are you going to compensate us for our forest? . . . Or are you going to compensate us for our great river—for her fish, her water, for the vegetables that grow along her banks, for the joy of living beside her? What is the price of this? . . . How are you compensating us for our fields either—we didn't buy this land; our forefathers cleared it and settled here. What price this land? Our gods, the support of those who are our kin—what price do you have for these? Our adivasi (tribal) life—what price do you put on it?[2]

Second, such a contextualization of the human person becomes empirically necessary, given the spatial boundaries—local, regional, national—that divide and define human communities. Just as territoriality gives rise to different sorts of claims to rights and citizenship in the case of refugees, migrants, and aliens, so too does the territorial boundedness of human populations as well as of certain natural resources, along with the rich variety of environmental practices, compel us to give recognition to location. This is particularly important in the context of biodiversity, where rival claims to biological resources are grounded in arguments about the specificity of their origin and location, on the one hand, and their universal and transnational character, on the other. The nongeneralizability of both categories—human beings and nature—follows.

The first part of this chapter frames the discussion of ethics and biodiversity by locating the major positions on biodiversity within two broad categories of arguments relating to the environment: arguments of an ecocentric nature and those of an anthropocentric nature.[3] The second part of the chapter identifies and problematizes, from the standpoint of ethics, three specific issues in the debate on biodiversity: those of national sovereignty, local communities, and future generations. The third and final section of the chapter raises the question of whether the language of property rights is the most appropriate language for the discussion of biodiversity.

The essential characteristic of all ecocentric arguments is the claim of intrinsic value, which encourages us to respect nature for its own sake, rather than instrumentally or for human purposes. Arguments of intrinsic value may derive intellectual sustenance from a variety of philosophical traditions. Hedonist utilitarian ethics, for instance, makes *sentience* the criterion for identifying the defensible boundaries of concern for the interests of others. A capacity for suffering—and its obverse, a capacity for enjoyment and happiness—gives a being a right to equal consideration, because it suggests that the being in question has *interests*. A stone cannot suffer, hence it does not have interests, and we have no moral obligation to concern ourselves about its welfare. There is, however, no moral justification for not taking into consideration the interests of beings who do suffer, though it may be necessary to establish that these interests are morally significant interests, or that the survival or realization of the "self" in question has moral value, apart from its importance in sustaining life.[4]

The moral significance of the interests of the biosphere can also be justified by an appeal to Kantian philosophy. Extending the basic Kantian principle of treating others as ends in themselves, and never only as a means to our ends, this is taken to imply that what is good for a person must be determined by that person's own nature and self-identity. Johnson has argued that, like individuals, species too have an interest in surviving well, in fulfilling their nature in an appropriate environment. A species is a living system, an ongoing coherent organic whole, with properties that are not simply the aggregate of those of the individual species members. If species are morally significant entities, further, the biosphere is a morally significant entity too.[5]

Deep ecology posits an intrinsic relation between things, such that the relation is itself constitutive of the definition of the things in question. It accords respect to ways and forms of life, on the premise that the equal right to live and blossom is a value axiom, and so provides the basis for the principle of biospherical egalitarianism, which forces us to recognize our close partnership and interdependence with other forms of life. It also urges a recognition of the principles of diversity and richness of forms and modes of life, which translates into a politics of decentralization and local autonomy.[6] Deep ecology is potentially hostile to the idea of asserting ownership—through, for instance, patenting—over other living beings and species. The case has even been made for incorporating some fundamental rights of nonhuman species in a constitutional bill of rights, to give the interests of such species a place in democratic decision making.[7]

For deep ecologists, therefore, species, ecological systems, and the biosphere as a whole are objects of value. They value human and nonhuman life on earth as having intrinsic value; they believe that the richness and diversity of life forms contributes to the realization of this value; and they are convinced, finally, that human beings have no right to diminish this richness and diversity except for the satisfaction of vital needs. This argument has been extended to assert that all organisms and entities in the ecosphere, being parts of an interrelated whole, are equal in their intrinsic worth. A common criticism of this perspective has been that, in privileging the worth of the whole ecosystem, it provides no grounds from which the value of individual plants or microorganisms might be determined. All organisms may be part of an interrelated whole, but this is insufficient to establish that they are all of intrinsic worth, let alone of equal intrinsic worth.[8] If we do extend our respect for sentient beings to nonsentient living things, on what basis do we differentiate between the greater or lesser worth of some, in situations where practical trade-offs have to be made?

Anthropocentrism is characterized by a certain ambivalence on the question of biodiversity conservation. There are two ways it can mandate and support conservation. The first is a weak anthropocentrism, which suggests that conservation is desirable either because human beings derive aesthetic satisfaction from the contemplation of natural diversity or because it satisfies human preferences in relation to nature in ways that welfare economics seeks to catalog and value such preferences. In the latter perspective, valuing biodiversity may lead to the creation of incentives for stakeholders to conserve it. A weak form of anthropocentrism may also be required by "moral expansionism,"[9] which proceeds on the principle that the only source of value is an evaluator, and since we have no way of determining the relative moral worth of the interests of different beings, we should not distinguish between them. As such, this approach seeks to expand outwards from a human-centered ethics to a fuller moral recognition of and protection for nonhuman animals and sentient life in general.[10]

A second, stronger form of anthropocentrism is provided by an ethic that appeals to the Aristotelian conception of well-being "in terms of a set of objective goods a person might possess, for example friends, the contemplation of what is beautiful and wonderful, the development of one's capacities, the ability to shape one's own life, and so on.[11] This approach suggests that we value things in the natural world for their own sake, as a component of a flourishing human existence, and not simply as an external means to our own satisfaction.

Despite their obvious differences, however, it is possible to see a convergence between ecocentric and anthropocentric arguments on biodiversity. Ecocentric arguments emphasize an essentially negative, leave-alone policy attitude toward biodiversity, *unless we choose to invoke the caveat regarding "the satisfaction of vital needs."* Anthropocentric arguments may positively value the conservation of biodiversity but may still endorse its exploitation for human purposes, as for instance in justifying the use of plants that possess medicinal value for fighting life-threatening diseases.

However, in environmental ethics, arguments about biodiversity appear to be arguments of a slightly different order than those relating to other elements of the natural environment. This is arguably one of the few dimensions of the natural environment in relation to which ethical claims are distinguished by a pervasive use of the language of property rights. This may be illustrated by taking a closer look at three issues that have acquired political salience in recent debates on biodiversity:

- The claim of national sovereignty
- The claim of local communities
- The claim of future generations

The debates on these issues present human beings as, variously, members and citizens of the biotic community, local communities, national communities, the global community, and a "transgenerational"[12] community. As such, though none of these claims is directly concerned with providing a moral justification for the rights of diverse biological resources, a recognition of the importance of biodiversity lies at the core of each. In each case, though in very different ways, the claim of autonomy generates a claim to rights. Thus, the claim of national sovereignty translates into the right of the sovereign nation to control and manage natural resources within its territorial boundaries as it wishes. Likewise, the claim of autonomy for local communities generates the idea of their rights—as collectivities—to the ecosystem that not only sustains them, but to the preservation and improvement of which they have historically contributed through creative practices now recognized as indigenous knowledge systems. Finally, the potential for agency and autonomy that is attributed to future generations (on the assumption of their continuity with the present) mandates the recognition of their rights to a natural world that is not destroyed and eroded either in the quantum of its resources or in their richness and diversity. Each of these claims, as we know, further translates from the ethical to the political realm, to yield definite policy prescriptions.

It is important also to recognize that each of these claims takes the particular form it does, namely, that of a *claim*, precisely because there are serious economic and political constraints on their fulfillment. These claims are underwritten not only by the awareness of depleting resources, but also by a sense of insecurity arising from the rival claims asserted by powerful states of the North as well as transnational corporations. Let us examine each of these issues, in terms of the moral claims they suggest, and in relation to the empirical context in which they have been articulated.

## Whose Resource? Negotiating Rival Claims on Biodiversity

### The Claim of National Sovereignty

The chief ethical issue underlying this claim is that of who should own and/or control natural resources, in this case resources rich in biodiversity, and, by extension, who should determine how these resources may be used. Do these resources belong to humankind in general, to nation-states within whose territory they happen to subsist, or to the peoples who live in and by them? On the answer to this first question depends the answer to the further questions of the limits and agents of such use. It is, for instance, possible to argue that such resources belong to humankind in general, and that there is therefore nothing wrong in countries exploiting, for commercial purposes, strains of plants or animals that have originated in other countries. In any event, it is clear that the argument of the "global commons" is not a conservationist argument, but an argument intended to justify access to biological resources regardless of their location, and to facilitate their unhindered exploitation.

This argument of the global commons[13] has the spurious attraction of being consistent with the rhetoric of global unity and the common endeavors of humankind in the field of "scientific progress." But its moral appeal is undermined by the fact that these uses are sought to be patented in the form of intellectual property rights in the context of a trade regime that is patently unequal as between nations. Indeed, it is odd that the rights claimed by Northern governments and transnational corporations to these resources are justified by arguments of the "common good of humanity," to be advanced by means of scientific and technological innovation, precisely that which environmentalists have long associated with environmental degradation and destruction.

On what moral grounds can a "commons" be transformed and brought under a private property regime? Do the developed nations that have the material and intellectual resources to experiment and devise adaptations in sophisticated laboratory conditions, also have the moral right to all future use of those genetic resources, regardless of the country of their origin, which is frequently a developing nation? This question is especially charged because many countries of the South are, by virtue of their location in tropical and subtropical zones, rich in germplasm, while Northern countries that have made great advances in technology in the field of genetics and genetic engineering have little or no germplasm.

It may be useful to consider these issues against the backdrop of the international conventions on this question. As is well known, the question of biodiversity conservation received its earliest expression in the form of the concern for the destruction of the tropical rainforests, articulated mostly by Northern NGOs like the World Wildlife Fund* (WWF) and the International Union for the Conservation of Nature and Natural Resources (IUCN). In collaboration with the World Resources Institute, the World Bank, and the United Nations Environmental Programme, it was these NGOs that drafted the original documents on biodiversity conservation. These drafts reflected a fairly traditional environmentalist approach, namely, that of resource management. It was only in 1991 that the Group of 77 countries (most of the developing countries) asked for the issue of biotechnology to be included with that of biodiversity. Through this, it was proposed that Northern access to the genetic resources of the South be linked to access, for the South, to biotechnology research in the North, and even to a share of the benefits arising from the commercial exploitation of genetic resources.[14] This proposal had thus far been strenuously resisted by the U.S. government, and the

* World Wildlife Fund is now called the World Wide Fund for Nature. The acronym WWF remains the same.

business community in America had also lobbied strongly against it. A month before the UN Conference on Environment and Development (the Earth Summit) held at Rio de Janeiro in June 1992, a document was drawn up at the seventh and final round of negotiations at Nairobi, which remained contentious even on the eve of the summit. Though it embodied substantial concessions for which the United States had driven a hard bargain—in not, for example, providing the South with the guarantees it had sought for the transfer of biotechnology or on the question of intellectual property rights—the United States refused to sign the Biodiversity Convention at Rio. It is significant that the reasons cited, by President George H. W. Bush, for this rejection was that the convention "threatens to retard biotechnology and undermine the protection of ideas."[15] The Convention on Biological Diversity, in force since December 1993, has however been ratified by over 150 countries.

The general declaration on environment and development that emerged from the Rio Summit recognizes both the intrinsic value and the anthropocentric value of biodiversity. It gives to nation-states the "sovereign right" to "exploit" their natural resources, in accordance with their own policies on environment and development. But its further recognition of a right to development—subject only to the minor caveat of the developmental and environmental needs of future generations—endorses the priority of development over environment, rather than the idea that development should take place only within environmental limits. Chatterjee and Finger have persuasively argued that the biodiversity convention exemplifies a "perversion" of the concern for the destruction of the world's biodiversity into a preoccupation with new scientific and biotechnological developments for economic growth. There are, in their view, three key arguments that hold this "perversion" together:

> First, the convention gives "nation-states the sovereign right to exploit their own resources pursuant to their environmental policies," thus *transforming biological diversity into a natural resource to be exploited and manipulated.* Then, the convention implicitly equates the diversity of life—animals and plants—to the diversity of genetic codes, for which read genetic resources. By doing so, *diversity becomes something modern science can manipulate.* Finally, the convention promotes *biotechnology as being "essential for the conservation and sustainable use of biodiversity."* (emphasis added)[16]

Consequently, biotechnology comes to be projected as something that is unambiguously good for the conservation of biodiversity, because it is conducive to the maintenance of genetic diversity, as well as for improving agricultural production, by making possible the increasing productivity of crops, livestock, and aquaculture species. Ironically, biotechnology, in this account, becomes the ideal instrument through which to secure not only material progress, but also biodiversity conservation itself!

It is, on the contrary, widely accepted that biotechnology contributes to the depletion, rather than the enhancement or even conservation, of biodiversity. The argument that only a small sample of genetic material is required for gene banks and laboratories ignores the extended consequences of these processes. Thus, for instance, laboratory research provides "improved" seeds, which promise higher yields and are mass produced by transnational corporations that market these to farmers everywhere, including in the South. The introduction of such seed varieties tends to erode traditional agricultural biodiversity, by discouraging the growing of traditional crops (as wheat and rice replace the local cereals that people traditionally consumed). The high-yielding varieties are also more susceptible to pests and disease, again requiring fresh infusions of genetic materials. Even where these are stored in gene banks, there is a degree of loss in storage. Meanwhile, the evolution of new varieties that tended to take place on farms has also stopped.[17]

For the countries of the South, the shift to high-technology agriculture increases their dependence on biotechnology patented and controlled by transnational corporations. Their vulnerability in the global economic system is compounded by their lack of technological ability to exploit or safeguard biological resources. Policy on biodiversity is thus inescapably driven by the economic interests of the industrialized world, for which conservation is an issue secondary to the primary concern for patents and profits.[18]

Thus, even as biotechnologists are projected as laboring painstakingly for the common future of humankind, there are clearly significant commercial interests—most notably in the pharmaceutical and seed industries—to be served by formalizing such arguments in international conventions. As biotechnology has moved out of the cloistered world of scientific laboratories into the market, the stakes have also risen. It has been estimated that in the next three decades, biotechnology will account for 60 to 70 percent of the global economy, in sectors as diverse as food, pharmaceuticals, energy, mining, and feedstock chemicals. It is also very likely that biotechnological production will be concentrated in the hands of the ten largest multinationals of the world.[19] These interests are supported and advanced by the governments of the North, through the patently unequal terms of international trade, even as they aver that they cannot make commitments on patents and technology transfer, on the grounds that these are in the hands of corporations. This makes it possible for Northern powers to remind the Third World of its debt, forgetting that debts have also been incurred, over the past two centuries, in the reverse direction in the form of not just natural resources, but also biological and genetic material. The question of whether biological and genetic resources, or knowledge and technology related to them, are appropriate objects of monopolistic rights of private ownership, remains an unresolved ethical issue.

Even within the states of the North, difficulties are arising from the tension between the ecocentric approach of the Convention on Biodiversity, and the decidedly anthropocentric manner of its implementation. Bosselmann illustrates this with reference to the German municipal law on Gene Technology, which—in accordance with Article 19 of the Convention—seeks to control biotechnological research, and so regulates the licensing of genetically modified products and their release into the environment. However, it is premised on the prior recognition and implicit acceptance of a fundamental right to engage in genetic engineering. As such, the burden of proof for risks involved rests entirely on the general public, rather than on the producer of the risk. Ultimately, it is the weighing of social costs and benefits that decides whether or not particular activities of genetic engineering are acceptable.[20]

In this contest for control between states of the North and the South, and transnational corporations, the original concern about the destruction of biodiversity, and what can be done to prevent it, is forgotten. While the Convention on Biodiversity's recognition of the right of sovereign governments (and, by implication, local industries) to use and manipulate their own natural resources was widely hailed as a victory for the G-77 countries and environmental NGOs, another important issue was given much less importance: the communities who depend on biodiversity for their sustenance, habitat, food, medicines, and even culture.

**The Claim of Local Communities**

In ethical terms, this question has two components: first, the argument that the local communities who live in and by these natural resources have a historically validated right to them; and, second, that these communities are living repositories of knowledge systems about these biological resources, on the basis of which they have creatively adapted and innovated in impressive ways that deserve recognition and protection.

In many countries of the South, including India, the recognition of the rights of local communities—including but not always indigenous peoples—has become an issue in recent years, chiefly in the context of involuntary displacement caused by development projects like big dams or mines. The claim of collective or group rights has often invoked similar claims made by cultural communities such as indigenous peoples in Canada and Australia. Establishing the admissibility of rights claims for collectivities, and not just for individual citizens, entails recognizing that communities, no less than individuals, possess a capacity for agency and autonomy. Though classical liberal theory attributed this capacity exclusively to individuals, the contemporary discourse on rights gives explicit recognition to such claims, which are seen as admissible even within liberal political philosophy—not just in communitarianism, which is more naturally hospitable to such claims.

If it is, then, in principle possible for collectivities and communities to have rights, what is the substance of such rights in relation to biodiversity? Do communities that have lived by the resources of biodiversity, for whom biodiversity has supplied not only all their material needs (food, fodder, fiber, energy, house-construction materials, medicines, and so on) but also their knowledge systems and cultural identity, have any rights over those resources? The basis of such a claim is chiefly historical, because it invokes the customary rights such communities have enjoyed over these resources. In India, such customary or communal rights range from exclusive rights of ownership and alienation to rights to control and manage resources, to simple rights of usufruct. There is, however, no necessary congruence between the right to own, control, manage, and use. Whatever the substance of the right, it is asserted by or on behalf of the community, and hence a collective. In opposition to this, attempts by the state to assume ownership and control over these resources is often expressed in terms of the doctrine of "eminent domain," by which the public purpose or national interest may justify overriding community rights. But does the fact of their having been historical inhabitants, users, and conservers of these resources give them a special lien on the resources, overriding the claims of the state and industry?[21]

Further, are the innovations brought about by communities or farmers through their own creativity and their intimate knowledge of the properties of various species deserving of protection? This is the basis for the further claim to recognition for the practices and creative innovations of such communities, as they have historically interacted with their natural environment.[22] It suggests that, as the authors of these innovations, they possess a sort of moral copyright over them, which must be respected against egregiously invasive attempts at intellectual piracy. The claim to collective or community rights thus extends into a claim to *collective intellectual property rights*. The normative superiority of this claim derives partly from historical priority and partly from the fact that it is innocent of the intention to commoditize and commercialize, which similar claims (based on the argument of intellectual-value addition) from biotechnologists and transnational corporations, supported by Northern governments, clearly are not. Thus, the use of biodiversity as a common property resource by communities is expressive of an anthropocentric ecological ethic that encompasses human beings, even recognizes the centrality of the human species on this planet and yet proposes a democratic, even communitarian, approach to nature and its diversity. As opposed to this, the attempt to patent an intellectual property right over any resource is an attempt to create value solely for profit. Indeed, it has been estimated that the cost of plant genetic screening, to a pharmaceutical corporation, gets reduced by anywhere between 50 and 90 percent where the corporation gets access to traditional knowledge about its medicinal properties.

While the sovereign rights of nation-states to biodiversity have received international recognition in the form of the Biodiversity Convention, these rights have yet to be adequately translated into the *prior* rights of the communities who have maintained and preserved it. The preamble to the Convention on Biodiversity recognizes the dependence of indigenous communities on biodiversity, and even considers it desirable that the benefits arising from the use of traditional knowledge, innovations, and practices should be equitably shared. However, the

Convention does not go so far as to recognize the *rights* of local communities over these resources, much less to discuss the mechanisms by which they may be compensated for sharing their knowledge.[23] Though Article 16 of the Convention recognizes the need to effectively protect intellectual property rights, its introduction of a caveat "subject to national legislation and international law" brings in an element of ambiguity in the formulation.

In the post-Rio era, the onus thus rests, to an alarming and even improbable degree, on governments to ensure that the diversity of species as well as of communities is protected. But government appears to be an unlikely agent of such an enlightened project because, in the Indian context at least, there is evidence to suggest that it is the centralization of government control over natural resources and agricultural and other development policies, which has historically been destructive of biodiversity. There is also evidence to show that areas rich in biodiversity are at the same time areas of considerable human poverty. The difficulty of devising legal mechanisms through which the objectives of the Biodiversity Convention—of equity in benefit sharing—could be secured, are rendered even more complex in the context of the World Trade Organization regime.[24]

In India, in fact, it is official attempts at conservation—blindly following the Western model—that have been responsible for recent conflicts between the parks and the people. This is because many national parks and biosphere reserves are inhabited by human populations,[25] and their conversion into protected areas requires that these people be displaced and deprived of their access to basic livelihood resources. Indeed, there are not just social but also ecological consequences that follow from such protection. In 1980, the Bharatpur bird sanctuary in northern India—a wetland with over 350 species of birds, including the migratory Siberian crane—was upgraded in status and renamed the Keoladeo Ghana National Park. Consequently, grazing in the park was banned, despite protests by the village folk in which seven people were killed. Meanwhile, a study by the Bombay Natural History Society showed that the ban on buffalo grazing had upset the ecosystem of the region, in which grazing had kept the wetland from turning into grassland. Grass cutting by the villagers has now been permitted.[26]

**The Claim of Future Generations**

This *normative* claim is, of course, generally justified by reference to the *fact* of the exhaustibility of natural resources, including biodiversity among others. Already, just three species—wheat, rice, and maize—provide half the world's food. If we add to this list potato, barley, sweet potato, and cassava, the total goes up to three-quarters. Such tremendous dependence on just a handful of crops is alarming because monocultural farming is susceptible to rapidly spreading diseases. Quite apart from this vulnerability to disease, it is arguable that we know very little at present about biodiversity. It has been estimated that only a fraction of 1 percent of the world's species have been studied for their potential value to humanity, whether for food, medicines, or raw material for industry, and that of a total of anywhere between 10 million and 80 million species, only 1.4 million have even been named.

Despite a variety of methodological problems in calculating the economic value of biodiversity, the projected loss of species in the twenty-first century has been estimated as being on the order of 20 to 50 percent of the world's totals, which means a rate between 1000 and 10,000 times the historical rate of extinction. Further, it is claimed that the rate of loss far exceeds the regenerative capacity of evolution to generate or evolve new species. Thus, extinction "outputs" far exceed the speciation "inputs."

> The implications of species depletion for the integrity of many vital ecosystems are far from clear. The possible existence of depletion thresholds, associated system collapse, and huge discontinuities in related social cost functions, are potentially the worst outcome in any reasonable human time horizon. Such scenarios are indicative of the links between ecosystem integrity and economic well-being. More immediately, the impoverishment of biological resources in many countries might also be regarded as an antecedent to a decline in community or cultural diversity, indices of which are provided in diet, medicine, language and social structure.[27]

The claim that biotechnology enhances the prospects of future generations through advances in the area of genetic engineering may be countered by arguing the value of preserving the natural world from further depredations and degradation so that future generations may enjoy it. The first argument is justified by reference to the Enlightenment view that science and material progress have the potential of solving most human problems and that this potential only increases as the frontiers of science expand. The second is justified by a view that is clearly ethico-philosophical as well as ecological.

Modern liberal philosophy endorses the principle that the beings who can have rights are precisely those who have or can have interests. Thus, plants and vegetables can be said to be capable of having a good, because they are living things with certain inherited biological propensities that determine the pattern of their natural growth, but is not that good distinct from having interests? For interests presuppose cognitive equipment, and are "compounded out of *desires* and *aims*, both of which presuppose something like *belief*, or cognitive awareness."[28] If the plant's need for nutrition or cultivation, its flourishing or languishing, suggest that it has interests, the interests that thrive when plants flourish are not plant interests, but human interests. However, though neither plants nor whole species can be said to have rights in the strict sense of the term, we may assert *duties* to protect threatened and endangered species, "not duties to the species themselves, but rather duties to future human beings, duties derived from our housekeeping role as temporary inhabitants of this planet."[29]

Any claim for the conservation of biodiversity that appeals to our concern for future generations is thus located in a human-centered ethic that invokes our belief that our descendants (will) have interests that are morally significant.[30] Whether it is the material interests of future generations that we wish to protect in the face of an alarming depletion of biodiversity, or a consideration for their aesthetic fulfillment that impels us to preserve the wilderness for their enjoyment, the rights we seek to protect for them are "contingent" rights—that is, the interests that they are sure to have when they come into being.[31]

## Transcending the Language of Ownership

The discussion so far suggests that these three claims are not necessarily compatible with each other, and there are tensions between them that are difficult to unequivocally resolve. Let us first take the claim made on behalf of the sovereign nation-state. This asserts the sovereign right of the nation-state to the biological resources found within its territorial boundaries, and seeks to formalize this ownership in instruments of international law. Developing countries have been only partially successful in their efforts to this end. While they have secured recognition of their sovereign rights to use and exploit (however repugnant that usage is to the environmentalist's sensibility) resources, and even to set their own terms and conditions on the transfer and sale of these resources, their attempts at negotiating a less inequitable international patents regime have not been as successful. For largely agrarian economies, this implies a new dependence on high-technology agriculture promoted by transnational seed corporations and a parallel process of biodiversity erosion.

Of greater consequence, however, is the fundamental issue of the moral status of such ownership, and of the putative rights of state elites to barter it away. It is, for instance, quite possible to conceive of a gene-rich but economically poor and environmentally insensitive Southern state, the government of which cheerfully permits the transfer and sale of genetic resources within its territory in a way that enriches the political elites but leads to irreplaceable biodiversity loss. How can we protect biodiversity against such possibilities without, on the other hand, falling into the trap of the "global commons" argument?

The sovereignty given to nation-states in this respect by the Biodiversity Convention may also not be a sufficient guarantee of the interests of local communities, because their rights to the biological resources by which they live remain unprotected from the depredations of the state as well as transnational and indigenous corporate interests. There is an inherent danger in vesting excessive faith in national governments, especially if the overall economic—and therefore developmental—strategy of Southern nations is imitative of the North. Indeed, in the negotiations that eventually culminated in the Convention on Biodiversity, the developing countries sought access to the biotechnologies of the North on preferential terms, as much as the developed countries sought access to, and protection for, the biological resources of the developing world. This tension also translates into an uncomfortable anomaly in the positions taken by environmental groups and NGOs. On the one hand, they must support national sovereignty in the face of external threats, whether from other states or from transnational corporations. On the other hand, to the extent that the rights of local communities are not adequately safeguarded by national governments, they must also oppose the latter.

In ethical terms, some of the intractability of these issues seems to arise from the fact that claims to rights in biodiversity—especially those of nation-states and local communities—typically invoke ideas of property and ownership. But is a right to property the best, most efficacious, and most morally satisfactory form of expressing such rights claims? What are the implications of appealing to the concept of property and seeking to alter only its rights holders? The political position that repeatedly emerges from environmental activism is that nation-states should have sovereign rights of ownership and control over their biological resources; that local communities should be recognized as the owners and creators of indigenous knowledge systems and indigenous innovations; and that future generations are, as the inheritors of this legacy, potential owners. All these claims strongly echo the imagery of property. The concept of property recalls not only the idea of commoditization, but also an attendant notion of inequity and lack of participation. Is it appropriate to couch such claims in the vocabulary of property rights?

The language of ownership and property seems to be critical to the appropriation of biodiversity for commercial purposes. Arguably, it facilitates biodiversity loss and possibly even its eventual destruction. This is because, in ways yet unknown to science, species are interrelated with each other through food chains and food webs in such a delicate balance that the extinction of one species could easily threaten the existence of another. Such an appropriation not only seeks to privatize a biological resource, it also seeks to privatize (in the form of intellectual property rights) knowledge about its properties, or a practice of its use, that has hitherto existed in the domain of the commons. Thus, is an emphatic assertion of biodiversity as *a common property resource* an adequate or satisfactory solution?[32] Such an assertion may possess the merit of protecting biological resources and traditional knowledge about them both from the state and from profit-driven efforts at privatization, but it would leave unresolved the question of how and under what conditions biodiversity loss (e.g., in the case of a plant needed to produce a lifesaving drug) may be permitted; in what form it should be compensated; and who the recipients of such compensation should be: states, communities as a whole, community leaders, or NGOs. Such a view remains, therefore, irredeemably burdened by its inability to comprehend the question of biodiversity outside of the vocabulary of property rights.

How then may the language of ownership and control be transcended? The claim (discussed in the first part of the chapter) that biological resources possess intrinsic value or moral significance, or a good of their own, should logically imply that no human being(s) can be said to "own" or "control" these resources. If we value nature, we approach it as an entity of equal moral status and value. As such, the language of property and ownership is clearly inappropriate in the context of biodiversity. We may not find it feasible to give representation to the natural world in processes of democratic decision making, but we can represent human beings as stewards and trustees of nature, for the present as well as for the future. In this sense, the ideas underlying the claim of future generations have greater potential in enabling us to unequivocally reject any argument for biodiversity that is couched in the vocabulary of property rights. An argument of stewardship or trusteeship echoes the convergence (noted in the first section) between ecocentric and anthropocentric arguments. It does this by invoking the caveat of vital needs in ecocentric arguments, or by endorsing the importance of conservation from an anthropocentric point of view, while making allowances for morally justifiable interventions such as lifesaving drugs. There is a parallel here between the rights claimed for humans and those claimed for nature, because infringements of both are seen equally to require moral justification.

Following from this, the rights that could be said to inhere in sovereign nation-states or local communities would also be contingent rights, like those conventionally assigned to future generations. In the liberal tradition—and most influentially, in the work of Robert Nozick—rights have long been conceptualized in terms of self-ownership.[33] In opposition to this, an alternative interpretation of rights as self-government[34] that incorporates principles of equal respect, symmetry, and universality may yield a more egalitarian, participatory, and democratic perspective, one that also has greater potential for a radical ecological agenda. However, both the language of self-ownership and, to a lesser degree, that of self-government tend to obscure the question of power, which is of paramount importance in relation to biodiversity. The gentle ethics of the environment must engage with the question of power if the multiple asymmetries—global, national, and local—that characterize the control and use of biodiversity are to be meaningfully addressed.

## Notes

1. Debates on equality and welfare are suggestive in another respect also. Take, for instance, Amartya Sen's argument about how the identification of objects of value specifies an evaluative space. Sen's "capability approach" expands the evaluative space to include, in a way that utilitarianism does not, the capability to lead different types of lives, and this capability depends on many factors, including personal characteristics and social arrangements. See Amartya Sen, "Capability and Well-Being," in *The Quality of Life*, ed. Martha C. Nussbaum and Amartya Sen (Oxford: Clarendon Press, 1993), 33.

2. Bava Mahalia, "Letter from a Tribal Village," *Lokayan Bulletin*, 11 (1994): 157–158.

3. It does not, however, deal separately with the different dimensions of biological diversity, such as genetic diversity, species diversity, and ecosystems diversity.

4. Peter Singer, *Practical Ethics* (Cambridge: Cambridge University Press, 1993), 283.

5. Lawrence E. Johnson, *A Morally Deep World: An Essay on Moral Significance and Environmental Ethics* (New York: Cambridge University Press, 1993), 265.

6. Arne Naess, "Deep Ecology," in *The Green Reader*, ed. Andrew Dobson (London: Andre Deutsch, 1991).

7. Robyn Eckersley, "Liberal Democracy and the Rights of Nature: The Struggle for Inclusion," in *Ecology and Democracy*, ed. Freya Mathews (London: Frank Cass, 1996), 181.

8. Singer, *Practical Ethics*, 282.

9. Holmes Rolston III, "Environmental Ethics: Values in and Duties to the Natural World," in *Applied Ethics: A Reader*, ed. Earl R. Winkler and Jerrold R. Coombs (Oxford: Blackwell, 1993).

10. It has been suggested that nature is "just the latest minority deserving a place in the sun of the American liberal tradition" (Roderick Nash, cited in Robyn Eckersley, "Liberal Democracy and the Rights of Nature," 125).

11. John O'Neill, *Ecology, Policy, and Politics* (London: Routledge, 1993).

12. Avner de-Shalit, *Why Posterity Matters: Environmental Policies and Future Generations* (London: Routledge, 1995).

13. Vandana Shiva has argued that biodiversity is not a global commons in the sense in which oceans or the atmosphere are, because biodiversity exists in specific countries and is used by specific communities. In this sense, biodiversity is and has always been a local common resource (Vandana Shiva, "Biodiversity Conservation, People's Knowledge and Intellectual Property Rights," in *Biodiversity Conservation: Whose Resource? Whose Knowledge?*, ed. Vandana Shiva (New Delhi: INTACH, 1994), 4). It may also be argued that the argument of the global commons cannot cut both ways. If genetic resources are a common

heritage, they cannot be privatized, and if they are sought to be privatized, they have to be acknowledged as the property of the Third World and paid for like any other resource. This is especially so because they tend to return to their places of origin in the form of a priced commodity.

14. Mukund Govind Rajan, *Global Environmental Politics: India and the North-South Politics of Global Environmental Issues* (Delhi: Oxford University Press, 1997), 206ff.

15. Fiona McConnell, "The Convention on Biological Diversity," in *The Way Forward: Beyond Agenda 21*, ed. Felix Dodd. (London: Earthscan, 1997), 51.

16. Pratap Chatterjee and Matthias Finger, *The Earth Brokers: power, politics and world development.* (London: Earthscan, 1995), 42.

17. Ashish Kothari, *Understanding Biodiversity: Life, Sustainability and Equity*, Tracts for the Times no. 11 (Delhi: Orient Longman, 1997), 55.

18. Marian A. L. Miller, "Sovereignty Reconfigured: Environmental Regimes and Third World States," in *The Greening of Sovereignty in World Politics*, ed. Karen T. Litfin. (Cambridge, MA: MIT Press, 1998), 189.

19. Suman Sahai, "Biotechnology: New Global Money-Spinner," *Economic and Political Weekly* 30, no. 46 (1994): 2916.

20. Klaus Bosselmann, "Human Rights and the Environment: Redefining Fundamental Principles?", in *Governing for the Environment: Global Problems, Ethics and Democracy*, ed. Brendan Gleeson and Nicholas Low (Hampshire: Palgrave, 2001), 131.

21. The polar opposite of this argument is, of course, explicitly developmental. It suggests that development, which is equated with economic growth, is hampered and obstructed by such claims by or on behalf of such communities. Another variation of this argument is that of "mainstreaming" tribal communities, on the premise that they are being excluded from, and are therefore deprived of, the benefits of progress and development by being kept caged within their traditional unchanging environs.

22. For a rich variety of examples, see Michael D. Warren, L. Jan Slikkerveer, and David Brokensha, eds., *The Cultural Dimension of Development: Indigenous Knowledge Systems* (London: Intermediate Technology Publications, 1995).

23. Environmental activists have argued that the formulation is weak and unclear, and will need sustained pressure to ensure that it works in the interests of biodiversity conservation as well as the equitable sharing of benefits.

24. In August 2001, the lower house of the Parliament of India passed the Protection of Plant Varieties and Farmers' Rights Bill, which gives some recognition to the rights of farmers who develop new strains through selection and breeding. This is apparently the first time in the world that farmers' and breeders' rights have received concurrent recognition. Nevertheless, farmers may find it hard to meet the criteria specified for registering their strains, and to muster the resources required to establish these. The benefit-sharing aspects of the bill are also ambiguous and are likely to advance the interests of commercial breeders.

25. A national-level survey by the Indian Institute of Public Administration showed that 69 percent of Protected Areas surveyed had human populations living inside them, and 64 percent had community rights leases or concessions inside them. The most common activities in the parks were grazing and the collection of nontimber produce (Kothari, *Understanding Biodiversity*, 27–28).

26. Kothari, *Understanding Biodiversity*, 32.

27. David Pearce and Dominic Morgan, *The Economic Value of Biodiversity* (London: Earthscan, 1994), 11.

28. Joel Feinberg, *Rights, Justice, and the Bounds of Liberty*. (Princeton, NJ: Princeton University Press, 1980), 168.

29. Feinberg, *Rights, Justice, and the Bounds of Liberty*, 172.

30. Avner de-Shalit extends communitarianism to develop the idea of a "transgenerational community" to which we can be said to have obligations consistent with the requirements of intergenerational justice. He believes that a communitarian model rescues us from first having to resolve the ontological problem of potential versus actual persons (*de-Shalit, Why Posterity Matters*, 127).

31. Feinberg, *Rights, Justice, and the Bounds of Liberty*, 182.

32. There are innumerable examples of communities traditionally engaged in biodiversity conservation, some religious in inspiration (like sacred groves) and others secular. See several case studies of South and Central Asia in *Communities and Conservation: Natural Resource Management in South and Central Asia*, ed. Ashish Kothari, Neema Pathak, R. V. Anuradha, and Bansuri Taneja (Delhi: UNESCO and Sage Publications, 1998). There are also instances, in a more modern political idiom, of collective action seeking to protect the right to save and exchange seeds noncommercially—for example, the "Seed Satyagraha" by farmers in Karnataka, India, in 1992, and the protest against the Dunkel Draft in 1993.

33. Robert Nozick, *Anarchy, State, and Utopia* (New York: Basic Books, 1974).

34. Attracta Ingram, *A Political Theory of Rights* (Oxford: Clarendon Press, 1994).

# Bibliography

Alston, Dana. "Moving beyond the Barriers." In *Proceedings: The First National People of Color Environmental Leadership Summit.* Edited by Charles Lee. New York: United Church of Christ Commission for Racial Justice, 1992.

Alston, Dana, ed. *We Speak for Ourselves: Social Justice, Race, and Environment.* Washington, DC: Panos Institute, 1991.

Alston, Philip. "Conjuring Up New Human Rights." *American Journal of International Law* 78 (1984): 607–621.

Anderson, Jack. "Danger to Children and Other Living Things." *Washington Post*, 1 August 1982, sec. B.

Anderson, Michael R. "Individual Rights to Environmental Protection in India.' In *Human Rights Approaches to Environmental Protection.* Edited by Alan Boyle and Michael R. Anderson. Oxford: Clarendon Press, 1996.

Anthony, Carl, Ben Chavis, Richard Moore, Vivien Li, Scott Douglas, and Winona LaDuke. "A Place at the Table: A Sierra Roundtable on Race, Justice, and the Environment." *Sierra*, May–June 1993.

Applegate, R. *Public Trusts: A New Approach to Environmental Protection.* Washington, DC: Exploratory Project for Economic Alternatives, 1976.

Archer, Jack H., et al. *The Public Trust Doctrine and the Management of America's Coasts.* Amherst: University of Massachusetts Press, 1994.

Arler, Finn. "Levn, levninger og brugt natur. Om forpligtelsen over for eftertiden [Relics, remnants, and used nature: On the responsibility toward posterity]." In *Naturminder—levnenes betydninger i tid og rum* [Nature memorials—The significance of relics in time and space.] Edited by J. Guldberg and M. Ranum. Odense: Odense University Press, 1997.

Arras, John D. "Getting Down to Cases: The Revival of Casuistry in Bioethics." *Journal of Philosophy and Medicine* 16 (1991): 29–51.

Arrow, Kenneth J., et al. "Intertemporal Equity, Discounting, and Economic Efficiency." In *Climate Change 1995: Economic and Social Dimensions of Climate*

Our sincere thanks and appreciation to Chris Hubbard at New York University for helping to compile this reference section—A. L. and A. de-S.

*Change*. Edited by James P. Bruce, Hoesung Lee, and Erik F. Haites. Cambridge: Cambridge University Press, 1995.

Attfield, Robin. *The Ethics of Environmental Concern*. 2nd ed. Athens: University of Georgia Press, 1991.

Attfield, Robin. "Intrinsic Value and Transgenic Animals." In *Animal Biotechnology and Ethics*, 172–189. Edited by Alan Holland and Andrew Johnson. London: Chapman and Hall, 1998.

Austin, Regina, and Michael Schill. "Black, Brown, Red, and Poisoned: Minority Grassroots Environmentalism and the Quest for Eco-Justice." *Kansas Journal of Law and Public Policy* 1 (1991).

Avery, Dennis T. "Why Biotechnology May Not Represent the Future in World Agriculture." In *World Food Security and Sustainability: The Impacts of Biotechnology and Industrial Consolidation*, NABC Report 11, pp. 97–109. Edited by Donald P. Weeks, Jane Baker Segelken and Ralph W. F. Hardy. Ithaca, NY: National Agricultural Biotechnology Council, 1999.

Babcock, H. "Has the U.S. Supreme Court Finally Drained the Swamp of Takings Jurisprudence?" *Harvard Environmental Law Review* 19 (1994).

Baier, Annette. "The Rights of Past and Future Persons." In *Responsibilities to Future Generations: Environmental Ethics*, 171–183. Edited by Ernest Partridge. Buffalo, NY: Prometheus Books, 1981.

Balzer, Philipp, Klaus Peter Rippe, and Peter Schaber. "Two Concepts of Dignity for Humans and Non-Human Organisms in the Context of Genetic Engineering." *Journal of Agricultural and Environmental Ethics* 13 (2000): 7–27.

Barringer, Felicity. "Debate over Lead in Gasoline Revs Up Again." *Washington Post*, 5 October 1981, sec. A.

Barry, Brian. "Justice between Generations." In *Law, Morality and Society*. Edited by P. M. S. Hacker and Joseph Raz. Oxford: Oxford University Press, 1977.

Barry, Brian. *Democracy, Power, and Justice*. Oxford: Oxford University Press, 1989.

Barry, Brian. "Sustainable and Intergenerational Justice." In *Fairness and Futurity: Essays on Environmental Sustainability and Social Justice*. Edited by Andrew Dobson. Oxford: Oxford University Press, 1999.

Barry, John *Rethinking Green Politics*. London: Sage, 1999.

Beachamp, Tom. "Principalism and Its Alleged Competitors." *Kennedy Institute Ethics Journal* 4, no. 3 (1994): 181–198.

Beachamp, Tom, and James Childress. *Principles of Medical Ethics*. Oxford: Oxford University Press, 1994.

Beckenstein, Alan R., et al. *Stakeholder Negotiations: Exercises in Sustainable Development*. Chicago: Irwin, 1996.

"Behind the Hubbub over Leaded Gas." *Changing Times* 25, no. 10 (1971): 6–9.

Bekoff, Gary, and Dale Jamieson. "Sport Hunting as Instinct—Another Evolutionary Just-So Story." *Environmental Ethics* 13, no. 4 (1991): 375–378.

Boch, C. "The Iroquois at the Kirchberg; or, some Naïve Remarks on the Status and Relevance of Direct Effect." *Jean Monnet Working Paper* no. 6, Harvard Law School, 1999.

Bockemühl, Jochen. "A Goethean View of Plants: Unconventional Approaches." In *Intrinsic Value and Integrity of Plants in the Context of Genetic Engineering*. Edited by David Heaf and Johnannes Wirz. Llanystumdwy, UK: International Forum for Genetic Engineering, 2001.

Bogert, George. *Trusts*. 6th ed. St. Paul, MN: West, 1987.

Bohman, James. "Democracy as Inquiry, Inquiry as Democratic: Pragmatism, Social Science, and the Cognitive Division of Labor." *American Journal of Political Science* 43, no. 2 (1999): 590–607.

Bookchin, Murray. *Ecology of Freedom*. Rev. ed. Montreal: Black Rose Books, 1991.

Borlaug, Norman E. "Ending World Hunger: The Promise of Biotechnology and the Threat of Anti-Science Zealotry." *Plant Physiology* 123 (2000): 487–490.

Bosselmann, Klaus. "Human Rights and the Environment: Redefining Fundamental Principles?" In *Governing for the Environment: Global Problems, Ethics and Democracy*. Edited by Brendan Gleeson and Nicholas Low. Hampshire: Palgrave, 2001.

Bowen, William M. *Environmental Justice through Research-Based Decision-Making*. New York: Garland, 2001.

Brandt, Allan. "Racism and Research: The Case of the Tuskegee Syphilis Study." *Hastings Center Report* 8, no. 6 (1978): 21–29.

Bratton, Susan Power. "Alternative Models of Ecosystem Restoration." In *Ecosystem Health: New Goals for Environmental Management*. Edited by Robert Costanza, Bryan G. Norton, and Benjamin D. Haskell. Washington, DC: Island Press, 1992.

Bretting, John, and Diane Michele Prindeville. "Environmental Justice and the Role of Indigenous Women Organizing Their Communities." In *Environmental Injustices, Political Struggles: Race, Class, and the Environment*. Edited by David Camacho. Durham, NC: Duke University Press, 1998.

Broome, John. *Counting the Costs of Global Warming*. Cambridge: White Horse Press, 1992.

Brown, Peter. "Climate Change and the Planetary Trust." *Energy Policy*, March 1992, pp. 208–222.

Brubaker, E. *Property Rights in the Defence of Nature*. London: Earthscan, 1995.

Bryant, Bunyan. *Environmental Justice: Issues, Policies, and Solutions*. Covelo, CA: Island Press, 1996.

Bryant, Bunyan, and Paul Mohai, eds. *Race and the Incidence of Environmental Hazards: A Time for Discourse*. Boulder, CO: Westview Press, 1992.

Bullard, Robert. *Dumping in Dixie: Race, Class, and Environmental Quality*. Boulder, CO: Westview Press, 1990.

Bullard, Robert, ed. *Confronting Environmental Racism: Voices from the Grassroots*. Boston: South End Press, 1993.

Bullard, Robert, ed. *People of Color Environmental Groups 1994–95 Directory*. Atlanta: Environmental Justice Resource Center, 1994.

Callicott, J. Baird. "Conservation Values and Ethics." In *Principle of Conservation Biology*. Edited by Gary K. Meffe. Sunderland, MA: Sinauer Associates, 1994.

Callicott, J. Baird. "Environmental Philosophy Is Environmental Activism: The Most Radical and Effective Kind." In *Environmental Philosophy and Environmental Activism*, 21. Edited by Don E. Marietta, Jr., and Lester Embree. Lanham, MD: Rowman and Littlefield, 1995: 19–36.

Callicott, J. Baird. "On Norton and the Failure of Monistic Inherentism." *Environmental Ethics* 18, no. 1 (1996): 219–221.

Callicott, J. Baird, and Eric T. Freyfogle, eds. *For the Health of the Land: Previously Unpublished Essays and Other Writings of Aldo Leopold*. Washington, DC: Island Press, 1999.

Callicott, J. Baird, and M. Nelson, eds. *The Great Wilderness Debate*. Athens: University of Georgia Press, 1998.

Camacho, David, ed. *Environmental Injustices, Political Struggles: Race, Class, and the Environment*. Durham, NC: Duke University Press, 1998.

Campaign for Responsible Technology. Available at http://www.svtc.org.

Capek, Sheila. "The 'Environmental Justice' Frame: A Conceptual Discussion and an Application." *Social Problems* 40, no. 1 (1993).

Caranta, R. "Governmental Liability after Francovich." *Cambridge Law Journal* 52 (1993): 272–297.

Care, Norman. "Future Generations, Public Policy, and the Motivation Problem." *Environmental Ethics* 4 (1982): 195–213.

Carpenter, Janet, E., and Leonard P. Gianessi. *Agricultural Biotechnology: Updated Benefit Estimates*. Washington, DC: National Center for Food and Agricultural Policy, 2001.

Cartmill, Matt. *A View to a Death in the Morning*. Cambridge, MA: Harvard University Press, 1993.

Charest, Pierre J. "Biotechnology in Forestry: Examples from the Forest Service." *Forestry Chronicle* 72, no. 1 (1996): 37–42.

Chatterjee, Pratap, and Matthias Finger. *The Earth Brokers: Power, Politics and World Development*. London: Earthscan, 1995.

Chavis, Benjamin, Jr. "Foreword." In *Confronting Environmental Racism*: Voices from the Grassroots. Edited by Robert D. Bullard. Boston: South End Press, 1993.

Cheney, Jim. "Postmodern Environmental Ethics: Ethics as Bioregional Narrative.' *Environmental Ethics* 11 (1989): 117–134.

Childress, James. *Practical Reasoning in Bioethics*. Bloomington: Indiana University Press, 1997.

Christensen, Norman L., et al. "The Report of the Ecological Society of America Committee on the Scientific Basis for Ecosystem Management." *Ecological Applications* 6, no. 3 (1996): 665–691.

Churchill, Robin. "Environmental Rights in Existing Human Rights Treaties." In *Human Rights Approaches to Environmental Protection*. Edited by Alan Boyle and Michael R. Anderson. Oxford: Clarendon Press, 1996.

Clark, C. W. "Profit Maximization and the Extinction of Animal Species." *Journal of Political Economy* 81 (1973): 950–961.

Cole, Luke W., and Sheila R. Foster. *From the Ground up: Environmental Racism and the Rise of the Environmental Justice Movement*. New York: New York University Press, 2001.

Collard, Andrée, and Joyce Contrucci. *Rape of the Wild*. London: The Women's Press, 1988.

Collins, Patricia Hill. *Fighting Words: Black Women and the Search for Justice*. Minneapolis: University of Minnesota Press, 1998.

Comstock, Gary L. *Vexing Nature: On the Ethical Case Against Agricultural Biotechnology*. Dordrecht: Kluwer, 2000.

Connolly, William. *Political Theory and Modernity*. 2nd ed. Ithaca, NY: Cornell University Press, 1993.

Cooper, David E. 'Intervention, Humility and Animal Integrity." In *Animal Biotechnology and Ethics*, 145–155. Edited by Alan Holland and Andrew Johnson. London: Chapman and Hall, 1998.

Countryside Alliance. *Hunting the Facts*. 1999. Available at http://www.countrysidealliance.org.

Craig, David. *On the Crofters' Trail*. London: Jonathan Cape, 1990.

Cranston, M. "Human Rights, Real and Supposed." In *Political Theory and the Rights of Man*. Edited by David D. Raphael. Bloomington: Indiana University Press, 1967.

Cronon, William, ed. *Uncommon Ground: Toward Reinventing Nature*. New York: Norton, 1995.

Curran, W. J. "The Tuskegee Syphilis Study." *New England Journal of Medicine* 289, no. 14 (1973): 730–731.

Daly, Herman E., and John B. Cobb, Jr. *For the Common Good*. Boston: Beacon Press, 1989.

Daniels, Norman. *Justice and Justification*. Cambridge: Cambridge University Press, 1996.

Daniels, Robert A. "Untested Assumptions: The Role of Canals in the Dispersal of Sea Lamprey, Alewife, and Other Fishes in the Eastern United States." *Environmental Biology of Fishes* 60, no. 4 (2000): 302–329.

Danish Economic Council. "Bæredygtighed, Balance mellem generationer" [Sustainability: Balance between generations]. In *Dansk økonomi*, 171–256. Copenhagen: Det økonomiske Råd, 1998.

Danish Ministry of Environment and Energy. *Statement on nature and environmental policy* 1995. Copenhagen: DanishMinistry of Environment and Energy, 1995.

Danish Ministry of Environment and Energy. *Denmark's Nature and Environment Policy 1995. Summary Report*. Copenhagen: Danish Ministry of Environment and Energy, 1995.

Danish Ministry of Environment and Energy. *Ecological utilization space—Where do we stand and where are we heading*? Copenhagen: Danish Ministry of Environment and Energy, 1998.

Danish Ministry of Environment and Energy. *Natur- og miljøpolitisk redegørelse 1999*. Copenhagen: Danish Ministry of Environment and Energy, 1999.

Deal, David T. "Mobile Source Fuels and Fuel Additives." In *Clean Air Law and Regulation*. Edited by Timothy A. Vandever. Washington, DC: Bureau of National Affairs, 1992.

De Certeau, Michel. *The Practice of Everyday Life*. Berkeley: University of California Press, 1984.

Deimann, Sven, and Bernard Dyssli, eds. *Environmental Rights: Law, Litigation and Access to Justice*. London: Cameron May, 1995.

Desgagné, R. "Integrating Environmental Values into the European Convention on Human Rights." *American Journal of International Law* 89 (1995).

de-Shalit, Avner. *Why Posterity Matters: Environmental Policies and Future Generations*. London: Routledge, 1995.

de-Shalit, Avner. *The Environment: Between Theory and Practice*. Oxford: Oxford University Press, 2000.

Devall, Bill, and George Sessions. *Deep Ecology: Living as if Nature Mattered*. Salt Lake City: Peregrine Press, 1985.

Dewey, John. "Philosophy and Democracy." In *John Dewey: The Political Writings*. Edited by Debra Morris and Ian Shapiro. Indianapolis: Hackett, 1993.

Di Chiro, Giovanna. "Defining Environmental Justice: Women's Voices and Grassroots Politics." *Socialist Review* 22, no. 4 (1992).

Dickens, Charles. *Little Dorrit*. Edited by Harvey Peter Sucksmith. Oxford: Clarendon Press, 1979.

Dickstein, Morris. "Introduction: Pragmatism Now and Then. In *The Revival of Pragmatism: New Essays on Social Thought, Law, and Culture*. Edited by Morris Dickstein. Durham, NC: Duke University Press, 1998.

Dickstein, Morris, ed. *The Revival of Pragmatism: New Essays on Social Thought, Law, and Culture*. Durham, NC: Duke University Press, 1998.

Dobson, Andrew. "Deep Ecology." *Cogito* (1989): 41–46.

Dobson, Andrew. *Green Political Thought*. London: HarperCollins Academic, 1990.

Dobson, Andrew. *Justice and the Environmental: Conceptions of Environmental Sustainability and Dimensions of Social Justice*. Oxford: Oxford University Press, 1998.

Dobson, Andrew, ed. *Fairness and Futurity: Essays on Environmental Sustainability and Social Justice*. Oxford: Oxford University Press, 1999.

Dombeck, Michael P. "Thinking Like a Mountain: BLM's Approach to Ecosystem Management." *Ecological Applications* 6, no. 3 (1996): 699–702.

Douglas-Scott, S. "Environmental Rights in the European Union—Participatory Democracy or Democratic Deficit?" In *Human Rights Approaches to Environmental Protection*. Edited by Alan Boyle and Michael R. Anderson. Oxford: Clarendon Press, 1996.

Dowie, Mark. *Losing Ground: American Environmentalism at the Close of the Twentieth Century*. Cambridge, MA: MIT Press, 1995.

Dryzek, John. *Deliberative Democracy and Beyond: Liberals, Critics, Contestations*. Oxford: Oxford University Press, 2000.

Dubgaard, Alex. "Bæredygtighed og forsigtighedsprincippet [Sustainability and the principle of precaution]." In *Fremtidens pris. Talmagi i miljøpolitikken* [The price of the future: Number magic in environmental policy]. Edited by Henning Schroll et al. Copenhagen: Mellemfolkeligt Samvirke/Det Økologiske Råd, 1999.

Du Bois, François. "Social Justice and the Judicial Enforcement of Environmental Rights and Duties." In *Human Rights Approaches to Environmental Protection*. Edited by Alan Boyle and Michael R. Anderson. Oxford: Clarendon Press, 1996.

Dumm, Thomas. "Strangers and Liberals." *Political Theory* 22, no. 1 (1994): 167–176.

Dunlap, Riley. "Public Opinion and Environmental Policy." In *Environmental Politics and Policy: Theories and Evidence*. Edited by James P. Lester. Durham, NC: Duke University Press, 1989.

Dworkin, Ronald. "What Is Equality? Part 2, Equality of Resources." *Philosophy and Public Affairs* 10, no. 4 (1981).

Eckersley, Robyn. "Greening Liberal Democracy: The Rights Discourse Revisited." In *Democracy and Green Political Thought: Sustainability, Rights and Citizenship*, 212–236. Edited by Brian Docherty and Marius de Geus. London: Routledge, 1996.

Eckersley, Robyn. "Liberal Democracy and the Rights of Nature: The Struggle for Inclusion." In *Ecology and Democracy*. Edited by Freya Mathews. London: Frank Cass, 1996.

Edward of Norwich, 2nd Duke of York. *The Master of the Game: The Oldest English Book on Hunting*. London: Chatto and Windus, 1909.

Edwards, Bob. "With Liberty and Environmental Justice for All: The Emergence and Challenge of Grassroots Environmentalism in the United States." In *Ecological Resistance Movements*. Edited by Bron Taylor. Albany, NY: SUNY Press, 1995.

Elias, Norbert. "An Essay on Sport and Violence." In *Quest for Excitement*. Edited by Norbert Elias and Eric Dunning. Oxford: Blackwell, 1986.

Ellen, Roy F. "The Cognitive Geometry of Nature: A Contextual Approach." In *Nature and Society: Anthropological Perspectives*. Edited by Philippe Descola and Gisli Palsson. London: Routledge, 1996.

Elliot, Robert. "Faking Nature." *Inquiry* 25 (1982): 81–93.

Elliot, Robert. *Faking Nature*. London: Routledge, 1997.

Elster, Jon. *Ulysses and the Sirens*. Rev. ed. Cambridge: Cambridge University Press, 1984.

Environmental Strategies, Inc. "Summary of Comments on the Environmental Protection Agency's February 22, 1982, Proposal on Regulation of Fuel and Fuel Additives (Lead Phasedown)." Prepared for the Office of Air, Noise and Radiation Enforcement, Environmental Protection Agency, 5 October 1982.

Epictetus. *The Moral Discourses of Epictetus*. Translated by Elizabeth Carter. London: Dent, 1910.

Epstein, Barbara. "The Environmental Justice/Toxics Movement: Politics of Race and Gender." *Capitalism, Nature, Socialism* 8, no. 3 (1997): 63–87.

ESI International. "Summary of Comments on the Environmental Protection Agency's August 2, 1984, Proposal on Regulation of Fuel and Fuel Additives (Lead Phasedown)." Prepared for the Office of Mobile Sources, Environmental Protection Agency, 1984.

Faber, Daniel, ed. *The Struggle for Ecological Democracy*. New York: Guilford, 1998.

Faucheux, Sylvie, David Pearce, and John Proops, eds. *Models of Sustainable Development*. Cheltenham: Elgar, 1996.

Feinberg, Joel. "The Rights of Animals and Unborn Generations." In *Philosophy and Environmental Crisis*. Edited by William Blackstone. Athens, GA: University of Georgia Press, 1974.

Feinberg, Joel. *Rights, Justice, and the Bounds of Liberty*. Princeton, NJ: Princeton University Press, 1980.

Festenstein, Matthew. *Pragmatism and Political Theory from Dewey to Rorty*. Chicago: University of Chicago Press, 1997.

Fiddes, Nick. *Meat: A Natural Symbol*. London: Routledge, 1993.

Fincham, J. R. S., and J. R. Ravetz. *Genetically Engineered Organisms: Benefits and Risks*. Toronto: University of Toronto Press, 1991.

Flader, Susan L. *Thinking Like a Mountain*. Madison: University of Wisconsin Press, 1974.

Fogel, Max L. "Warning: Auto Fumes May Lower Your Kid's IQ." *Psychology Today* 13 (1980).

Foucault, Michel. "Interview with Herodote." In *Power/Knowledge: Selected Interviews and Other Writings 1972–1977*, p. 77. Edited by C. Gordon. Brighton: Harvester, 1980.

Foucault, Michel. "The Subject and Power.' In *Beyond Structuralism and Hermeneutics*, 208–226. Edited by Hubert Dreyfus and Paul Rabinow. Chicago: University of Chicago Press, 1982.

Foucault, Michel. "Governmentality." In *The Foucault Effect: Studies in Governmentality*. Edited by G. Burchill, C. Gordon, and P. Miller. Sussex: Harvester, 1993.

Fox, Warwick. "Deep Ecology: A New Philosophy of our Time?" *Ecologist* 14, nos. 5–6 (1984).

Fox, Warwick. *Toward a Transpersonal Ecology*. Totnes: Green Books, 1995.

Franklin, A. "On Fox-Hunting and Angling: Norbert Elias and the 'Sportisation' Process." *Journal of Historical Sociology* 9, no. 4 (1996).

Franklin, Jerry F. "Ecosystem Management: An Overview." In *Ecosystem Management: Applications for Sustainable Forest and Wildlife Resources*. Edited by Alan Haney and Mark S. Boyce. New Haven, CT: Yale University Press, 1997.

Fraser, Nancy. *Justice Interruptus: Critical Reflections on the "Postsocialist" Condition*. New York: Routledge, 1997.

Fraser, Nancy. "Social Justice in the Age of Identity Politics: Redistribution, Recognition, and Participation." In *The Tanner Lectures on Human Values* 19 (Salt Lake City: University of Utah Press, 1998).

Frasz, Geoffrey B. "Environmental Virtue Ethics: A New Direction for Environmental Ethics." *Environmental Ethics* 15 (1993): 259–274.

Frazier, Claude A. "Suffer Little Children." *Saturday Evening Post* 252, no. 7 (October 1980).

Freeden, Michael. *Ideologies and Political Theory*. Oxford: Clarendon Press, 1996.

Freeden, Michael. "Ideologies as Communal Resources." *Journal of Political Ideologies* 4 (1999): 411–417.

Freeden, Michael. "Practising Ideology and Ideological Practices." *Political Studies* 48 (2000): 302–322.

Freudenberg, Nicholas, and Carol Steinsapir. "Not in Our Backyards: The Grassroots Environmental Movement." In *American Environmentalism: The U.S. Environmental Movement, 1970–1990*. Edited by Riley E. Dunlap and Angela G. Mertig. Philadelphia: Taylor & Francis, 1992.

Friedman, Mira. *Hunting Scenes in the Art of the Middle Ages and the Renaissance*. 2 vols. Tel Aviv: Tel Aviv University Press, 1978.

Frodeman, Robert. "Radical Environmentalism and the Political Roots of Postmodernism." *Environmental Ethics* 14, no. 4 (1992): 307–320.

Furze, Brian. "Ecologically Sustainable Rural Development and the Difficulty of Social Change." *Environmental Values* 2, no. 2 (1992): 141–156.

Gaard, Greta. *Ecological Politics: Ecofeminists and the Greens*. Philadelphia: Temple University Press, 1998.

Gasset, José Ortegay. *Meditations on Hunting* (New York: Charles Scribner's Sons, 1972).

Ghiselin, Michael. "Species Concepts, Individuality, and Objectivity." *Biology and Philosophy* 2 (1987): 127–143.

Gibbs, Lois. *Love Canal: My Story*. Albany, NY: SUNY Press, 1982.

Gobster, Paul, and Bruce Hull, eds. *Restoring Nature: Perspectives from the Social Sciences and Humanities* (Washington, DC: Island Press, 2000).

Goldman, Benjamin A. "What Is the Future of Environmental Justice?" *Antipode* 28, no. 2 (1996): 122–141.

Goldsmith, Edward. *The Way: An Ecological Worldview*. Totnes: Green Books, 1996.

Gonsalves, Dennis. 'Control of Papaya Ringspot Virus in Papaya: A Case Study." *Annual Review of Phytopathology* 36 (1998): 415–437.

Goodin, Robert. *Protecting the Vulnerable*. Chicago: University of Chicago Press, 1985.

Goodin, Robert. *Green Political Theory*. Cambridge: Polity Press, 1992.

Goodman, Robert. "Ensuring the Scientific Foundations for Agriculture's Future." In *Visions of American Agriculture*, 187–204. Edited by William Lockeretz. Ames: Iowa State University Press, 1997.

Goodman, Sherri W. "Ecosystem Management at the Department of Defense." *Ecological Applications* 6, no. 3 (1996): 706–707.

Gottlieb, Robert S. *Forcing the Spring: The Transformation of the American Environmental Movement*. Washington, DC: Island Press, 1993.

Gould, Carol. "Diversity and Democracy: Representing Differences." In *Democracy and Difference*. Edited by Seyla Benhabib. Princeton, NJ: Princeton University Press, 1996.

Gould, Kenneth, Allan Schnaiberg, and Adam Weinberg. *Local Environmental Struggles: Citizen Activism in the Treadmill of Production*. Cambridge: Cambridge University Press, 1996.

Gravelle, R. K. "Enforcing the Elusive: Environmental Rights in East European Constitutions." *Virginia Environmental Law Journal* 16, no. 4 (1997): 633–660.

Griffith, William B. "Protecting and Providing for Future Generations: The Present Generation as Trustee." In *Deciding for the Future: Balancing Risks and Benefits Fairly across Generations: Issue Papers*. Edited by B. Catrou. Washington, DC: National Academy of Public Administration/Battelle Institute, 1994.

Griffith, William B. "Public Lands, Property Rights, and Intergenerational Justice." Paper presented at the Seventh Annual Meeting of the Society for Philosophy in the Contemporary World, Fort Estes, CO, July 2000.

Griffith, William B. "The Reach of Property Rights When Environmental Effects Are Not Local." In *Values in an Age of Globalization: Selected Proceedings of the 28th Conference on Value Inquiry, April 2000*. Edited by K. Dobson. Forthcoming.

Grosz, S. "Access to Environmental Justice in Public Law." In *Public Interest Perspectives in Environmental Law*. Edited by David Robinson and John Dunkley. Chichester: Chancery Law Publishing, 1995.

Gunn, Alastair. "The Restoration of Species and Natural Environments." *Environmental Ethics* 13, no. 3 (1991): 291–309.

Habermas, Jürgen. *Faktizität und Geltung*. Frankfurt am Main: Suhrkamp, 1992.

Hamilton, Cynthia. "Coping with Industrial Exploitation." In *Confronting Environmental Racism: Voices from the Grassroots*. Edited by Robert Bullard. Boston: South End Press, 1993.

Hamilton, Cynthia. "Concerned Citizens of South Central Los Angeles." In *Unequal Protection: Environmental Justice and Communities of Color*. Edited by Robert Bullard. San Francisco: Sierra Club Books, 1994.

Hammond, P. B. "Lead Poisoning: An Old Problem with a New Dimension." In Senate Committee on Public Works and Committee on Commerce, *Joint Hearings before the Subcommittee on Air and Water Pollution of the Committee on Public Works and the Committee on Commerce: Air Pollution—1970*, 91st Cong., 2d sess., 24 and 25 March 1970, 1151–1177.

Handl, G. "Human Rights and the Protection of the Environmental: A Mildly Revisionist View." In *Human Rights, Sustainable Development and the Environment*. Edited by E. D. Weiss et al. Brazil: Instituto Interamericano de Derechos Humanos, 1992.

Hargrove, Eugene C. "Environmental Therapeutic Nihilism." In *Ecosystem Health: New Goals for Environmental Management*. Edited by Robert Costanza, Bryan G. Norton, and Benjamin D. Haskell. Washington, DC: Island Press, 1992.

Hayward, Tim. *Ecological Thought: An Introduction*. Cambridge: Polity Press, 1995.

Hayward, Tim. "Anthropocentrism: A Misunderstood Problem." *Environmental Values* 6, no. 1 (1997): 49–64.

Hayward, Tim. *Political Theory and Ecological Values*. Cambridge: Polity Press, 1998.

Hegel, G. W. F. *Hegel's Philosophy of Right*. Oxford: Oxford University Press, [1821] 1967.

Hill, Thomas E., Jr. "Ideals of Human Excellence and Preserving Natural Environments." *Environmental Ethics* 5 (1983): 211–224.

Hoffer, Peter Charles. *The Law's Conscience: Equitable Constitutionalism in America*. Chapel Hill: University of North Carolina Press, 1990.

Hofrichter, Richard, ed. *Toxic Struggles: The Theory and Practice of Environmental Justice*. Philadelphia: New Society, 1993.

Holland, Alan. "Sustainability: Should We Start from Here?" In *Fairness and Futurity*. Edited by Andrew Dobson. Oxford: Oxford University Press, 1999.

Holland, Alan, and Kate Rawles. *The Ethics of Conservation*. Report presented to The Countryside Council for Wales. Thingmount Series no. 1. Lancaster: Department of Philosophy, Lancaster University, 1994.

Honneth, Axel. "Integrity and Disrespect: Principles of Morality Based on the Theory of Recognition." *Political Theory* 20 (1995): 187–201.

Honneth, Axel. *The Struggle for Recognition: The Moral Grammar of Social Conflicts*. Cambridge, MA: MIT Press, 1995.

Honore, A. M. "Ownership." In *Oxford Essays in Jurisprudence*. Edited by Anthony Guest. Oxford: Oxford University Press, 1968.

Hood, Robert L. "Ecosystem Health: A Critical Analysis." Unpublished doctoral dissertation, Bowling Green State University, 1998.

Huffman, J. "A Fish out of Water: The Public Trust Doctrine in a Constitutional Democracy." *Environmental Law Review* 19 (1989).

Hughes, D. "Analysis of Duddridge Case." *Journal of Environmental Law* 7, no. 2 (1995): 238–244.

Hunold, Christian, and Iris Marion Young, "Justice, Democracy, and Hazardous Siting." *Political Studies* 46 (1998).

Hunt, Murray. *Using Human Rights Law in English Courts*. Oxford: Hart, 1998.

*Illinois Central R. R. v. Illinois*. 146 U. S. 387 (1892).

Ingram, Attracta. *A Political Theory of Rights*. Oxford: Clarendon Press, 1994.

Itzkowitz, D. *Peculiar Privilege: A Social History of English Fox-Hunting 1753–1885*. Brighton: Harvester, 1977.

Jacoby, K. "Slaves by Nature? Domestic Animals and Human Slaves." *Slavery and Abolition* 15, no. 1 (1984): 89–99.

Jamieson, Dale. "Discourse and Moral Responsibility in Biotechnical Communication." *Science and Engineering Ethics* 6 (2000): 265–273.

Jasanoff, Sheila. "Acceptable Evidence in a Pluralistic Society." In *Acceptable Evidence: Science and Values in Risk Management*. Edited by Deborah G. Mayo and Rachelle D. Hollander. New York: Oxford University Press, 1991.

Johnson, Lawrence E. *A Morally Deep World: An Essay on Moral Significance and Environmental Ethics*. New York: Cambridge University Press, 1993.

Johnson, Trebbe. "Native Intelligence." *Amicus Journal* 14, no. 4 (1993).

Jonas, Hans. "Technology and Responsibility: The Ethics of an Endangered Future." In *Responsibilities to Future Generations*. Edited by Ernest Partridge. Buffalo, NY: Prometheus Books, 1981.

Jonas, Hans. *The Imperative of Responsibility: In Search of Ethics for the Technological Age*. Chicago: University of Chicago Press, 1984.

Jones, James. "The Tuskegee Legacy: AIDS and the Black Community." *Hastings Center Report* 22, no. 6 (1992): 38–40.

Jones, Michael. *Finland: Daughter of the Sea*. Folkestone: Dawson, 1977.

Jonsen, Albert R. *The Birth of Bioethics*. New York: Oxford University Press, 1998.

Jonsen, Albert R., Mark Siegler, and William J. Winslade. *Clinical Ethics: A Practical Approach to Ethical Decisions in Clinical Medicine*. 4th ed. New York: McGraw-Hill Health Professions Division, 1998.

Jonsen, Albert R., and Stephen Toulmin. *The Abuse of Casuistry: A History of Moral Reasoning*. Berkeley: University of California Press, 1998.

Jutila, H. "The Seed Bank of Ballast Area in Reposaari, SW Finland." Unpublished ms.

Kant, Immanuel. "On the Common Saying: 'This May Be True in Theory, But It Does Not Apply in Practice.'" In *Kant: Political Writings*. Edited by H. Reiss. Cambridge: Cambridge University Press, 1991.

Katz, Eric. *Nature as Subject: Human Obligation and Natural Community*. Lanham, MD: Rowman and Littlefield, 1997.

Katz, Eric. "A Pragmatic Reconsideration of Anthropocentrism." *Environmental Ethics* 21 (1999): 377–390.

Kelso, J. R. M., and K. I. Cullis. "The Linkage among Ecosystem Perturbations, Remediation, and the Success of the Nipigon Bay Fishery." *Canadian Journal of Fisheries and Aquatic Sciences* 53 (1996): 67–78.

King, Roger J. H. "Environmental Ethics and the Case for Hunting." *Environmental Ethics* 13, no. 1 (1991): 59–85.

Kitchell, James F., et al. "Sustainability of the Lake Superior Fish Community: Interactions in a Food Web Context." *Ecosystems* 3, no. 6 (2000): 545–560.

Klauer, Irene. "The Implicit Practice of Environmental Philosophy." In *Environmental Philosophy and Environmental Action*. Edited by Don E. Marietta, Jr., and Lester Embree. Lanham, MD: Rowman and Littlefield, 1998: 149–168.

Kothari, Ashish. *Understanding Biodiversity: Life, Sustainability and Equity*. Tracts for the Times no. 11. Delhi: Orient Longman, 1997.

Kothari, Ashish, Neema Pathak, R. V. Anuradha, and Bansuri Taneja, eds. *Communities and Conservation: Natural Resource Management in South and Central Asia*. Delhi: UNESCO and Sage Publications, 1998.

Kozlowski, Richard G. "Revisiting the Lead Phasedown." *EPA Journal* 13 (October 1987): 28.

Krauss, Celene. "Women of Color on the Front Line." In *Unequal Protection: Environmental Justice and Communities of Color*. Edited by Robert Bullard. San Francisco: Sierra Club Books, 1994.

Kripke, Saul. *Naming and Necessity.* Oxford: Blackwell, 1980.

Ksentini, Fatma. *Final Report of the UN Sub-Commission on Human Rights and the Environment.* UN Doc. E/CN.4/Sub.2/1994/9.

Lake, Robert. "Volunteers, NIMBYS, and Environmental Justice: Dilemmas of Democratic Practice." *Antipode* 28, no. 2 (1996).

Lau, Martin. "Islam and Judicial Activism." In *Human Rights Approaches to Environmental Protection.* Edited by Alan Boyle and Michael R. Anderson. Oxford, Clarendon Press, 1996.

Lavelle, Marianne, and Marcia Coyle. "Unequal Protection." *National Law Journal* 14 (1992): A16.

Lazarus, R. "Changing Conceptions of Property and Sovereignty in Natural Resources: Questioning the Public Trust Doctrine." *Iowa Law Review* 71 (1986).

Lee, Charles. "Beyond Toxic Wastes and Race." In *Confronting Environmental Racism: Voices from the Grassroots.* Edited by Robert Bullard. Boston: South End Press, 1993.

Lee, Charles, ed. *Proceedings: The First National People of Color Environmental Leadership Summit.* New York: United Church of Christ Commission for Racial Justice, 1992.

Leopold, Aldo. "Report to the American Game Conference on an American Game Polity." In *The River of the Mother of God and Other Essays by Aldo Leopold.* Edited by Susan L. Flader and J. Baird Callicott. Madison: University of Wisconsin Press, 1991.

Leopold, Aldo. *A Sand County Almanac*, rev. ed. New York: Oxford University Press, [1949] 1993.

Lester, James, and David Allen. "Environmental Justice in the U.S.: Myths and Realities." Paper presented at the annual meeting of the Western Political Science Association, Seattle, 1999.

Lewis, C. J. "The Timid Approach of the Federal Courts to the Public Trust Doctrine." *Public Land and Resources Law Review* 19 (1998).

Lewis, Jack. "Lead Poisoning: A Historical Perspective." *EPA Journal* 4 (1985): 15.

Lewontin, Richard. "Genes in the Food!" *New York Review of Books* 48, no. 10 (2001): 81–84.

Light, Andrew. "Ecological Restoration and the Culture of Nature: A Pragmatic Perspective." In *Restoring Nature: Perspectives from the Social Sciences and Humanities*, 49–70. Edited by Paul Gobster and Bruce Hull. Washington, DC: Island Press, 2001.

Light, Andrew. "Restoration, the Value of Participation, and the Risks of Professionalization." In *Restoring Nature: Perspectives from the Social Sciences and Humanities*, 163–184. Edited by Paul Gobster and Bruce Hull. Washington, DC: Island Press, 2000.

Light, Andrew. "Taking Environmental Ethics Public." In *Environmental Ethics: What Really Matters? What Really Works?* Edited by David Schmidtz and Elizabeth Willott. Oxford: Oxford University Press, 2001: 556–566.

Light, Andrew. "The Case for a Practical Pluralism." In *Environmental Ethics: An Anthology*. Edited by Andrew Light and Holmes Rolston III. Cambridge, MA: Blackwell, 2002.

Light, Andrew. "Contemporary Environmental Ethics: From Metaethics to Public Philosophy." Forthcoming in *Metaphilosophy* (2002).

Light, Andrew. "'Faking Nature' Revisited." In *The Beauty around Us: Environmental Aesthetics in the Scenic Landscape and Beyond*. Edited by Diane Michelfelder and Bill Wilcox. Albany, NY: SUNY Press, forthcoming, 2002.

Light, Andrew. "Restoring Ecological Citizenship." In *Democracy and the Claims of Nature*. Edited by Ben Minteer and Bob Pepperman-Taylor. Lanham, MD: Rowman and Littlefield, 2002: 153–172.

Light, Andrew, and Eric Higgs. "The Politics of Ecological Restoration." *Environmental Ethics* 18 (1996): 227–248.

Light, Andrew, and Eric Katz. *Environmental Pragmatism*. New York: Routledge, 1996.

Light, Andrew, and Eric Katz. "Environmental Pragmatism and Environmental Ethics as Contested Terrain." In *Environmental Pragmatism*. Edited by Andrew Light and Eric Katz. London: Routledge, 1996.

Lindblom, Charles. *Inquiry and Change: The Troubled Attempt to Understand and Shape Society*. New Haven, CT: Yale University Press, 1990.

List, C. J. "Is Hunting a Right Thing?" *Environmental Ethics* 19, no. 4 (1997): 405–416.

Littlewood, David. "The Wilderness Years: A Critical Discussion of the Role of Prescribed Newness in Environmental Ethics." Unpublished doctoral dissertation, Lancaster University (UK), 2001.

Losey, J. E., S. Raynor, and M. E. Carter. "Transgenic Pollen Harms Monarch Larvae." *Nature* 399 (1999): 214.

Low, Nicholas, and Brendan Gleeson. *Justice, Society and Nature: An Exploration of Political Ecology*. London: Routledge, 1998.

*Lucas v. So. Carolina Coastal Council*, 505 U.S. 1003 (1992).

Luke, Brian. "A Critical Analysis of Hunters' Ethics." *Environmental Ethics* 19, no. 1 (1997): 25–44.

Lupoi, Maurizio. *Trusts: A Comparative Study*. Translated by S. Dix. Cambridge: Cambridge University Press, 2001.

Macklin, Ruth. "Can Future Generations Correctly Be Said to Have Rights?" In *Responsibilities to Future Generations: Environmental Ethics*, ed. Ernest Partridge (Buffalo, NY: Prometheus Books, 1981).

Macleod, Colin. *Collected Essays*. Oxford: Clarendon Press, 1983.

Madison, Isaiah, Vernice Miller, and Charles Lee, "The Principles of Environmental Justice: Formation and Meaning." In *Proceedings: The First National People of Color Environmental Leadership Summit*. Edited by Charles Lee. New York: United Church of Christ Commission for Racial Justice, 1992.

Madison, James, Alexander Hamilton, and John Jay. *The Federalist Papers*. Harmondsworth: Penguin, 1987.

Mahalia, Bava. "Letter from a Tribal Village." *Lokayan Bulletin* 11 (1994): 157–158.

Mansbridge, Jane. "Public Spirit in Political Systems." In *Values and Public Policy*. Edited by Henry J. Aaron, Thomas E. Mann, and Timothy Taylor. Washington, DC: Brookings Institution, 1994.

Marvier, Michelle. "Ecology of Transgenic Crops." *American Scientist* 89 (2001): 160–167.

Mayr. Ernst. "The Ontological Status of Species." *Biology and Philosophy* 2 (1987).

McConnell, Fiona. "The Convention on Biological Diversity." In *The Way Forward: Beyond Agenda 21*. Edited by Felix Dodd. London: Earthscan, 1997.

McHoul, Alec, and Wendy Grace. *A Foucault Primer: Discourse, Power and the Subject*. Melbourne: Melbourne University Press, 1993.

McNeill, J. R. *Something New under the Sun: An Environmental History of the Twentieth Century World*. New York: Norton, 2000.

Meine, Curt. *Aldo Leopold: His Life and Work*. Madison: University of Wisconsin Press, 1988.

Meyers, G. "Variations on a Theme: Expanding the Public Trust Doctrine to Include Protection of Wildlife." *Environmental Law* 19 (1989).

Miles, Irene, William C. Sollivan, and Frances E. Kuo. "Psychological Benefits of Volunteering for Restoration Projects." *Ecological Restoration* 18, no. 4 (2000): 218–227.

Mill, J. S. *On Liberty*. London: Dent, [1859] 1910.

Miller, C. "Environmental Rights: European Fact or English Fiction?" *Journal of Law and Society* 22, no. 3 (1995): 374–397.

Miller, David. *Social Justice*. Oxford: Clarendon Press, 1976.

Miller, Marian A. L. "Sovereignty Reconfigured: Environmental Regimes and Third World States." In *The Greening of Sovereignty in World Politics*. Edited by Karen T. Litfin. Cambridge, MA: MIT Press, 1998.

Moore, G. E. *Ethics*. Oxford: Oxford University Press, [1911] 1966.

Moriarty, P. V., and Mark Woods. "Hunting Does Not Equal Predation." *Environmental Ethics* 19, no. 4 (1997): 391–404.

Munasinghe, M., et al. "Applicability of Techniques of Cost-Benefit Analysis to Climate Change." In *Climate Change 1995: Economic and Social Dimensions of Climate Change*, ed. James P. Bruce, Hoesung Lee and Erik F. Haites. Cambridge: Cambridge University Press.

Munksgaard, J., and A. Larsen. "Miljømæssigt råderum—et vildskud? [Environmental utilization space—An aberration?]." *Samfundsøkonomen* 1999, 1.

Naess, Arne. "The World of Concrete Contents." *Inquiry* 28 (1986): 417–428.

Naess, Arne. *Ecology, Community and Lifestyle*. Edited and translated by David Rothenberg. Cambridge: Cambridge University Press, 1989.

Naess, Arne. "The Way." *Ecologist* 19, no. 5 (1989): 196–197.

Naess, Arne. "Deep Ecology." In *The Green Reader*. Edited by Andrew Dobson. London: Andre Deutsch, 1991.

Nagel, Thomas. *The Last Word*. Oxford: Oxford University Press, 1997.

Narby, Jeremy. "Shamans and Scientists." In *Intrinsic Value and Integrity of Plants in the Context of Genetic Engineering*. Edited by David Heaf and Johnannes Wirz. Llanystumdwy, UK: International Forum for Genetic Engineering, 2001.

National Oil Refinery Action Network. Available at http://www.igc.org/cbesf/index_frame.html.

National Research Council. *Ecological Risks of Transgenic Crops*. Washington, DC: National Academy of Sciences, 2002.

Nature Conservancy Council. *Guidelines for Selection of Biological Sites of Special Scientific Interest*. Peterborough: Nature Conservancy Council, 1989.

Naturrådet. *Dansk naturpolitik. Visioner og anbefalinger* [Danish nature politics: Visions and recommendations]. Copenhagen: Naturrådet, 2000.

Netherlands Scientific Council for Government Policy. *Sustained Risks: A Lasting Phenomenon*. Report to the Government no. 44. The Hague: Netherlands Scientific Council for Government Policy, 1995.

Newton, David. *Environmental Justice*. Santa Barbara, CA: ABC-Clio, 1996.

Newton, Lisa, and Catherine Dillingham. *Watersheds II: Ten Cases in Environmental Ethics*. New York: Wadsworth, 1996.

Nickle, James. "The Right to a Safe Environment." *Yale Journal of International Law* 18, 1 (1993): 281–295.

Nietzsche, Friedrich. "Unzeitgemässe Betrachtungen, Zweites Stück, Vom Nutzen und Nachtheil der Historie für das Leben." In *Werke I*. Edited by K. Schlechta. Frankfurt am Main: Ullstein, [1874] 1976.

Norton, Bryan. *Toward Unity among Environmentalists*. Oxford: Oxford University Press, 1991.

Norton, Bryan. "Why I Am Not a Nonanthropocentrist: Callicott and the Failure of Monistic Inherentism." *Environmental Ethics* 17, no. 4 (1995): 341–358.

Norton, Bryan. "Review of Holmes Rolston III's *Conserving Natural Value*." *Environmental Ethics* 18 (1996): 209–214.

Norton, Bryan, and Bruce Hannon. "Environmental Values: A Place-Based Approach." *Environmental Ethics* 19, no. 2 (1997): 227–245.

Nowell-Smith, P. H. *Ethics*. Harmondsworth: Penguin, 1954.

Nozick, Robert. *Anarchy, State, and Utopia*. New York: Basic Books, 1974.

Nussbaum, Barry D. "Phasing Down Lead in Gasoline through Economic Instruments." *Journal of Energy Engineering* 117, no. 3 (1991): 115–124.

O'Neill, John. *Ecology, Policy, and Politics*. London: Routledge, 1993.

O'Neill, John. *The Market: Ethics, Knowledge and Politics*. London: Routledge, 1998.

Oelschlaeger, Max. "Earth-Talk: Conservation and the Ecology of Language." In *Wild Ideas*. Edited by David Rothenberg. Minneapolis: University of Minnesota Press, 1995.

Oelschlaeger, Max, ed. *Post-Modern Environmental Ethics*. Albany, NY: SUNY Press, 1995.

Olwig, Kenneth. "Reinventing Common Nature: Yosemite and Mt. Rushmore—A Meandering Tale of a Double Nature." In *Uncommon Ground: Toward Reinventing Nature*. Edited by William Cronon. New York: Norton, 1995.

Palmer, Clare. "'Taming the Wild Profusion of Existing Things?' A Study of Foucault, Power and Animals." *Environmental Ethics* 23, no. 4 (2001): 339–358.

Palumbi, Stephen R. "The High-Stakes Battle over Brute-Force Genetic Engineering." *Chronicle of Higher Education*, April 13, 2001, pp. B7–B9.

Pardo, Mary. "Mexican American Women Grassroots Community Activists: 'Mothers of East Los Angeles.'" *Frontiers* 11, no. 1 (1990).

Parfit, Derek. *Reasons and Persons*. 3rd ed. Oxford: Clarendon Press, 1987.

Parker, Kelly A. "Pragmatism and Environmental Thought." In *Environmental Pragmatism*. Edited by Andrew Light and Eric Katz. London: Routledge, 1996.

Passmore, John. *Man's Responsibility for Nature*. New York: Scribner, 1974.

Patton, Paul. "Taylor and Foucault on Power and Freedom." *Political Studies* 37 (1989): 260–276.

Pearce, David. *Environmental Values and the Natural World*. London: Earthscan, 1993.

Pearce, David, and Dominic Morgan. *The Economic Value of Biodiversity*. London: Earthscan, 1994.

Pearce, David, and R. Kerry Turner. *Economics of Natural Resources and the Environment*. New York: Harvester Wheatsheaf, 1990.

Peña, Devon. "Nos Encercaron: A Theoretical Exegesis on the Politics of Place in the Intermountain West." Paper presented at the New West Conference, Flagstaff AZ, 1999.

Peña, Devon, ed. *Chicano Culture, Ecology, Politics: Subversive Kin.* Tucson: University of Arizona Press, 1998.

Perkins, John. *Insects, Experts and the Insecticide Crisis.* New York: Plenum Press, 1982.

Peterson, Cass. "How the EPA Reversed around the Gas Pumps." *Washington Post*, 1 August 1984, sec. A.

Pew Initiative on Biotechnology. *Harvest on the Horizon: Future Uses of Agricultural Biotechnology.* Washington, DC: Pew Initiative on Biotechnology, 2001.

Pinchot, Gifford. *The Fight for Conservation.* New York: Doubleday, Page and Company, 1910.

Pinchot, Gifford. *The Training of a Forester.* Philadelphia: Lippincott, 1914.

Pinchot, Gifford. *Breaking New Ground.* New York: Harcourt, Brace and Company, 1947.

Pitkin, Hannah. "Obligation and Consent." In *Philosophy, Politics and Society*, 4th ed., pp. 45–85. Edited by Peter Laslett, W. G. Runciman, and Quentin Skinner. Oxford: Blackwell, 1972.

Plater, Z., R. Abrams, and W. Goldfarb. *Environmental Law and Policy.* 2nd ed. St. Paul, MN: West, 1998.

Popovic, N. "In Pursuit of Environmental Human Rights: Commentary on the Draft Declaration of Principles on Human Rights and the Environment." *Columbia Human Rights Law Review* 27, no. 3 (1996): 497.

Preston, Christopher J. "Epistemology and Intrinsic Values: Norton and Callicott's Critiques of Rolston." *Environmental Ethics* 20 (1998): 409–428.

Pulido, Laura. *Environmentalism and Economic Justice: Two Chicano Struggles in the Southwest.* Tucson: University of Arizona Press, 1996.

Quarles, John. *Cleaning Up America: An Insider's View of the Environmental Protection Agency.* Boston: Houghton Mifflin, 1976.

Quigley, Peter. "Rethinking Resistance: Environmentalism, Literature, and Poststructural Theory." *Environmental Ethics* 14, no. 4 (1992): 291–306.

Raboy, V. "Accumulation and Storage of Phosphate and Minerals." In *Cellular and Molecular Biology of Plant Seed Development*, 441–477. Edited by B. A. Larkins and I. K. Vasil. Dordrecht: Kluwer, 1997.

Rajan, Mukund Govind. *Global Environmental Politics: India and the North-South Politics of Global Environmental Issues.* Delhi: Oxford University Press, 1997.

Rapport, David. "What Is Clinical Ecology?" In *Ecosystem Health: New Goals for Environmental Management.* Edited by Robert Costanza, Bryan G. Norton, and Benjamin D. Haskell. Washington, DC: Island Press, 1992.

Rawles, Kate. "The Missing Shade of Green." In *Environmental Philosophy and Environmental Activism*, Edited by Don E. Marietta, Jr., and Lester Embree. Lanham, MD: Rowman and Littlefield, 1995: 149–168.

Rawls, John. *A Theory of Justice.* Oxford: Oxford University Press, 1971.

Rees, William E., and Mathis Wackernagel. *Our Ecological Footprint: Reducing Human Impact on the Earth.* Philadelphia: New Society Publishers, 1996.

Rehling, David. "Legal Standing for Environmental Groups within the Administrative System—The Danish Experience and the Need for an International Charter on Environmental Rights." In *Participation and Litigation Rights of Environmental Associations in Europe*, 151–156. Edited by M. Führ and C. Roller. Frankfurt am Main: Peter Lang, 1991.

Rest, A. "Europe—Improved Environmental Protection through an Expanded Concept of Human Rights?" *Environmental Policy and Law* 27, no. 3 (1997).

Richardson, Henry S. *Practical Reasoning about Final Ends.* Cambridge: Cambridge University Press, 1997.

Ricoeur, Paul. *Interpretation Theory: Discourse and the Surplus of Meaning.* Fort Worth: Texas Christian University Press, 1976.

Rieser, A. "Ecological Preservation as a Public Property Right: An Emerging Doctrine in Search of a Theory." *Harvard Environmental Law Review* 15 (1991).

Rippe, Klaus Peter. "Dignity of Living Beings and the Possibility of a Non-Egalitarian Biocentrism." In *Intrinsic Value and Integrity of Plants in the Context of Genetic Engineering*, pp. 12–14. Edited by David Heaf and Johnannes Wirz. Llanystumdwy, UK: International Forum for Genetic Engineering, 2001.

Rissler, Jane, and Margaret Mellon. *The Ecological Risks of Transgenic Crops.* Cambridge, MA: MIT Press, 1997.

Rogoff, M., and S. M. Rawlins. "Food Security: A Technological Alternative." *BioScience* 37 (1987): 800–807.

Rolston, Holmes III. *Environmental Ethics: Duties to and Values in the Natural World.* Philadelphia: Temple University Press, 1988.

Rolston, Holmes III. "Environmental Ethics: Values in and Duties to the Natural World." In *Applied Ethics: A Reader.* Edited by Earl R. Winkler and Jerrold R. Coombs. Oxford: Blackwell, 1993.

Rolston, Holmes III. *Conserving Natural Value.* New York: Columbia University Press, 1994.

Rolston, Holmes III. "Nature for Real: Is Nature a Social Construct?" Public lecture, Mansfield College, Oxford, 1996.

Roman, A. J. "Locus Standi: A Cure in Search of a Disease?" In *Environmental Rights in Canada.* Edited by John Swaigen. Toronto: Butterworths, 1981.

Rooney, Anne. *Hunting in Middle English Literature.* Cambridge: Brewer, 1993.

Rorty, Richard. "Pragmatism, Relativism, and Irrationalism." In *Consequences of Pragmatism: Essays: 1972–1980.* Minneapolis: University of Minnesota Press, 1982.

Rose, C. "Joseph Sax and the Idea of the Public Trust." *Ecology Law Quarterly* 25 (1998): 351–362.

Rosenthal, Sandra B., and Rogene A. Buchholz. "How Pragmatism *Is* an Environmental Ethic." In *Environmental Pragmatism.* Edited by Andrew Light and Eric Katz. London: Routledge, 1996.

Rosman, K. J. R., et al. "Isotopic Evidence for the Source of Lead in Greenland Snows Since the Late 1960s." *Nature* 362 (1993): 333–335.

Rosner, David, and Gerald Markowitz. "A 'Gift of God'?: The Public Health Controversy over Leaded Gasoline during the 1920s." *American Journal of Public Health* 75, no. 4 (1985).

Ross, W. D. *The Right and the Good.* Oxford: Clarendon Press, 1930.

Rossignol, Anne, and Phillipe A. Rossignol. "A Rift in the Rift Valley." *Reflections: Newsletter of the Program for Ethics, Science, and the Environment, Oregon State University* 5, no. 2 (1998): 2, 7.

Rothenberg, Stephen J., et al. "Blood Lead Levels in Children in South Central Los Angeles." *Archives of Environmental Health* 5, no. 51 (1996): 383.

Ruffins, Paul. "Defining a Movement and a Community." *Crossroads/Forward Motion* 11, no. 2 (1992).

Ruhl, J. B. "An Environmental Rights Amendment: Good Message, Bad Idea." *Natural Resources and Environment* 11, no. 3 (1997): 46–49.

Runte, A. *National Parks: The American Experience.* 2nd ed. Lincoln: University of Nebraska Press, 1987.

Russow, Lilly-Marlene. "Why Do Species Matter?" *Environmental Ethics* 3 (1981): 101–112.

Ryan, Alan. *John Dewey and the High Tide of American Liberalism.* New York: Norton, 1995.

Sagoff, Mark. *The Economy of the Earth: Philosophy, Law, and the Environment.* Cambridge: Cambridge University Press, 1988.

Sahai, Suman. "Biotechnology: New Global Money-Spinner." *Economic and Political Weekly* 35 (1994): 2916.

Sandel, Michael J. *Liberalism and the Limits of Justice.* Cambridge: Cambridge University Press, 1982.

Sax, Joseph. "The Public Trust Doctrine in Natural Resource Law." *Michigan Law Review* 68 (1970).

Sax, Joseph. "The Search for Environmental Rights." *Journal of Land Use and Environmental Law* 6 (1990): 93—105.

Sayer, Kenneth M. "An Alternative View of Environmental Ethics." *Environmental Ethics* 13 (1991): 195–213.

Scherer, Donald. "Evolution, Human Living, and the Practice of Ecological Restoration." *Environmental Ethics* 17, no. 3 (1995): 359–380.

Scherer, Donald, ed. *Upstream/Downstream.* Philadelphia: Temple University Press, 1990.

Schlosberg, David. "Challenging Pluralism: Environmental Justice and the Evolution of Pluralist Practice." In *The Ecological Community: Environmental Challenges for Philosophy, Politics, and Morality.* Edited by Roger Gottlieb. London: Routledge, 1997.

Schlosberg, David. *Environmental Justice and the New Pluralism: The Chal-lenge of Difference for Environmentalism.* Oxford: Oxford University Press, 1999.

Schlosberg, David. "Networks and Mobile Arrangements: Organizational Innovation in the U.S. Environmental Justice Movement." *Environmental Politics* 6, no. 1 (1999): 122–148.

Schmidtz, David. "The Problem with Preservation." *Environmental Values* 6 (1997): 327–340.

Schneider, C. P., et al. "Predation by Sea Lamprey (Petromyzon Marinus) on Lake Trout (Salvelinus Namaycush) in Southern Lake Ontario, 1982–1992." *Canadian Journal of Fisheries and Aquatic Sciences* 53 (1996): 1921–1932.

Schwartz, H. "In Defense of Aiming High: Why Social and Economic Rights Belong in the New Post-Communist Constitutions of Europe." *East European Constitutional Review* 1 (1992): 25–29.

Scott, G. R. "The Expanding Public Trust Doctrine: A Warning to Environmentalists and Policy Makers." *Fordham Law Review* 10 (1998).

Scruton, Roger. "From a View to a Death: Culture, Nature and the Huntsman's Art." *Environmental Values* 6, no. 4 (1997): 471–482.

Selvin, Molly. *This Tender and Delicate Business: The Public Trust Doctrine in American Law and Economic Policy 1789–1920.* New York: Garland, 1987.

Sen, Amartya. "Capability and Well-Being." In *The Quality of Life.* Edited by Martha C. Nussbaum and Amartya Sen. Oxford: Clarendon Press, 1993.

Shaffer, Thomas L., and Carol Ann Mooney. *The Planning and Drafting of Wills and Trusts.* 3rd ed. Westbury, NY: Foundation Press, 1991.

Shapiro, Ian. *Political Criticism.* Berkeley: University of California Press, 1990.

Shelton, D. "Environmental Rights in the European Community." *Hastings International and Comparative Law Review* 16 (1993).

Shiva, Vandana. "Biodiversity Conservation, People's Knowledge and Intellectual Property Rights." In *Biodiversity Conservation: Whose Resource? Whose Knowledge?* Edited by Vandana Shiva. New Delhi: INTACH, 1994.

Shrader-Frechette, Kristin. *Burying Uncertainty: Risk and the Case against Geological Disposal of Nuclear Waste.* Berkeley: University of California Press, 1993.

Shrader-Frechette, Kristin, and Earl D. McCoy. *Method in Ecology: Strategies for Conservation.* New York: Cambridge University Press, 1993.

Shue, Henry. *Basic Rights: Subsistence, Affluence, and US Foreign Policy.* Princeton, NJ: Princeton University Press, 1980.

Singer, Peter. *Practical Ethics.* Cambridge: Cambridge University Press, 1993.

Singh, R. *The Future of Human Rights in the United Kingdom: Essays on Law and Practice.* Oxford: Hart, 1997.

Sitar, Shawn P., et al. "Lake Trout Mortality and Abundance in Southern Lake Huron." *North American Journal of Fisheries Management* 19, no. 4 (1999): 881–900.

Smith, B. R., and J. J. Tibbles. "Sea Lamprey in Lakes Huron, Michigan, and Superior: History of Invasion and Control, 1936–1978." *Canadian Journal of Fisheries and Aquatic Sciences* 37 (1980): 1780–1801.

Sobotka and Company. "An Analysis of the Factors Leading to the Use of Leaded Gasoline in Automobiles Requiring Unleaded Gasoline." In United States House Committee on Interstate and Foreign Commerce, *Hearings before the Subcommittee on Oversight and Investigations: Environmental Effect of the Gasoline Tilt Rule.* 96th Cong., 1st sess., 12 and 13 March 1979.

Solow, Robert M. "Sustainability: An Economist's Perspective." In *Economics of the Environment: Selected Readings*, 181–182. Edited by Robert Dorfman and Nancy S. Dorfman. New York: Norton, 1993.

Somerville, C. R., and B. Dario. "Plants as Factories for Technical Materials." *Plant Physiology* 125 (2001): 168–171.

Soule, Edward. "Assessing the Precautionary Principle." *Public Affairs Quarterly* 14 (2000): 309–328.

Soulé, Michael E., ed. *Conservation Biology: The Science of Scarcity and Diversity.* Sunderland, MA: Sinauer Associates, 1986.

Southwest Organizing Project (SWOP). *Intel Inside New Mexico: A Case Study of Environmental and Economic Injustice.* Albuquerque: SWOP, 1995.

Spangler, G. R., and J. J. Collins. "Response of Lake Whitefish (Coregonus Clupeaformis) to the Control of Sea Lamprey (Petromyzon Marinus) in Lake Huron." *Canadian Journal of Fisheries and Aquatic Sciences* 37, no. 11 (1980): 2039–2046.

Stein, P. "A Specialist Environmental Court: An Australian Experience." In *Public Interest Perspectives in Environmental Law.* Edited by D. Robinson and J. Dunkley. London: Wiley Chancery, 1995.

Stevens, William K. *Miracle under the Oaks.* New York: Pocket Books, 1995.

Stevenson, C. P. "A New Perspective on Environmental Rights after the Charter." *Osgoode Hall Law Journal* 21, no. 3 (1983): 390–421.

Stewart, C. Neal, Jr., Harold A. Richards IV, and Matthew D. Halfhill. "Transgenic Plants and Biosafety: Science, Misconceptions, and Public Perceptions." *BioTechniques: The Journal of Laboratory Technology for Bioresearch* 29 (2000): 832–843.

Stone, Deborah A. *Policy Paradox: The Art of Political Decision Making.* New York: Norton, 1997.

Subrin, S. "How Equity Conquered Common Law." *University of Pennsylvania Law Review* 135 (1987): 918–920.

Sugawara, Sandra. "EPA Trying to Ease Out of a Leaden Box." *Washington Post*, 19 May 1982, sec. A.

Sunstein, Cass. "Against Positive Rights." *East European Constitutional Review* 2 (1993): 37.

Sunstein, Cass, and E. Ullmann-Margalit. "Second-Order Decisions." *Ethics* 110 (1999): 5–31.

Suomin, J. "The Grain Immigrant Flora of Finland." *Acta Botannica Fennica* 111 (1979): 1–108.

Swink, William D. "Effectiveness of an Electrical Barrier in Blocking a Sea Lamprey Spawning Migration on the Jordan River, Michigan." *North American Journal of Fisheries Management* 19, no. 2 (1999): 397–405.

Szasz, Andrew. *EcoPopulism: Toxic Waste and the Movement for Environmental Justice.* Minneapolis: University of Minnesota Press, 1994.

Szasz, Andrew, and Michael Meuser. "Environmental Inequalities: Literature Review and Proposals for New Directions in Research and Theory." *Current Sociology* 45, no. 3 (1997): 99–120.

Taylor, Charles. *Multiculturalism.* Princeton, NJ: Princeton University Press, 1994.

Taylor, Dorceta. "Can the Environmental Movement Attract and Maintain the Support of Minorities?" In *Race and the Incidence of Environmental Hazards: A Time for Discourse.* Edited by Bunyan Bryant and Paul Mohai. Boulder, CO: Westview Press, 1992.

Taylor, Dorceta. "Environmentalism and the Politics of Exclusion." In *Confronting Environmental Racism: Voices from the Grassroots.* Edited by Robert Bullard. Boston: South End Press, 1993.

Taylor, Paul. *Respect for Nature.* Princeton, NJ: Princeton University Press, 1986.

Tesh, Sylvia, and Bruce Williams. "Identity Politics, Disinterested Politics, and Environmental Justice." *Polity* 18, no. 3 (1996): 285–305.

Thompson, Paul B. "Uncertainty Arguments in Environmental Issues." *Environmental Ethics* 8, no. 1 (1986): 59–75.

Thompson, Paul B. "Risk, Ethics and Agriculture." *Journal of Environmental Systems* 13 (1993): 137–155.

Thompson, Paul B. "Pragmatism and Policy: The Case of Water." In *Environmental Pragmatism.* Edited by Andrew Light and Eric Katz. London: Routledge, 1996.

Thompson, Paul B. *Food Biotechnology in Ethical Perspective.* London: Chapman and Hall, 1997.

Thompson, Paul B. "Food Biotechnology's Challenge to Cultural Integrity and Individual Consent." *Hastings Center Report* 27, no. 4 (1997): 34–38.

Paul B. Thompson. "Science Policy and Moral Purity: The Case of Animal Biotechnology." *Agriculture and Human Values* 14 (1997): 11–27.

Throop, William. "The Rationale for Environmental Restoration." In *The Ecological Community*, 39–55. Edited by Roger Gottlieb. London: Routledge, 1997.

Toulmin, Stephen. *An Examination of the Place of Reason in Ethics.* Cambridge: Cambridge University Press, 1958.

Tuan, Yi-Fu. *Dominance and Affection: The Making of Pets.* New Haven, CT: Yale University Press, 1984.

Turner, Bryan S. *The Body and Society.* London: Sage, 1993.

Tzfira, T., A. Zuker, and A. Altman. "Forest-Tree Biotechnology: Genetic Transformation and Its Application to Future Forests." *Trends in Biotechnology* 16 (1998): 439–446.

United Church of Christ. *Toxic Wastes and Race in the United States: A National Report on the Racial and Socio-Economic Characteristics of Communities with Hazardous Waste Sites.* New York: United Church of Christ, 1987.

United Kingdom House of Commons Hayward Debates. November 29, 1997. Part 3.

United States Environmental Protection Agency. "Regulation of Fuels and Fuel Additives." *Federal Register* 37, no. 36 (23 February 1972): 3882.

United States Environmental Protection Agency. *EPA's Position on the Health Implications of Environmental Lead.* Washington, DC: U.S. Environmental Protection Agency, 1973.

United States Environmental Protection Agency. "Regulation of Fuels and Fuel Additives." *Federal Register* 38, no. 6 (10 January 1973): 1258.

United States Environmental Protection Agency. "Regulation of Fuels and Fuel Additives." *Federal Register* 47, no. 27 (August 1982): 38078.

United States Environmental Protection Agency. *Motor Vehicle Tampering Survey.* Denver: National Enforcement Investigations Center, 1983.

United States Environmental Protection Agency. *Costs and Benefits of Reducing Lead in Gasoline: Final Regulatory Impact Analysis.* Washington DC: U.S. Environmental Protection Agency, 1985.

United States Environmental Protection Agency. "Regulation of Fuels and Fuel Additives." *Federal Register* 50, no. 45 (7 March 1985); 9386.

United States Environmental Protection Agency. *Motor Vehicle Tampering Survey 1990.* Washington, DC: Office of Air and Radiation, 1993.

United States Environmental Protection Agency. *National Air Quality and Emissions Trends Report, 1992.* Research Triangle Park, NC: Office of Air Quality Planning and Standards, 1993.

United States Environmental Protection Agency. *The Model Plan for Public Participation.* Washington, DC: Office of Environmental Justice, 1996.

United States Environmental Protection Agency. *National Air Pollutant Emission Trends, 1900 to 1996.* Research Triangle Park, NC: Office of Air Quality Planning and Standards, 1977.

United States General Accounting Office. *Siting of Hazardous Waste Landfills and Their Correlation with Racial and Economic Status of Surrounding Communities.* Washington, DC: Government Printing Office, 1983.

United States House Committee on Government Operations. *Hearing before a Subcommittee of the Committee on Government Operations: Lead in Gasoline: Public Health Dangers.* 97th Cong., 2d sess., 14 April 1982, 83.

United States House Committee on Ways and Means. *Hearings before the Committee on Ways and Means on the Subject of the Tax Recommendations of the President.* 91st Cong., 2d sess., 9, 10, 14, 15, 16, and 17 September 1970.

United States Senate Committee on Energy and Natural Resources. *Hearing before the Committee on Energy and Natural Resources: Price Differential between Leaded and Unleaded Gasoline.* 95th Cong., 2d sess., 11 December 1978.

United States Senate Committee on Environment and Public Works. *Hearing before the Committee on Environment and Public Works: Airborne Lead Reduction Act of 1984.* 98th Cong., 2d sess., 22 June 1984, 203, 204, 206.

United States Senate Committee on Environment and Public Works. *A Legislative History of the Clean Air Act Amendments of 1990.* Washington, DC: Government Printing Office, 1993.

United States Senate Committee on Public Works and Committee on Commerce. *Joint Hearings before the Subcommittee on Air and Water Pollution of the Committee on Public Works and Committee on Commerce: Air Pollution—1970, Part* 3.91st Cong., 2d sess., 24 and 25 March 1970.

VanDeVeer, Donald, and Christine Pierce. *The Environmental Ethics and Policy Book.* Belmont, CA: Wadsworth, 1994.

Verhoog, Henk. "The Intrinsic Value of Animals: Its Implementation in Governmental Regulations in the Netherlands and Its Implication for Plants." In *Intrinsic Value and Integrity of Plants in the Context of Genetic Engineering,* 15–18. Edited by David Heaf and Johnannes Wirz. Llanystumdwy, UK: International Forum for Genetic Engineering, 2001.

Vitali, T. "Sport Hunting—Moral or Immoral?" *Environmental Ethics* 12, no. 1 (1990): 69–82.

Wackernagel, Mathis. *How Big Is Our Ecological Footprint?* Vancouver: The UBC Task Force on Healthy and Sustainable Communities.

Waddington C. H. *The Evolution of an Evolutionist.* Edinburgh: Edinburgh University Press, 1975.

Waldron, Jeremy. "A Rights-Based Critique of Constitutional Rights." *Oxford Journal of Legal Studies* 13 (1993): 18–51.

Walsh, Michael P. "Phasing Lead Out of Gasoline: The Experience with Different Policy Approaches in Different Countries." Draft Issue Paper written for the United Nations Environment Programme and the Organisation for Economic Cooperation and Development, 1998.

Walzer, Michael. *Spheres of Justice.* Oxford: Blackwell, 1983.

Walzer, Michael. *Interpretation and Social Criticism.* Cambridge, MA: Harvard University Press, 1987.

Walzer, Michael. *Thick and Thin: Moral Argument at Home and Abroad.* South Bend, IN: University of Notre Dame Press, 1994.

Warren, Michael D., L. Jan Slikkerveer, and David Brokensha, eds. *The Cultural Dimension of Development: Indigenous Knowledge Systems.* London: Intermediate Technology Publications, 1995.

Weber, S. "Environmental Information and the European Convention on Human Rights." *Human Rights Law Journal* 12, no. 5 (1991): 177–185.

Weiss, Edith Brown. "Our Rights and Obligations to Future Generations for the Environment." *American Journal of International Law* 84 (1990).

Welchman, Jennifer. "The Virtues of Stewardship." *Environmental Ethics* 19, no. 4 (1999): 411–423.

Wenz, Peter. *Environmental Justice.* Albany, NY: SUNY Press, 1988.

Westra, Laura. "A Transgenic Dinner: Social and Ethical Issues in Biotechnology and Agriculture." *Journal of Social Philosophy* 24, no. 3 (1993): 213–232.

Westra, Laura. *Living in Integrity.* Totowa, NJ: Rowman and Littlefield, 1997.

Weterings, R., and J. B. Opshoor. *The Ecocapacity as a Challenge to Technological Development.* Publikatie RMNO no 74A. Rijswijk: Advisory Council for Research on Nature and Environment, 1992.

Wilkinson, C. "The Headwaters of the Public Trust." *Environmental Law* 19 (1989).

Will, George. "The Poison Poor Children Breathe." *Washington Post,* 16 September 1982, sec. A.

Williams, Bernard. *Ethics and the Limits of Philosophy.* Cambridge, MA: Harvard University Press, 1985.

Williams, Bruce A., and Albert R. Matheny. *Democracy, Dialogue and Environmental Disputes: The Contested Languages of Social Regulation.* New Haven, CT: Yale University Press, 1995.

Wilson, Hugh. "Gene Flow in Squash Species." *Bioscience* 40 (1990): 49–55.

Wolch, Jennifer R., and Jody Emel. *Animal Geographies: Place, Politics, and Identity in the Nature-Culture Borderlands.* New York: Verso, 1998.

Wolfenbarger, L. L., and P. R. Phifer. "The Ecological Risks and Benefits of Genetically Engineered Plants." *Science* 290 (2000): 2088–2093.

World Commission on Environment and Development. *Our Common Future.* Oxford: Oxford University Press, 1987.

Wuppertal Institute. *Toward a Sustainable Europe.* Copenhagen: NOAH/Friends of the Earth Denmark. 1996.

Yaffee, Steven L., et al., *Ecosystem Management in the United States: An Assessment of Current Experience.* Washington, DC: Island Press, 1996.

Young, Iris Marion. *Justice and the Politics of Difference.* London: Routledge, 1990.

# About the Contributors

**Finn Arler** is Associate Professor in the Department of Development and Planning, Division of Technology Environment and Society, Aalborg University, Denmark. His Ph.D. is in philosophy. He has edited several books and authored over fifty articles (in Danish, English, and German), most of them within the field of environmental ethics and human ecology. E-mail ⟨arler@i4.auc.dk⟩

**Avner de-Shalit** is Associate Professor of Political Science at the Hebrew University of Jerusalem, and Associate Fellow at the Oxford Centre for Environment, Ethics and Society, Oxford, UK. He is the author of *The Environment: Between Theory and Practice* and *Why Posterity Matters*. He has coedited four books and has published numerous articles in political theory, environmental philosophy, and environmental politics. E-mail ⟨msads@mscc.huji.ac.il.⟩

**Michael Freeden** is Professor of Politics at Oxford University and Professorial Fellow of Mansfield College. Among his books are *The New Liberalism: An Ideology of Social Reform, Liberalism Divided: A Study in British Political Thought (1914–1939)*, and *Ideologies and Political Theory: A Conceptual Approach*. He is the founder-editor of the *Journal of Political Ideologies*.

**William B. Griffith** is Professor of Philosophy and Director of the Graduate Program in Philosophy and Social Policy at George Washington University in Washington, D.C. He has written numerous papers on philosophical aspects of economics, of law, and issues of public policy. E-mail ⟨wbg@gwu.edu⟩

**Tim Hayward** is Reader in Politics at the University of Edinburgh. He is the author of *Ecological Thought: An Introduction* and *Political Theory and Ecological Values*. E-mail ⟨tim.hayward@ed.ac.uk⟩

**Alan Holland** is Professor of Applied Philosophy at Lancaster University (UK). He has coedited collections of articles on global sustainability and animal biotechnology, and writes widely on issues in environmental and applied philosophy. E-mail ⟨a.holland@lancaster.ac.uk⟩

**Robert Hood** is Assistant Professor of Philosophy at Middle Tennessee State University. He has published several essays on environmental ethics and ecosystem health and is currently working on a book on the relationship between bioethics and environmental ethics. E-mail ⟨rhood@mtsu.edu⟩

**Mathew Humphrey** is Lecturer in Political Theory at the University of Nottingham. He is editor of *Political Theory and the Environment: A*

*Reassessment* and has published various articles on environmental ethics and deep ecology. E-mail ⟨mathew.humphrey@nottingham.ac.uk⟩

**Niraja Gopal Jayal** is Professor of Law and Governance at the Jawaharlal Nehru University, New Delhi. Her publications include *Democracy and the State: Welfare, Secularism and Development in Contemporary India*, an edited volume *Democracy in India*, and essays on applied ethics. E-mail ⟨jayal@ndb.vsnl.net.in⟩

**Andrew Light** is Assistant Professor of Environmental Philosophy at New York University, and Research Fellow at the Institute for Environment, Philosophy and Public Policy at Lancaster University (UK). Light is the author of over fifty articles and book chapters on environmental ethics, philosophy of technology, and philosophy of film, and has edited or coedited thirteen books, including *Environmental Pragmatism*; *Technology and the Good Life?*, and *Beneath the Surface: Critical Essays on the Philosophy of Deep Ecology*. He is currently a Harrington Faculty Fellow in Architecture, Philosophy, and Geography at The University of Texas at Austin. E-mail ⟨andrew.light@nyu.edu.⟩

**Francis O'Gorman** is Lecturer in Victorian Literature in the School of English at the University of Leeds. He is the author of books on John Ruskin and the Victorian novel, as well as a number of articles on Victorian conceptions of the natural, including the relationship between manliness and hunting.

**John O'Neill** is Professor of Philosophy at Lancaster University (UK). He has written widely on the philosophy of economics, political theory, environmental philosophy, and the philosophy of science. His books include *The Market: Ethics, Knowledge and Politics* and *Ecology, Policy and Politics: Human Well-Being and the Natural World*. He is a recent Hallsworth Senior Research Fellow at Manchester University. E-mail ⟨j.oneill@lancaster.ac.uk⟩

**Clare Palmer** is Senior Lecturer in Philosophy at Lancaster University (UK). She is the author of *Environmental Ethics and Process Thinking* and *Environmental Ethics*. She is also editor of the journal *Worldviews: Environment, Culture, Religion*. E-mail ⟨c.palmer@lancaster.ac.uk⟩

**David Schlosberg** is Associate Professor of Political Science at Northern Arizona University, where he teaches political theory and environmental politics. He is the author of *Environmental Justice and the New Pluralism* and coeditor (with John Dryzek) of *Debating the Earth*. E-mail ⟨david.schlosberg@nau.edu⟩

**Paul B. Thompson** holds the Joyce and Edward E. Brewer Chair of Applied Ethics at Purdue University. He has published over 100 papers and book chapters on topics in agricultural and environmental ethics, and is the author or editor of seven books, including *The Agrarian Roots of Pragmatism*. E-mail ⟨pault@purdue.edu⟩

**Vivian E. Thomson** is Assistant Professor in the Department of Environmental Sciences and the Department of Politics at the University of Virginia. She codirects Virginia's interdisciplinary program in Environmental Thought and Practice and is the recent Fulbright Chair of American Studies at the University of Southern Denmark, Odense. E-mail ⟨vthomson@virginia.edu⟩

# Index